Handbook of Isotopes in the Cosmos
Hydrogen to Gallium

Each naturally occurring isotope has a tale to tell about the history of matter, and each has its own special place in cosmic evolution. This volume aims to grasp the origins of our material world by looking at the abundance of the elements and their isotopes, and how this is interpreted within the theory of nucleosynthesis. Each isotope of elements from hydrogen to gallium is covered in detail. For each, there is an historical and chemical introduction, and a table of those isotopes that are abundant in the natural world. Information given on each isotope includes its nuclear properties, solar-system abundance, nucleosynthesis in stars, astronomical observations, and isotopic anomalies in presolar grains and solar-system solids. Focussing on current scientific knowledge, this *Handbook of Isotopes in the Cosmos* provides a unique information resource for scientists wishing to learn about the isotopes and their place in the cosmos. The book is suitable for astronomers, physicists, chemists, geologists and planetary scientists, and contains a glossary of essential technical terms.

DONALD CLAYTON obtained his Ph.D. at Caltech in 1962, studying nuclear reactions in stars. He became Andrew Hays Professor of Astrophysics at Rice University, Texas. In 1989 he moved to Clemson University, South Carolina, where he became Centennial Professor of Physics and Astronomy in 1996. Clayton has received numerous awards for his work, including the Leonard Medal of the Meteoritical Society in 1991, the NASA Headquarters Exceptional Scientific Achievement Medal in 1992, and the Jesse Beams Award of the American Physical Society in 1998. Clayton is a fellow of the American Academy of Arts and Sciences. He has published extensively in the primary scientific literature, and has written four previous books and published on the web his Photo Archive for the History of Astrophysics.

Handbook of Isotopes in the Cosmos

Hydrogen to Gallium

DONALD CLAYTON

Clemson University

CAMBRIDGE
UNIVERSITY PRESS

PUBLISHED BY THE PRESS SYNDICATE OF THE UNIVERSITY OF CAMBRIDGE
The Pitt Building, Trumpington Street, Cambridge, United Kingdom

CAMBRIDGE UNIVERSITY PRESS
The Edinburgh Building, Cambridge CB2 2RU, UK
40 West 20th Street, New York, NY 10011-4211, USA
477 Williamstown Road, Port Melbourne, VIC 3207, Australia
Ruiz de Alarcón 13, 28014 Madrid, Spain
Dock House, The Waterfront, Cape Town 8001, South Africa

http://www.cambridge.org

First published 2003

Printed in the United Kingdom at the University Press, Cambridge

Typeface Quadraat Regular 10/14 pt   *System* LaTeX 2_ε [TB]

A catalogue record for this book is available from the British Library

Library of Congress Cataloging in Publication data

Clayton, Donald D.
Handbook of isotopes in the cosmos / Donald Clayton.
    p.   cm.
Includes bibliographical references and index.
Contents: v. 1. Hydrogen to Gallium.
ISBN 0 521 82381 1 (v. 1 : hardback)
1. Cosmochemistry.   2. Isotopes.   3. Nucleosynthesis.   I. Title.
QB450.C55   2003
523′.02–dc21   2002041702

ISBN 0 521 82381 1 hardback

Contents

Contents

Contents

Illustrations

Atomic number, Z , plotted against neutron number, N , for:

Stable natural isotopes are outlined solidly, whereas those few radioactive isotopes which have natural observable abundances in astrophysics have dashed outlines.

Preface

This book concerns the common atoms of our natural world. How many of each element exist, and why? What variations are found in the relative numbers of the isotopes of each element, and how are those variations interpreted? If I could write an epic poem, I would lyricize over the history of the universe writ small by their natural abundances. I would rhapsodize over the puzzling arrangements at different times and places of the thousand or so different isotopes of some ninety chemical elements. These different arrangements speak of distant past events.

My more prosaic approach is to consider the elements one by one. For each chemical element I first introduce some properties, perhaps chemical, perhaps poetic, perhaps cultural. Few today know the elements, and fewer can choose to read a chemistry textbook to find out. Each element introduction is followed by an account, isotope by isotope, of the isotopic abundance and its measured variations, how these may be accounted for on the basis of the nucleosynthesis theory, and of the cosmochemical implications for interstellar dust and for the origin of the solar system. These have inspired my scientific life. So penetrating are their clues that the proliferation of isotopic connections smacks of a hard rain on the face – bracing, daunting, overwhelming, refreshing. That is how I want the reader to experience the isotopes, because that is what they are to me.

This element by element consideration is preceded by an introductory essay styled less technically for general readership. At the end of the book I place a **Glossary**. It describes the meanings of concepts that are used repeatedly in the story of the isotopes. In the **Glossary** I have taken pains to be technically correct without any burden of appearing to be overly technical. I try to explain these building blocks of the science as I would in conversation with any science-educated person. One might read the **Glossary** prior to reading about a specific isotope – but not necessarily so. My goal is a communication that can be opened at any point and simply read. I envision readers who will, as the spirit moves, open the book to any point and be able to read of a wondrous world. The reader will decide which concepts of the **Glossary** s/he needs.

Some may criticize my omission of references to the research literature; their inclusion is so traditional for scientists. But to include them would detract from my goal. I imagine instead a conversation between learned people. If on a dining occasion I relate, to a physician, say, my astonishment that one can collect from the meteorites huge numbers of small rocks that are older than the Earth, he will usually be hooked by curiosity. He will want to know how I can say that, but he will not want to hear, "You must read Zinner *et al.* in the 1996 *Astrophysical Journal*. I can give you the reference." Intellectual discourse relies on the ability to relate discoveries and the reasons for their interpretation. Nor would things be different if my conversation were with a nuclear chemist. Our identities might conveniently allow us to assume some common knowledge, but he still will want to hear, "What is known? How is it known? Why should I care?"

This book is therefore not offered as a handbook of technical detail. Such a book, admirable though it would be, would be replete with references to the journal papers that have discovered and interpreted the facts. So rich are the isotopic phenomena that even practicing isotope scientists are hard pressed to commit their riches to instant recall. Four decades of research leave me totally preoccupied with the natural manifestations of the abundances of the isotopes. My aim is to share my own fascination at those discoveries. To lend appreciation of the immense fabric of natural philosophy is the goal. Technical details appear throughout because understanding requires them. Readers will appreciate the issues hanging on the isotopic abundances by reading of them. Clues to the origins of nuclei lie hidden in the manifestations of their abundances. My aim is to grasp the origins of our material world. This might be likened to the viewing of a great painting, which is clearly much more than the countless technical details of the brushstrokes. My topic is the painting, not its brushstrokes.

Many scientists see need for a readable companion to the isotopes. One wants not so much the nuclear data characterizing each isotope, for which large data bases of nuclear physics exist, and for which web sites will allow you to download more than one ever wants to know. One often wants just to experience directly the natural history that the fossil clues within isotopic abundances reveal. The scientific literature describing the highlights related here fills thousands of published technical papers. No attempt is made herein to provide attributions to them. For a scientist, finding the reference is the easy part; it is finding the idea that is hard. This book is intended to be not a source of references but of scientific ideas and related phenomena.

For anyone wanting to read more, monographs are more accessible than the research-journal literature. Insofar as understanding of the conceptual issues is concerned, almost all of those concepts can be drawn from six books, which contain ample references to the scientific journals.

Principles of Stellar Evolution and Nucleosynthesis, Donald D. Clayton (McGraw-Hill: New York, 1968; University of Chicago Press: Chicago, 1983)

The Evolution and Explosion of Massive Stars II. Explosive Hydrodynamics and Nucleosynthesis, Stanford E. Woosley and Thomas A. Weaver, *Astrophysical Journal Supplement*, **101**, 181 (1995)

Supernovae and Nucleosynthesis, W. D. Arnett (Princeton University Press: Princeton, 1996)

Nucleosynthesis and Chemical Evolution of Galaxies, B. E. J. Pagel (Cambridge University Press: Cambridge, 1997)

Astrophysical Implications of the Laboratory Study of Presolar Materials, T. Bernatowicz and
    E. Zinner, eds. (American Institute of Physics: New York, 1997)
Meteorites and the Early Solar System, J. F. Kerridge and M. S. Mathews, eds. (University
    of Arizona Press: Tucson, 1988)

I was fortunate to have become involved early in the questions of nucleosynthesis in stars, when tenacious questioning could expose key nuclear and astrophysical issues. Hoyle's sweeping canvas was inspiring, but significantly incomplete; and
some processes were misleadingly formulated or not envisioned in the epochal 1957
exposition by Burbidge, Burbidge, Fowler and Hoyle (commonly cited as B^2FH). The
opportunity fell to me with Caltech colleagues to formulate mathematical solutions
for the s process and the r process of heavy-element nucleosynthesis, to discover the
quasiequilibrium nature of silicon burning, to reformulate the e process for radioactive
nickel rather than iron, and, with my Rice colleagues, to show how explosive oxygen
and silicon burning can lead to an alternative quasiequilibrium known as "alpha-rich
freezeout" if the peak temperature is high enough, and how a neutron-rich version of
that alpha-rich quasiequilibrium accounted for many neutron-rich isotopes. Motivation to demonstrate the correctness of this post-B^2FH picture presented the chance to
predict astronomical tests for gamma-ray-line astronomy and to predict presolar grains
of outlandish isotopic compositions. I experienced the joy of asking these previously
unasked questions, and to see their dramatic observational confirmations. These gave
excitement to my scientific life and account for my eagerness to share a naturalist's
stories, as they are found within the chart of the nuclides.

I have not tried to address all scientific areas in which isotopes play a role. To
do so would vastly overreach my goal. Medical research lies beyond my qualifications;
and those applications of isotope tracers are man-made rather than natural. I choose
to emphasize the natural manifestations, those that occurred through nature's laws
rather than through human technology. Even so I have omitted those natural small
isotopic fractionations that living things display and that are generated by the natural
laws of biochemical evolution. Except for a few culturally related remarks I have largely
omitted geology. Isotopic tracers are of very great significance to human understanding
of geologic science. My choice, however, is to omit those isotopic abundance signals
that inform of the natural evolution of the Earth. Because I am interested in those
isotopes that maintain natural abundances over long times, I am not addressing the
huge numbers of radioactive isotopes that have so many applications by man. Only
naturally occurring radioactivity is essential to the cosmic origin and evolution of
matter. And even in the cosmic applications I omit many that the reader may wish
to be aware of. In the case of cosmic rays I mention only a few specifics of their
abundances, giving short change to most of the isotopic alterations that occur within
their abundances as they collide with interstellar atoms. For interstellar molecules,
as observed so brilliantly by radio astronomers, I give only inklings of the clues to

interstellar chemistry that are provided, limiting my reporting to several dramatic instances of isotopic selectivity in interstellar chemistry. I have tried, in the interests of brevity and of focus, to concentrate on the origin and evolution of matter in the universe.

Finally, it is important to not misrepresent the true nature of science. Science is neither a collection of facts nor of their interpretations; and this book is largely a collection of facts and interpretations. Science itself is a mosaic of methods, hypotheses, ideas, criticism and above all, *skepticism*. Scientific knowledge is never really known to be true, unless it is so by human definition. The skeptical challenges to conventional wisdoms have always provided great scientific rewards. So in writing of the interpretations that mankind has placed on the peculiar circumstances surrounding each isotope, I write of them as if they were incontrovertible. I may hint at but do not document the uncertainties, the competing interpretations, the controversies that are the lifeblood of knowledge. These the reader can imagine for himself or herself.

Donald Clayton

Introduction

Nuclei are nature's primal species. Most came into being before life emerged on Earth. Each atomic nucleus has its own characteristics: its mass, its number of constituent elementary particles, its spin rate, its magnetic strength, its electric charge, its multitude of excited forms, each form with its own set of properties. Unlike life forms, where modest variations exist within a given species, all nuclei of a given type are identical – exactly the same as far as any measurement has ever revealed. The exactness of the replicas is remarkable, totally outside of common human experience. They are totally indistinguishable, one from another. They are perfect clones of a master form understood latterly by mankind in terms of the quantum physics of particles. A fundamental goal of physics, maybe one should say "a dream" of physics, is to understand each species in terms of a fundamental set of laws governing the indivisible particles of which each is assembled. Despite unprecedented progress in understanding, this goal still eludes mankind. It may ever elude us. Nonetheless, a formidable description of the physics of the atomic nucleus now exists. The community of nuclear physics gains each year more sophistication in its formulation of this realm of quantum mechanics. This realm extends from the elementary quarks, from which it seems that constituent nuclear particles – neutrons, protons, mesons – are constructed, to the fluidlike phenomena observed when one nucleus containing many protons and neutrons collides with another. This is the subject of textbooks, monographs, and popular books on nuclear and elementary particle physics; and it is not the goal of this book.

We experience these nuclei in their lowest energy states, called their "ground states," nuclei "unheated," without any extra energy of excitation, through to a myriad of distinct energized forms, or "excited states." We do not experience the atomic nuclei in daily life, however. But we can locate them and study them resting at the centers of atoms. The atoms are much larger than these central nuclei, about 100 000 times larger, owing to the orbits of electrons that revolve around each nucleus. It is the whole atom, specifically these orbiting electrons, that gives each nucleus its chemical properties. The entire science of chemistry concerns itself with how the electron orbits of one nucleus interact with the electron orbits of another. The orbiting electrons make of each nucleus an atom of a chemical element. Each atom is itself electrically neutral, because their several electrons each carry identical negative electric charge, that, taken together, in sum, exactly cancel the positive electric charge that resides in the nucleus at the atomic center, a positive charge that is one of the distinct properties of each nucleus. Every nucleus of a given chemical element carries exactly the same value for the positive charge on its nucleus. And the chemical properties of that element are determined by that charge.

This book concerns the relative numbers of the nuclei. It considers their properties and the numbers of each kind that are found to exist naturally, that is to say, that are found in natural settings. The attempt here is to describe the applications to mankind's knowledge of the world that can be discerned from the populations of the isotopes found to occur in nature. Many demonstrations concerning the history of the universe derive from the numbers of each species, which differ among astronomical objects.

When humans speak of a species, we normally think of living species, of tigers, of sea gulls, of oak trees, or of dandelions. The rich beauty of our experience reflects our appreciation of their various forms and numbers. The sea gull is far more numerous than the tiger. We speak normally of their population, rather than of their abundance; but it is the same idea. We are now accustomed to interpret the living populations not as the will of a Creator, but in terms of a dynamic ecological balance. Each species feeds off others. Each is devoured by predators. Each copes with the environment. The populations of species reflect aspects of their fitness for this struggle, and for their parallel struggle with the changing face of the Earth. So the populations represent not precisely fitness, for in some sense all are fit, but rather the numbers that can coexist within this ecological balance. Populations come into balance with their food supplies, and also with their risks. Mankind is a part of this balance, although we have difficulty in perceiving ourselves as part of a balanced fabric.

Nuclei too have populations. They come into being by being assembled from others. They are destroyed by transmutations into others. Their populations are determined by a balance within the universe that might poetically be described as ecological. Within that balance the total number of constituent nuclear particles is a fixed constant, or believed to be so in terms of the Big-Bang paradigm for the origin of matter and the evolution of the early universe. But the fixed number of nucleons (protons plus neutrons) does not fix the relative numbers of the diverse nuclear species into which the nucleons can be assembled. The populations of nuclear species record ancient events in the universe. The properties of each nucleus endow it with a different kind of "fitness" than that evinced by life; and that fitness plays a role in determining their present populations. Scientists commonly call the population numbers for nuclei their *abundances*. The distinct nuclear species vary hugely in their abundances, just as do the life species on Earth. Iron is millions of times more abundant than gold; just as are sea gulls in relation to tigers.

It is worthwhile to momentarily consider this subject in relation to the philosophical history of western ideas and culture. Long before atoms were known the Greeks developed philosophical ideas that remain part of our everyday thinking, even if not justifiable. Aristotle and Plato argued that although individuals within a species die and perish, the ideal form for their species is fixed and eternal. Differences among individuals could be seen as deviations from the ideal form. These ideas were strongly influenced by living forms. When Charles Darwin revisited this subject, he presented a revolutionary picture that replaced Aristotle's concept of a species as an eternal ideal form with concepts of groups of individuals making up competitive populations. To him the species became the abstraction and the groups of individuals the hard truths. Groups of individuals made up the populations whose numbers and characteristics reflected and determined their fitness. The *Origin of Species* described Darwin's idea of natural selection in analogy to the selections of plants and animals practiced intentionally by human breeding. Of this revolt Ernst Mayr has said, "No two ways of looking at nature could be more different." Lingering public belief after 2000 years of acceptance of the "ideal form" concept accounts for the public sense of horror at intentionally replacing a gene in one species by a gene from another species. Of this tension between the holistic ideal and the material reductionist way of thinking, Keith Davies has written: "This tension maintains openness and is progressive. For science to have a healthy future, the balance between these approaches must never become dogmatic. Our imagination gives our guesses a holistic basis, our reductive experiments a way to falsify them. The confrontation is essential."

Darwin's scientific reductionism won the day. And yet an irony occurs when we consider the populations of nuclei instead of the populations of living species. Each proton is identical, conforming in perfection to the Platonic and Aristotelian ideal. So too is each ^{12}C nucleus identical to all others. Plato and Aristotle surely would have embraced these examples of perfect replication of the eternal form, perhaps finding some ultimate good that allowed them their perfect adherence to the universal form. Today we have invented the principles of quantum mechanics to make good this Aristotelian victory. No genetic evolution accompanies the competition among populations of nuclei during evolution. But Darwinian ideas nonetheless find their place in the history of the isotopes. The natural environments of the interiors of stars find one nucleus more fit than another, and hence its ultimate population becomes greater. Some irony surely lies in finding that this history embraces aspects of both the Aristotelian and Darwinian pictures in that ancient philosophical debate.

The fascination with the abundances of the atomic nuclei is that they inform of ancient events. The events that are recorded in their populations depend upon the material sample in question. In the crust of the Earth, they record its geologic evolution. Silicon in that crust is much more abundant than iron, for example, because the Earth's crust is sandy, whereas its iron sank to the Earth's core during its early molten state. In the Earth's oceans the elemental abundances reflect their solubilities in water. In the Earth's atmosphere, their numbers reflect their volatilities. And so it goes. Such abundance-sets reflect and record the geophysical history of the Earth and the chemical properties of the chemical elements. Atmospheric carbon dioxide (CO_2) and methane (CH_4) record an extra wrinkle, the impact of human beings on the Earth's atmosphere.

That the populations of nuclei depend on the sample being examined holds true throughout the universe. In differing samples the bulk abundances have differing physical significances. One may seek the total number of each specific kind of atom in the solar system, for example. Today these reside overwhelmingly within the Sun owing to its dominant mass, although complemented to lesser small degrees by the planets. One speaks of this set of abundances as "solar abundances." Determination of their values is itself a lengthy scientific quest, and it is not over. New NASA space missions greeting the millenium will continue this quest to know the solar abundances. Historically the solar abundances are regarded as those that existed 4.6 billion years ago in the interstellar cloud of gas and dust from which our solar system was soon to be born. Human experience has been limited to these abundances throughout our lengthy evolution, and only very recently has mankind sampled the stuff of other worlds and compared it to ours. That comparison pulses with scientific excitement.

At other times and other places in our universe the sets of elemental and isotopic abundances differ from the solar abundances. Astronomers first noticed this in other stars. The abundances of atoms in stars can be inferred from the strength of the atomic light that arrives at Earth from each star. Superimposed on their continuous distribution of light wavelengths, or colors, are the unmistakable atomic lines that identify the chemical elements in those stars as being the same chemical elements that exist on Earth. *Atomic lines* are light with exactly specified wavelengths, the fingerprints of the chemical element. Indeed, this sameness agrees with the interpretation of quantum mechanics and with the belief that the laws of physics should be the same throughout the universe. As far as one understands, the entire universe not only does contain but must contain only the same elements that we know here on Earth. Because physics is universal, so too are the elements universal. It is their populations that are not universal, but vary from sample to sample.

But the relative strengths of atomic lines differ from star to star. The confluence of atomic physics, of quantum mechanics, and of statistical mechanics has allowed astronomers to understand these variations in detail. These issues were at the heart of the revolution that was 20th-century physics; but today they are understood. The net result is that other stars have different abundances of the elements than does our own. Perhaps one should say "modestly different." The broad comparisons between the elements remain valid – iron is quite abundant, vanadium is rather rare. That remains true; but many stars have many fewer of each. A few have more of each. This was a great discovery of 20th-century astronomy, because it established the *nucleosynthesis* of the elements as an observational science. Astronomers also learned how old the stars are, for there do exist telltale signs of a star's age. The oldest stars are found to have many fewer of all chemical elements (except the three lightest elements) than does the Sun. These came to be called metal-poor stars, because the heavy elements were lumped together under the term "metals" by astronomers. It may seem paradoxical that the oldest stars have the fewest metals; but the key is that the abundances within each star record the abundance of metals in the gas from which the star formed. The most metal-poor stars in our Milky Way galaxy of stars have only 1/10 000th of the iron atoms (in comparison with hydrogen) as the Sun. These are also the oldest stars, probably among the first to be born in the infancy of the Milky Way.

During the period 1960–80 it became increasingly clear that the earliest stars to form in our Galaxy did so almost entirely from hydrogen (H) and helium (He), the two lightest elements (charge $Z = 1$ and $Z = 2$, also called "atomic number"). It became clear too that stars forming later had more of the heavier elements in relation to hydrogen and helium. Stars came to be characterized by their ratio of iron to hydrogen, written Fe/H by use of the chemical symbols for the elements iron and hydrogen; and this is called the *metallicity* of the star. The first stars inherited gas having the lowest metallicities, and those forming later, that is "younger" stars, inherited gas having increasingly higher metallicity. The metallicity of the Galaxy's gas has increased with time. This empirical base substantiated the theory of nucleosynthesis in stars, an idea that had arisen theoretically.

The theory of nucleosynthesis in stars was set forward by Englishman Fred Hoyle, first in 1946 and in improved detail in 1954. It had previously been thought possible that all of the chemical elements had been created at the beginning of the universe. Some attributed this to God. Theories of cosmic creation were invented that utilized a much more dense early epoch of the universe, one in which the heavy elements might possibly have been created by action of nuclear physics. This wave was fueled by Hubble's characterization of the expansion of the universe, observed by astronomers, and by the birth of cosmology based on Einstein's relativistic theory of gravity, which replaced Newton's theory. These rationalized the decisive fact of the expanding universe. They also required an early epoch that was increasingly more dense and hot as one looks backward to the beginning of time. This is called the "Big Bang." And the modern nuclear theory of the Big Bang showed that the ashes of that dense hot early universe would be the three lightest elements (H, $Z = 1$; He, $Z = 2$; and Li (lithium), $Z = 3$). Furthermore, the relative abundances of those elements agreed with the values found in stars. For the first time mankind understood that the universe *should have begun* with hydrogen and helium comprising more than 99% of all atoms, and in the ratio H/He $= 10/1$, just as observed in old stars. This was a very great triumph of human natural philosophy, combining parts of nuclear physics, the relativity theory of gravity, particle physics, and quantum statistical mechanics – all buttressed by observational astronomy.

In the 1950s and 1960s Hoyle's interpretation, that the heavier nuclei were synthesized from lighter nuclei within the interiors of stars, took hold. More detailed formulations of pieces of Hoyle's theory were set forth by others, especially A. G. W. Cameron and W. A. Fowler but also by this writer and others. These were buttressed by a great increase in the knowledge of the relative abundances of the chemical elements that was being provided at the same time by geochemists studying meteorites.

Geochemists, especially V. M. Goldschmidt, Hans Suess, and H. C. Urey, argued in a 1956 review paper that the meteorites gave a good solar-system sample of the heavy elements that had not been fractionated by geochemistry. They were not only able to provide a reasonably accurate compilation of the abundances of the elements, but they also saw in rough outline how those heavier than iron might have been assembled by the capture of neutrons by lighter elements. These ideas were reinforced in 1957 by Burbidge, Burbidge, Fowler and Hoyle in a famous review paper (which came to be referred to by an acronym built from the authors's names, B^2FH). I joined this quest in 1957 and participated in improved quantitative formulations of the process involving slow capture of neutrons in stars (the s process) and of the process involving rapid capture of neutrons in stars (the r process). In the late 1960s Hoyle's e process for synthesizing ^{56}Fe was replaced by quasiequilibrium synthesis of ^{56}Ni instead. These years created a compelling picture of the origin of almost all of the nuclei within the interiors of the stars as they aged, contracted, heated, and finally exploded. But the problems were by no means solved, for new knowledge invariably increases ignorance as well, because previously unasked questions emerge. Not only does that process of discovery continue today with refinements of the stellar settings and of the nuclear reaction rates, but meteoritics and cosmochemistry have made equally great strides delineating the abundances of the elements and their isotopes.

Only the five lightest elements owe their abundances to origins outside the stars – the first three in the Big Bang and the fourth and fifth (beryllium and boron) by cosmic-ray interactions with interstellar atoms. From the stars came all the rest. From atomic number $Z = 6$ (carbon) to atomic number $Z = 94$ (plutonium), we look to the stars. This range of atomic numbers includes all of the common elements of human experience on Earth, save for the hydrogen within the water that blesses the Earth's surface. Textbooks provided standard learning vehicles for the stellar nucleosynthesis theory for generations of astrophysicists, who cemented the theory with hundreds of brilliant papers on an incredibly large number of observable manifestations of the theory and of astronomical tests of it. The poet Walt Whitman divined, in *Leaves of Grass*, this mystic hope:

> I believe that a leaf of grass is no less
> than the journeywork of the stars.

This theory was launched with renewed fervor by discovery in the 1970s of solid samples carrying isotopic ratios that differ from those on Earth. These solid samples came from the meteorites. Meteorites as a whole are rubble piles of stones made in the early solar system; but some of those stones "remembered" the unusual presolar isotopic compositions of the grains from which they had been assembled. Isolated somewhat later were pieces of stardust, refractory dust grains that had condensed during cooling of the gases ejected from specific single stars, and which recorded the isotopic compositions of each of those stars. These were first characterized in the laboratory in 1987 with the aid of isotopic predictions for them made a dozen years earlier. These solid fragments of stars could be studied in laboratories here on Earth! They were diamonds, silicon carbides, graphite, silicon nitride, aluminum oxide, and others. Suddenly mankind possessed measurements of stellar isotopic ratios that were accurate to 0.1%, far better than any astronomer of conventional type could hope for. These presolar grains could be sorted into stardust families, and their stellar parents identified. Instead of that one isotopic composition that we on Earth inherited from our birthing molecular cloud, we have thousands of accurate measurements of other isotopic compositions. The new challenge to nucleosynthesis theory was immediate and invigorating. During this same period radio astronomers began measuring the isotopic ratios of elements within interstellar molecules. These too gave a diverse and fascinating account of an interstellar gas that had not completely mixed, and that had evolved in time to new and changing isotopic structures. Our solar system could be clearly seen for the first time as but one point within space and time, that huge framework of astrophysics and cosmochemistry. This could be understood, if at all, only with the aid of the theory of synthesis of elements in stars.

To share nucleosynthesis theory is, however, also not the primary goal of this book. Books on the theory already exist. Rather it is to introduce and summarize fascinating and varied aspects of the abundances of the elements and their isotopes. Those abundances, and how they are interpreted within the theory of nucleosynthesis, are my real topics. Each isotope of each element has a far-reaching tale to tell. And if you do not share the technical interest of the scientist, you may share the impulse of the poet. In material both dry and technical from one point of view one uncovers a canvas of brushstrokes no less than that of Turner's Venetian sunbursts. Those brushstrokes too are dry and technical; and yet they give song to the human spirit. This book invites the reader on my journey of four decades, coming to know each isotope intimately. Each one of them, some 286 that exist naturally on the Earth and in the universe, has its own personality within cosmic evolution. For example, the personalities of the mass-181 isotope of tantalum, written ^{181}Ta, and of the mass-56 isotope of iron, ^{56}Fe, differ as dramatically as those of the sea gull and the tiger.

One can not progress far toward this goal without more specific consideration of the isotopes of the elements. The distinct isotopes of a given chemical element differ in their masses and associated properties of their nuclei. Each isotope of a given chemical element has the same nuclear charge, has therefore the same number of orbiting electrons in its structure, and has therefore the same chemical properties. Each isotope of oxygen behaves chemically as oxygen. The distinct isotopes of oxygen have distinct nuclear masses because their nuclei contain differing numbers of neutrons, the uncharged nucleon. By contrast, the number of positively charged protons determines the nuclear charge and the chemical identity of the element. The extra neutrons just go along for the ride insofar as chemistry is concerned. One speaks of each nucleus by the numbers of these constituent nucleons: Z for the number of protons, the so-called "atomic number"; N for the number of neutrons; and A for their total. That is, $A = Z + N$, where A is the "mass number" (nucleon number). To again use the example of oxygen, each isotope thereof has the same number of protons, $Z = 8$. But there exist three stable isotopes, $A = 16$, $A = 17$, and $A = 18$. From the sum of nucleons it is evident that these contain respectively $N = 8$, $N = 9$, and $N = 10$ neutrons within their nuclei, so that they produce the three mass numbers for oxygen. As a matter of notation these are conventionally written with the mass number as a preceding superscript: ^{16}O, ^{17}O and ^{18}O.

Cosmic portraits of the isotopes are the main burden of this book. For each element there is an historical and chemical introduction, followed by a table of those isotopes of that element that have observed abundance in the natural world. Then a section on each isotope describes its nuclear characteristics, its abundance in the solar system relative to that of silicon, the means of nucleosynthesis of that isotope, astronomical observations of it, and its participation in isotopic patterns differing from those known on Earth, primarily in presolar grains extracted from the meteorites.

ELEMENTS

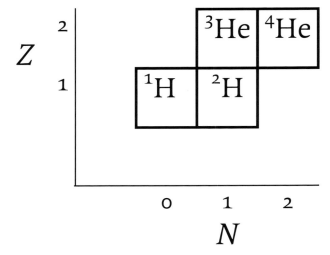

1 Hydrogen (H)

Hydrogen is a very special element. It takes its name from the fact that its combustion yields water (*hydra-gene*). The universe is primarily composed of hydrogen. How should man understand this – that material existence is 90% hydrogen? This is thought to be the result of a "Big Bang" beginning to the universe. Chemically, hydrogen is the first and simplest element. Its mass is the lightest of all elements. Deciphering its atomic spectrum observed in the laboratory gave birth in physics to quantum mechanics, the single greatest intellectual revolution of the 20th century. The proton and electron provide the simplest atomic electric dipole, resulting in series of lines of light characterized by

$$\text{frequency of light} = k(1/n^2 - 1/m^2),$$

where n and m are two integers and k is a constant. This amazing formula deduced by Balmer, with its apparently mystical appeal to numerology, was in fact explained by the development of quantum mechanics. Astronomical observations of hydrogen's emission and absorption lines provide the best data about the distant, early, universe, its motions and its clustering in space. If n and m are large integers, these frequencies lie in the radio band, so that radio astronomers use them to learn physical conditions in interstellar gaseous nebulae. But the proton and electron also are magnetic dipoles, giving hydrogen an important magnetic radioemission line. The flip of the spin of the electron in its ground state yields an electromagnetic photon having a wavelength of 21 cm, making that radio transition the dominant tool used by astronomers to measure the rotational velocity and structure of galaxies.

Hydrogen's fusion into helium by thermonuclear reactions provides power to keep the stars hot. It is the only element whose nucleus consists of a simple nucleon, the proton. It is the only element whose electronic structure consists of but a single electron. It is the dominant constituent of the most abundant phase of matter in the universe, plasma, or ionized gas, which is the natural state of the great mass of matter contained within the interiors of the stars in the universe. Hydrogen's second isotope, ^{2}H, or deuterium (D), also plays a key role in the picture of the universe, giving the best indication of the density of ordinary matter in the universe. Its third isotope, ^{3}H, or tritium (T), is radioactive with halflife 12.33 yr, and is therefore very rare in nature, being produced currently by cosmic rays striking the Earth. But it is produced in bulk in nuclear reactors from the fission of Li, and it must be stored in very safe facilities owing to its radioactivity. Both heavy isotopes, D and T, were key ingredients of the thermonuclear "H bomb." Deuterium in "heavy water," D_2O, is a superb detector of neutrinos from the Sun, so that the Sudbury Neutrino Observatory in Canada was constructed with a kiloton (million kilogram) ball of D_2O placed at its center.

Hydrogen's list of distinctions goes on and on.

All hydrogen nuclei (H or D, its mass-2 isotope) have charge +1 electronic unit, so that a single electron orbits the nucleus of the neutral atom. Its electronic configuration is designated 1s in the shorthand of quantum mechanics. Although the pre-quantum-mechanics interpretation by Bohr imagined the electron orbiting the proton, that description is not correct for the 1s electron of the H ground state. The electron moves radially back and forth through the proton, being equally likely to be moving in any radial direction. It is a sobering reflection on quantum mechanics that the electron moves easily through the great mass of the proton without sensing its existence, save for its electric attraction. In chemical compounds, H has the distinction of being both a donor and an acceptor of an electron, so it sits ambiguously either at the top of Group IA (valence +1) of the periodic table or at the top of Group VIIA (valence −1). In Group IA it resembles Li, Na and K (as in HCl); whereas in Group VIIA it resembles F, Cl and Br (as in NaH). No other element has this chameleon-like property. Not surprisingly, its own molecule, H_2, is the most abundant molecule in the universe, although difficult to detect in the cold clouds of galaxies where it resides.

Although Robert Boyle had done experiments by 1671 to generate this highly flammable gas from acid and iron filings, it was not isolated and shown to be a chemical element until Henry Cavendish did so a century later. Its most common commerical application is in the preparation of commercial acids. Its most common molecule on Earth, H_2O, provides the water basis of life. It also gave the element its name. But because most hydrogen was lost into space from the early hot Earth, hydrogen is today only the tenth most abundant element on the Earth, being found primarily in the water in the Earth's thin layer of oceans and in hydrated rocks.

Hydrogen was not always known to be the most abundant element in the universe. This long oversight in astronomy occurred because its spectral lines do not dominate the spectrum of light arriving from the Sun. This is now understood as the result of the relatively low surface temperature (for stars) of the Sun, which at 5700 K is not quite hot enough to populate profusely the excited states of H; and it is only those absorption wavelengths produced by the excited states, the Balmer series, that are visible to the human eye. Therefore, one of mankind's greatest discoveries about the universe was that it is composed primarily (90% of its atoms) of hydrogen. This fact was not accepted by leading astronomers when it was first discovered in the early 1920s at Harvard College Observatory by a young woman graduate student from England,

Cecelia Payne. She provided in her 1925 Ph.D. thesis the spectral analysis of stars that made clear that it was so. Rather unjustly, Cecelia Payne-Gaposhkin is usually not listed among the great pioneers of human knowledge; but it can be argued that her discovery dominates man's picture of his universe. It is as important as Hubble's discovery of the expansion of the universe and Penzias and Wilson's discovery of the 2.7 K temperature of the universe. Because stars initially were formed from hydrogen, its is from hydrogen that the heavy elements must have been assembled during the thermonuclear fusion within stars.

Natural isotopes of hydrogen and their solar abundances

A	Solar percent	Solar abundance per 10^6 Si atoms
1	99.997	2.79×10^{10}
2	0.0034	9.49×10^5

¹H $Z = 1, N = 0$

Spin, parity: $J, \pi = 1/2+$;   **Mass excess:** $\Delta = 7.289$ MeV

H is the only stable nucleus to contain no neutron. The isolated neutron is itself unstable, decaying to a proton. A mass-2 nucleus having only two protons is unbound and exists only fleetingly, although the nucleus having one neutron and one proton is bound and stable (see ²H below). Despite this glaring difference, the force between two nucleons is almost independent of their identity as protons or neutrons.

Abundance

From the isotopic decomposition of normal H one finds that the mass-1 isotope, ¹H, is overwhelmingly its most abundant isotope. It is 99.985% of all H isotopes in the H_2O (water) of the oceans. From astronomical observations it is known that 71% of all of the nucleons in the universe are in its H atoms; that is, when elements are compared by total mass rather than by numbers, 71% of the mass (visible mass, that is) of the universe is hydrogen. If the comparison is made by numbers of atoms instead of by the relative masses of the elements, hydrogen atoms are 90% of all atoms in the universe!

To set an abundance scale for listing the abundances of the elements, astronomers usually set $H = 10^{12}$ atoms for the size of the sample. Other elemental abundances in stars are then given by their numbers per thousand billion (10^{12}) H atoms. For the heavy elements more reliable information about relative abundances comes from the primitive classes of meteorites, which are dominated by silicon. Thus the scale frequently used for geochemistry and for stellar nucleosynthesis takes a sample containing one million Si atoms, so that abundances of the elements are then their numbers per million Si atoms. Since hydrogen in the universe is observed by astronomers to be 27 900 times more abundant than silicon,

$$\text{solar abundance of H} = 2.79 \times 10^{10} \text{ per million Si atoms.}$$

Nucleosynthesis origin

Even after it became known that the most abundant atom is H, the cause of that fact received little attention. It seemed to many natural that the universe might have begun with matter in that form. Did not God create water? After modern cosmology began with the superposition of three things – the expansion of the universe, Einstein's theory of gravity, and the discovery that the universe is filled with thermal radiation having a temperature of but 2.7 K – it was realized that the universe was once quite dense and hot and that the ratio of photons to nucleons is roughly known. This enabled a realistic calculation of the fate of nuclear matter as it cooled off from the intial fireball, now commonly called "the *Big Bang*" (see **Glossary**). The intense thermal bath of photons and neutrinos ensured that matter passed through a cooling below 1 billion (10^9) degrees during the expansion when the nucleon composition was about 88% free protons and 12% free neutrons. This ratio was established slightly earlier by the intense bath of neutrinos and antineutrinos at temperatures above 3 billion degrees. At even earlier time and hotter temperatures the nucleons themselves would not have been stable, but rather destined to emerge later from the ragingly hot sea of quarks and gluons that then dominated matter. During the expansion, the rapidly falling density of neutrinos becomes unable to convert neutrons to protons below about 3 billion degrees, so that the neutron/proton ratio "freezes" at a value near 1/7. This occurs at about 1 second after the beginning! As matter cools to 1 billion degrees, these neutrons are almost completely assembled into ^{4}He nuclei before many neutrons can decay. At the ratio 1/7, sixteen nucleons would consist of two neutrons and 14 protons, so that when the two neutrons have been captured to form one ^{4}He nucleus, the ratio would clearly be ^{4}He/^{1}H $= 1/12$. Detailed calculations confirm this picture, showing that the Big-Bang universe should have frozen all nuclear reactions with about 75% hydrogen by mass, or about 92% by numbers of atoms. The bulk of the remainder should have been ^{4}He. Amazingly, this is what astronomers see in the oldest stars. In later stars like the Sun the previous nucleosynthesis in stars has increased the ^{4}He nuclei to an abundance of 10% of H.

The relative numbers of H and of He established during the first three minutes of the Big Bang can be viewed as firstly a competition set up by the neutrino-absorbing reactions between protons and neutrons. Secondly, whatsoever neutrons exist after sufficient expansion until neutrinos can no longer be absorbed faster than nuclear reactions occur will then combine with protons in the assembly of ^{4}He nuclei; therefore the number of ^{4}He nuclei is half the number of neutrons, and the rest are protons, which later find an electron and settle down to life as H atoms. The final H/He ratio could have been different than it is, however; it depends on the universe having very high entropy, which can be regarded as a large ratio between the number of final photons and the number of final nuclei. Big Bang with a different entropy would yield different answers. That aspect of the Big Bang still must be accepted as a "just so" story; but a convincing reason may someday emerge.

For the first time a universe composed of H and He has a scientific explanation. Only ^{1}H, ^{2}H, ^{3}He, ^{4}He and much ^{7}Li seem to have inherited their abundances as ashes of that Big Bang. The stars manufactured the remainder of the elements, except for small abundances created by cosmic-ray collisions. Stars in fact slowly destroy ^{1}H by fusing it into He. Cecelia Payne's discovery in stellar spectra that H dominates the abundances in stars became a prime fact of the cosmology of the universe. The origin of both isotopes of hydrogen, the first and seventh most abundant nuclei in the universe, as well as of helium, in the initial fireball is one of the great achievements of that theory.

The abundances of the isotopes are also governed by their destruction rates in stars. Hydrogen is destroyed by fusion into helium nuclei in stars. Most H spends too little time in such circumstances, however, so that stars have not been able to materially reduce the H abundance. Stars have to date only reduced the initial H abundance from 75% of all mass to about 71%. But the issue remains as to how hydrogen is able to fuse in stars to provide their power. The problem is the lack of a stable nucleus formed with either another proton or with a helium nucleus. It was Hans Bethe who calculated that when one proton approaches another in stars there is a small chance of beta decay changing one of them into a neutron at the time of their closest encounter. In those rare events the proton and new neutron can remain bound as ^{2}H. Although the collision probability for this starting reaction in the Sun cannot be measured on Earth, it can fortunately be calculated with high reliability. Its bulk rate can also be measured in the Sun by detecting the neutrinos emitted. Its rate is of high importance to stars.

Anomalous isotopic abundance

This isotope of H does not always occur in all natural samples in its usual proportion to D (see ^{2}H).

^{2}H $\vert$ $Z = 1, N = 1$

Spin, parity: $J, \pi = 1+$;

Mass excess: $\Delta = 13.136$ MeV; $\qquad S_n = 2.224$ MeV

^{2}H is the lightest stable isotope to contain a neutron and is very abundant in comparison with the abundances of most heavier elements; but, paradoxically, that neutron is less bound than in any other nucleus that owes its abundance to either the Big Bang or to stellar nucleosynthesis. Only stable ^{9}Be, which is a fragment of cosmic-ray collisions, has smaller value of S_n. The small binding endows ^{2}H with a large and sloppy size, larger than a helium nucleus. Its nuclear spin $J = 1$ shows that the nuclear force is somewhat greater when proton and neutron spins are aligned. Deuterium in "heavy water," D_2O, is a superb detector of neutrinos from the Sun because the neutron in D can absorb electron neutrinos by the so-called "charged W^- current," liberating detectable particles that can be counted. For that reason the Sudbury Neutrino Observatory in Canada was constructed with a million-kilogram ball of D_2O placed at its center. These observations changed neutrino physics. Curiously, this neutrino absorption reverses the solar reaction that is responsible for most of the neutrinos coming from the Sun; namely, $H + H \rightarrow D + neutrino + e^+$.

Abundance

From the isotopic decomposition of normal H one finds that the mass-2 isotope, ^{2}H, or D (for deuterium) as it is also written, is relatively rare. On Earth it constitutes only 0.015% of all H isotopes. This makes it 6670 times less abundant than ^{1}H. This information comes from the isotopic analysis of sea water; however, deuterium is even more rare in the universe! Modern observations of the interstellar gas reveal it to be ten times less abundant relative to H than it is in sea water. This makes the deuterium abundance of Earth the first great isotopic anomaly; namely, that D in sea water has been enriched tenfold by the historical processes by which the Earth's oceans were formed from the initial interstellar matter from which the solar system was built.

Using the total abundance of elemental H = 2.79×10^{10} per million silicon atoms in solar-system matter and an initial isotope ratio in the Sun of D/H = 1.5×10^{-5}, as determined in today's interstellar medium (ISM), the D isotope has

solar abundance of D = 4.2×10^5 per million silicon atoms,

that is, D/Si = 0.42. That deuterium is almost as abundant as silicon in the universe shows that it is not a rare atom at all – just dwarfed by the huge abundance of ^{1}H. In fact, D is quite abundant. It would have been the seventh most abundant nucleus in the universe had not destruction in stars lowered its abundance by more than half. In the ISM today D/H = 1.5×10^{-5}, almost exactly ten times smaller than in sea water. But observations of hydrogen absorption in the distant, therefore early, universe indicate D/H = 3.3×10^{-5}, more than twice as great as in our interstellar gas today. Apparently incorporation into stars in our Galaxy followed by return of the gas later, but without

D, to the interstellar medium has lowered our ISM value to its present level, just less than half of what had been initially present following the Big Bang.

A NASA space far-ultraviolet telescope (FUSE) has shown that D/O = 0.038 in the interstellar gas in the "local bubble," the low-density region within about 100 light-years of the Sun. Taking the correct oxygen abundance to be O/H = 500×10^{-6} would translate into D/H = 1.9×10^{-5} in the local bubble, slightly greater than the ratio D/H = 1.5×10^{-5} in the nearby interstellar medium. This surprises, since the D in the local bubble might be expected to be less than in the ISM generally owing to the supernova ejecta that exist in the local bubble (see ^{60}Fe). The observation by FUSE may indicate a local O abundance slightly less than 500×10^{-6} H. (See ^{16}Oxygen (O) for the controversy over the O abundance.)

Unfortunately the D/H ratio in the initial Sun cannot be measured. It is gone. The convective (boiling) motions of the solar surface take atoms down to depths where temperatures are a million degrees, far hotter than is needed for D to be destroyed by interacting with protons according to D + H $\rightarrow$ ^{3}He. Thus any initial solar D is now ^{3}He. Return of such matter from other stars has similarly depleted the D/H ratio of the ISM by a factor near two. An interesting correlation is that the ^{3}He concentration in today's Sun exceeds what was there initially, which is recorded in planetary materials.

Nucleosynthesis origin

Hans Bethe won the 1968 Nobel Prize in physics for his two papers analyzing how the Sun and stars might achieve the power required to preserve their internal temperatures from the fusion of hydrogen into helium. In low-mass stars like the Sun, that pathway proceeds through deuterium. Although nuclear reactions within stars create even more D than helium, the final product of the fusion cycle, very little of that D can long survive. The hot protons (near 14 million degrees) convert D into ^{3}He as fast as D can be produced. The paradoxical result is that the huge solar D production results in only a miniscule isotopic ratio in the Sun's core, near D/H = 10^{-17}. Stars are destroyers of deuterium. Even before a new star settles to the Main Sequence, the convective motions have dragged any initial D downward to temperatures of a million degrees or so, where it is destroyed before core fusion can even commence. The Sun is therefore also devoid of D, and before this was fully known the famous nuclear astrophysicist Willy Fowler quipped in a bet that he would personally "eat all of the D in the Sun."

Considering that D is the seventh most abundant nucleus in the universe, a very prolific source for it must have existed. Long hidden from view, that source was the early universe itself. The principles of general relativity enable one to know the temperature in the past from its value today (2.7 K). Picture the universe when it was once dense and very hot, say one billion degrees. At earlier hotter temperatures, any D made by nuclear reactions in the early Big Bang is immediately disintegrated by the hot photons, because the binding of the proton and neutron in the deuteron is but S_n = 2.224 MeV, too low for survival above a billion degrees. Only the small steady ratio $D/H = 10^{-12}$ can exist in balance between creation and destruction in such heat. Abundant free neutrons did indeed exist at that time owing to the absorption of antineutrinos by free protons. The early neutron-to-proton ratio $n/p = 1/7$ at temperature $(T) = 3$ billion degrees is prelude to the conversion of those free neutrons to neutrons bound in ^{4}He nuclei (see ^{1}H, **Nucleosynthesis origin** above). But owing to the rapid expansion of the universe causing its density to fall rapidly, not all of those neutrons can complete the assmbly into ^{4}He nuclei. A small fraction make it only to ^{2}H nuclei and to ^{3}He nuclei, with insufficient collisons to finish the process. Thus both D and ^{3}He nuclei exist at a level near several $\times$ 10^{-5} of H as the temperature falls below a billion degrees, where the D can survive subsequent expansion and cooling. Calculations show that the amount of D that survived that Big Bang is near $D/H = $ few $\times$ 10^{-5} provided that the present-day baryon density of the universe is near 10^{-30} g/cm^3. This density today, about one proton per cubic meter, is comparable to the mean value estimated from the number of visible galaxies in the universe. The numbers fit. The Big Bang seems to be the origin of both isotopes of hydrogen, both very abundant. The nucleosynthesis of D is seen to have been even more strange than mankind had imagined. It is a nuclear ash of the fireball that began our universe.

Anomalous isotopic abundance

This isotope of H does not always occur in all natural samples in its usual proportion. A major datum is that D/H in sea water is about ten times greater than it is in the interstellar medium. Since the Earth and solar system formed from that interstellar medium, the constitution of sea water is a large isotopic anomaly. It is inescapable that when the Earth formed, it formed from initial materials in which the D/H ratio was enriched about tenfold. In no other element (except those modified by radioactive decay) does the isotopic composition on Earth differ so greatly from its starting materials. That this is true is confirmed by even larger enrichments in D/H ratio found in other samples in the solar system. On Mars the D/H ratio has been measured to be six times greater than that on Earth; and on Venus it is 120 times greater. These amazing values have much to say about the formation and evolution of all three planetary atmospheres. The molecules H_2O and HCN have been analyzed in comets and found to give $D/H = 3 \times 10^{-4}$ and 2×10^{-3} respectively, both greater than the value 1.5×10^{-4} in sea water. In the meteorite Renazzo, the value $D/H = 2 \times 10^{-3}$ has been found. But the record enhancement within a sample measured in the laboratory, 20 times greater than the ocean value, has been measured in a cluster of interplanetary dust particles that was collected in the Earth's stratosphere as the cluster drifted downward toward Earth. The variability of these D/H ratios and the fact that all are much much greater than the interstellar value ($D/H = 1.5 \times 10^{-5}$) confirms that chemical memory of the interstellar molecules has occurred in the formation of planetary solids. The writer has called such explanations "cosmic chemical memory."

Astronomers have measured even larger interstellar D enrichments by radio-spectroscopic measurements of the D/H ratio in several interstellar molecular clouds. They measure ratios much greater than the D/H ratio of the average interstellar abundance. Numbers as great as DCN/HCN = 0.02 have been found in the HCN molecule using radio and microwave spectroscopy. Other molecules show similar enrichments of D. The question naturally arises, "Why is D so enriched in interstellar molecules?" The most studied answer recognizes the very cold conditions in molecular clouds (10–50 K), so that molecules move very slowly and only reactions between ions and atoms are expected to occur. The molecules, once formed, vibrate; but the vibration energy of the D-enriched molecule is less than that of the H-bearing version. The H or D atom may be likened to the mass on the end of a spring, which vibrates more slowly if it is more massive. This means that the DCN molecule has lower vibrational energy than the HCN molecule. In the cold conditions the lower energy makes DCN more stable than HCN. Other mechanisms of isotopic fractionation have also been put forward. The bottom line is that in all models this D enrichment represents a chemical consequence of it being more massive than H. Accordingly, these dramatic enrichments are known as "isotopic fractionation" rather than as a sample from a different pool of atoms that is enriched in D.

2 | Helium (He)

Helium is a very special element, second only to hydrogen in cosmologic importance. It is by a large margin the second most abundant element in the universe. This is thought to be the result of a "Big Bang" beginning to the universe.

All helium nuclei (^{3}He or ^{4}He) have charge +2 electronic units, so that two electrons orbit the nucleus of the neutral atom. Its electronic configuration is $(1s)^2$. Both electrons speed back and forth through the nucleus, as in hydrogen. Because only two electrons are allowed in the 1s orbital by the exclusion principle, that shell is full and helium is chemically a noble gas. It neither wants nor donates electrons. So tightly bound are those two electrons (25 eV first ionization potential) that no stable compounds of helium have ever been produced, a degree of nonreactivity that has been breached for its heavier noble sisters Ar, Kr and Xe. But helium has successfully bound a bare proton, resulting in the HeH$^+$ ion. Helium is extremely volatile for related reasons, boiling at -268.934 °C, just 4 deg. above absolute zero. Helium is the only element that cannot be cooled to solid form unless put under high pressure. At 26 times atmospheric pressure, it solidifies below -272.2 °C.

Helium has the distinction of being the smallest of all atoms. Both 1s electrons are pulled toward the nucleus more than in hydrogen; and atoms of larger Z must place electrons in shells farther from the center. A practical consequence is in the technology of leak detectors for vacuum systems, because helium atoms entering through the smallest leaks can be counted and thereby locate the leak.

Because helium forms no compounds and is almost absent in the Earth's atmosphere, it was unknown for a long time. The first clue leading to its discovery was an unidentified yellow emission line in the solar chromospheric spectrum observed by French astronomer Pierre Janssen during an eclipse of the Sun in 1868. Lockyer named the unknown element *helium* for the Greek sun god, *helios*. Subsequently it was discovered to be rather abundant in radioactive rocks, where it is trapped after emission from uranium series alpha decays. Ramsay and Soddy showed that the alpha rays were helium atoms whose electrons had been stripped away. In his biography of Lord Rutherford, A. S. Eve wrote:

> About this time (1904) Rutherford, walking in the campus with a small black rock in his hand, met the professor of Geology; "Adams," he said, "how old is the Earth supposed to be?" The answer was that various methods led to an estimate of

100 million years. "I know," said Rutherford quietly, "that this piece of pitchblende is 700 million years old."

This was the first occasion when so large a value was given, based too on evidence of a reliable character; for Rutherford had determined the amount of uranium and radium in the rock, calulated the annual output of alpha particles, was confident that these were helium, measured the amount of helium in the rock and by simple division found the period during which the rock had existed in a compacted form.

The first large supply of terrestrial helium on Earth was discovered in 1903 when drillers opened a huge pocket of natural gas near Dexter, Kansas. A quarter million cubic meters of gas spewed forth daily. Residents were amazed when it extinguished a flaming bale of hay rather than producing a huge torch. Two University of Kansas chemistry professors identified it to contain abundant helium.

Space science produced many surprising evidences of helium. Perhaps the most dramatic was revealed by measurements on the samples of the Moon returned to Earth by the *Apollo 11* astronauts. They were loaded with helium! The surface of the Moon contains fine dust particles that are occasionally overturned by collisions with the Moon – *gardening of the soil* it was called. While on the surface, each grain was exposed to the solar wind, a plasma of solar composition that flows out from the Sun and strikes the Moon with a speed of 400 km/s. The abundant He in the solar wind was driven into the grains, accounting for their large He content. Similar effects are seen in meteorites, especially those known as gas-rich meteorites. Their trapped gas reveals the initial ^{3}He/^{4}He ratio. The regolith of the meteorite's parent body was similarly exposed to the solar wind until around a million years ago, when a collision sent a rocky fragment our way. These samples look like rocks, but they are actually compacted breccias of fine-grained material.

How surprising it must have been for astronomers to later discover that He is the second most abundant element in the universe! This long oversight in astronomy occurred, as it had for hydrogen, because the helium spectral lines from starlight are almost absent in the spectrum of the Sun and were in any case poorly understood in other stars. This is now understood as the result of the relatively low surface temperature (for stars) of the Sun, which at 5700 K is not hot enough to energize the excited states of He; and it is only those excited-state absorption wavelengths that are visible to the human eye. But Big-Bang cosmology depends on the dominance of He among the abundances other than of H.

Most commercial helium is extracted from natural gas. The natural gas contains a significant amount of helium because large amounts of helium are trapped in the Earth by the radioactive decays of uranium. By cooling the natural gas, the methane is drained off as a liquid. At somewhat lower temperature the trapped nitrogen gas is also drained off as a liquid. Then only helium remains. Helium became famous as the gas for lighter-than-air balloons. It is now very important for low-temperature research, because it remains liquid at very low temperatures. Its superfluid properties amazed physicists, providing one of the great discoveries of the quantum world.

Natural isotopes of helium and their solar abundances

A	Solar percent	Solar abundance per 10^6 Si atoms
3	0.0142	3.86×10^5
4	99.99	2.72×10^9

^{3}He | $Z = 2$, $N = 1$
Spin, parity: J, $\pi = 1/2+$;
Mass excess: $\Delta = 14.931$ MeV; $\quad S_n = 7.718$ MeV

> *If a neutron is removed from ^{3}He the entire nucleus breaks apart. ^{3}He is the only stable nucleus having more protons than neutrons except for the proton itself. The addition of the second proton makes the binding of the single neutron 5.5 MeV stronger than in ^{2}H. Interestingly, ^{3}He cannot bind another proton, which renders it immune to proton destruction in stars. Having Z/A $= 2/3$, ^{3}He has a larger charge-to-mass ratio than any other stable nucleus except H.*

Abundance

From the isotopic decomposition of normal He one finds that the mass-3 isotope, ^{3}He, is quite rare relative to ^{4}He. It is 0.0142% of all helium in solar gases at the time the Sun was forming, as recorded in "planetary gases" trapped within gas-rich meteorites that formed in the primitive solar disk. It is somewhat larger, 0.0166% in the helium in Jupiter's atmosphere, which could be a better measure of initial ^{3}He/^{4}He. But ^{3}He is about 100 times more rare in the Earth's atmosphere (relative to ^{4}He) because the history of radioactive decay of uranium in the Earth (see the Rutherford anecdote above) has enriched our atmosphere in daughter ^{4}He.

To set an abundance scale for listing the abundances of the elements, astronomers usually set H $= 10^{12}$ atoms. Other elemental abundances in stars are then given by their numbers per thousand billion H atoms. In the Sun the ratio H/He $= 10$. For ^{3}He more reliable information about relative He isotopic abundances comes from the primitive classes of meteorites, which are dominated by silicon. Thus the scale frequently used for geochemistry and for stellar nucleosynthesis takes a sample containing one million Si atoms, so that abundances of the elements are then their numbers per million Si atoms. Since helium in the Sun is observed by astronomers to be 2720 times more abundant than silicon, the He total solar abundance is therefore

He $= 2.72 \times 10^9$ on the scale Si $= 10^6$. On that same scale the ^{3}He isotope has an initial solar-system abundance (using 0.0142%) of ^{3}He/H $= 1.42 \times 10^{-5}$, or

$$^3\text{He} = 3.86 \times 10^5 \text{ per million silicon atoms (initial solar)},$$

that is ^{3}He/Si $= 0.386$. The value in the interstellar medium is observed to be almost identical. That ^{3}He is almost as abundant as silicon shows that it is not a rare atom at all – just dwarfed by the huge abundance of ^{4}He. In fact, ^{3}He is the twelfth most abundant nucleus in the universe. Although the rarer isotopes of H and He seem rare on Earth, abundances between seventh ranked deuterium (initially) and twelfth ranked ^{3}He include the major isotopes of silicon, magnesium, iron and sulfur, elements that dominate the chemistry of the Earth. ^{3}He is somewhat more abundant in the Sun, as recorded in the solar wind, because convective (boiling) motions of the solar surface take atoms down to depths where temperatures are a million degrees, far hotter than is needed for solar D to have been destroyed by interacting with protons according to D $+$ H $\rightarrow$ ^{3}He. Thus the initial solar D is now ^{3}He. The ^{3}He concentration in today's Sun is measured in the solar wind and is known to exceed what was there initially, because the initial solar value has been recorded in meteorites where cosmochemists measure trapped helium that had never been in the Sun.

Astronomers are able to measure the ^{3}He concentration in ionized nebulae by a spin-flip transition of singly ionized ^{3}He$^+$. When ^{3}He loses one of its electrons, the remaining electron can make a magnetic transition by flipping its spin, in exact analogy to the famous 21-cm spin-flip transition of neutral hydrogen. This transition has a wavelength of 3.46 cm. First detections concentrated on just finding the line, so the large concentrations in planetary nebulae were focussed on and seemed to suggest that stars produce ample ^{3}He. But subsequent evidence in other ionized nebulae have not borne this out. Many HII regions have also been studied, and show no evidence of an increased ^{3}He as part of the chemical evolution of the Galaxy. There is no decline in the ratio ^{3}He/H with galactocentric distance, for example, as stellar production would require. A common plateau suggests that after small contributions for stellar production are accounted for, ^{3}He/H $= 1.9 \times 10^{-5}$. One HII region at large galactocentric radius has revealed ^{3}He/H $= 1.1 \times 10^{-5}$, however; so the true value of Big-Bang ^{3}He may lie between those limits.

The main reaction for destroying ^{3}He in stars is its capture by a ^{4}He nucleus, since protons cannot attack it: ^{3}He $+ ^4$He $\rightarrow$ ^{7}Be $+$ gamma ray. The large cross section for that reaction was very important in stimulating the solar neutrino experiments. Today one of the challenges is to detect specifically the number of neutrinos arriving from the Sun owing to the decay of the radioactive ^{7}Be made in this way. The reason for the meagre survival rate of ^{3}He in low-mass stars appears to be deep surface mixing, as attested to by carbon isotope ratios. For this reason the initial ^{3}He/H ratio can be used to help define the properties of the Big Bang.

Nucleosynthesis origin

Some ^{3}He has resulted from stellar nucleosynthesis and some has been created in the Big Bang. Hans Bethe won the 1968 Nobel Prize in physics for his two papers analyzing how the Sun and stars might achieve the power required to preserve their internal temperatures from the fusion of hydrogen into helium. In low-mass stars like the Sun, that pathway proceeds through deuterium and ^{3}He. Although reactions in such stars create both D and ^{3}He at a rapid rate, almost none of that D and little of that ^{3}He can survive. The hot protons (near 14 million degrees at solar center) convert D into ^{3}He as fast as D can be produced; so the D abundance can not build up. At the solar center the ^{3}He is also destroyed rapidly by collisions with other He nuclei in its conversion into ^{4}He, the end product of Bethe's proton–proton chain. But at intermediate distances from the solar center the ^{3}He remains very abundant. Temperatures of several million degrees are sufficient to create deuterium and turn it quickly into ^{3}He, but not sufficient to finish the proton–proton chain by converting the ^{3}He into ^{4}He. When stars similar to the Sun lose their outer layers, those exterior to the ultimate white-dwarf stars that their cores become, that lost mass is rich in ^{3}He. Astronomers have observed this as ^{3}He-rich planetary nebulae. Ratios ^{3}He/H = 2–5 $\times 10^{-4}$ are observed. The action of many such stars over cosmic history has been responsible for a part of the ^{3}He found in the universe today. But it seems unlikely that it has created the lion's share of ^{3}He (See **Abundance** above).

Considering that ^{3}He is the twelfth most abundant nucleus in the universe, a very prolific source for it must have existed. Long hidden from view, that source was the early universe itself. The early neutron-to-proton ratio n/p= 1/7 at T = 3 billion degrees in the Big Bang is prelude to the conversion of those free neutrons to neutrons bound in ^{4}He nuclei (see **^{1}H Nucleosynthesis origin** above). But owing to the rapid expansion of the universe causing its density to fall rapidly, not all of those neutrons can complete the assembly into ^{4}He nuclei. Some make it only to ^{2}H nuclei and to ^{3}He nuclei, with insufficient collisions to finish the process. Thus both D and ^{3}He nuclei survive the Big Bang at a level near several $\times 10^{-5}$ of H as the temperature falls. Calculations show that the amount of ^{3}He that survived that Big Bang is near ^{3}He/H = 10^{-5} provided that the present-day baryon density of the universe is near 10^{-30} g/cm^3. This density today, about one proton per cubic meter, is comparable to the mean value estimated from the number of visible galaxies in the universe. For both the ratio and the baryon density to be true, the entropy of the universe must be large, about two billion photons for every nucleon. The numbers fit, just as they did for D/H. The Big Bang seems to be the origin of both isotopes of hydrogen and of helium.

Skeptics of the correctness of the Big Bang, especially the famed cosmologists Fred Hoyle and Geoffrey Burbidge, envision "smaller bangs" occurring as new matter is created in the universe even today. Those are the ^{3}He sources in that picture.

Anomalous isotopic abundance

^{3}He does not occur in all natural samples in its usual fraction of helium. Most astonishing enrichments of ^{3}He relative to ^{4}He are found in atoms accelerated by solar flares. The charge-to-mass ratio $Z/A = 2/3$ is greater than any other nucleus save ^{1}H. This fact favors its acceleration by the hydromagnetic phenomenon of solar flares, or ^{3}He-rich flares. Observations of solar-flare particles at 385 keV/nucleon made from NASA's *Advanced Composition Explorer (ACE)* in January 2000 revealed an abundance ratio ^{3}He/ ^{4}He = 33, a million times greater than the mean of that ratio in the Sun! Abundant flares of this type may have been responsible for one of the most significant of the *extinct radioactivities*, ^{26}Al, in the early solar system through collisions of ^{3}He ions accelerated by those early solar flares with the abundant ^{24}Mg nuclei already contained in the solids in the solar disk.

In plumes of terrestrial gases from the Earth's mantle, gases with isotopic ratios near ^{3}He/ ^{4}He = 20–30 times the atmospheric ratio are recorded, showing that those mantle gases are ancient and have not mixed with today's atmosphere, which is known to be rich in ^{4}He as a result of uranium decay within the Earth followed by volcanic outgassing.

Some meteorites have ^{3}He/ ^{4}He ratios smaller than the solar ratio because they trapped primordial solar gases before the solar D had been converted to ^{3}He, as has happened within the present Sun. But in yet other meteorite samples, ^{3}He and other nuclei created by collisions of cosmic rays with atoms in the meteorite are found in excess; consequently the magnitudes of their excesses can be used to determine the length of time that the meteorite fragment was exposed to cosmic rays in its planetary orbit following ejection from the parent body by a collision. These studies show that the meteorites mostly spend 1–10 million years in orbit before crashing to Earth. But the cosmic rays themselves are enriched in ^{3}He owing to collisions of cosmic rays with interstellar atoms. In general it can be said that the isotopic compositions of helium and other noble gases record many effects that illuminate geologic and cosmic history.

⁴He | $Z = 2, N = 2$
 | **Spin, parity:** $J, \pi = 0+$;
 | **Mass excess:** $\Delta = 2.425$ MeV; $\qquad S_n = 20.577$ MeV

The neutron binding energy S_n of ⁴He is greater than that of any other stable nucleus. The same is true for protons. For this reason ⁴He is emitted intact from certain radioactive nuclei that reduce their excess mass in this way. In that guise the ⁴He nucleus is called an alpha particle, initially an "alpha ray."

What a curiosity it is that the nucleus from which it is hardest to eject a nucleon is also the atom from which it is hardest to eject an electron!

Because the ⁴He nucleus does not bind to either proton or neutron, it cannot be destroyed by interacting with either in stars. Another mechanism, the so-called "triple alpha reaction" allows ⁴He to fuse slowly into ¹²C at the cores of red-giant stars.

Abundance

From the isotopic decomposition of normal He one finds that the mass-4 isotope, ⁴He, is 99.986% of all helium. It is the second most abundant nucleus in the universe! Modern observations of the interstellar gas reveal it to be 10.3 times less abundant than hydrogen. The elemental abundance is He = 2.72×10^9 per million silicon atoms in solar-system matter.

Nucleosynthesis origin

Although stars manufacture helium, it is not enough to account for the present-day abundance of He in the Galaxy. Hans Bethe won the 1968 Nobel Prize in physics for his two papers analyzing how the Sun and stars might achieve the power required to preserve their internal temperatures from the fusion of hydrogen into helium. But that helium is either left bound forever in a white-dwarf star at the end of stellar life of low-mass stars or most of it is processsed to heavier elements in the cores of massive stars prior to their explosions. Calculations show that only several percent of matter can be turned into helium in this way. But stars are found with as much as 28% helium by mass and as little as 24%. Even the oldest known stars, those that formed before stellar nucleosynthesis had enriched the gas, have about 24% helium by mass. This suggests that it was the Big Bang (see **Glossary**) that created that large mass of helium.

Considering that 24% of all mass was from the beginning ⁴He, the second most abundant nucleus in the universe, about 1/12th of hydrogen by number, a very prolific source for it must have existed. Long hidden from view, that source was the early universe itself. The early neutron-to-proton ratio n/p= 1/7 at T = 3 billion degrees (about 1 second into the Big Bang) is prelude to the conversion of those free neutrons to neutrons bound in ⁴He nuclei (see **¹H** *Nucleosynthesis origin* above for more details). The nucleons of the Big Bang end up overwhelmingly as either H or within ⁴He nuclei, and with a neutron-to-proton ratio n/p = 1/7, that final ratio is ⁴He/H = 1/12, which is very near the smallest values of that abundance ratio to be observed anywhere in the universe today. Crucial to the high abundance of ^{4}He is the nuclear fact that it can not form bound nuclei by reacting with either protons or other ^{4}He nuclei. Thus ^{4}He survives the Big Bang easily.

After it first became known that the second most abundant nucleus is ^{4}He, the origin of that fact received little attention. It seemed to many natural that the universe might have begun with matter in that form. After modern Big-Bang cosmology was formulated it was realized that the universe was once quite dense and hot and that the ratio of photons to nucleons could be measured. So the first calculation of that origin for ^{4}He did not occur until the mid-1960s. Today this astonishing paradigm works amazingly well. The origin of both isotopes of both hydrogen and helium in the initial fireball of the Big-Bang picture is one of the great achievements of that theory.

Skeptics of the correctness of the Big Bang, especially the famed cosmologists Fred Hoyle and Geoffrey Burbidge, envision "smaller bangs" occurring as new matter is created in the universe even today. Those are the He sources in that picture.

Anomalous isotopic abundance

The isotopes of He do not always occur in all natural samples in their usual proportion. Because helium has only two stable isotopes, variations in their abundance ratio are usually attributed to ^{3}He. But in cases where radioactive alpha decays have enriched ^{4}He, that reason for ^{4}He richness is usually fairly obvious. One example is He in rocks containing uranium. The ^{4}He/^{3}He ratio is about 100 times greater than solar in the Earth's atmosphere because the history of radioactive decay of uranium in the Earth (Rutherford!) and its outgassing has enriched our atmosphere in daughter ^{4}He.

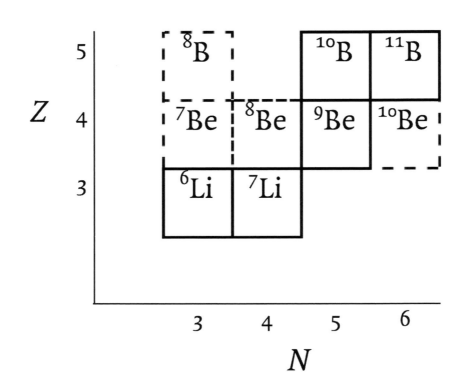

3 | Lithium (Li)

All lithium nuclei contain three protons and therefore +3 electronic units, so that three electrons orbit the nucleus of the neutral atom. Its electronic configuration is $1s^2 2s$. This can be abbreviated as an inner core of inert helium (a noble gas) plus a single s electron: (He)2s, which locates it immediately under hydrogen in Group I of the periodic table. Lithium therefore has valence + 1. It is one of the simplest of all chemical elements. In pure form it is a silvery metal that melts at 180 °C. Its small mass number makes lithium the lightest of all the alkali metals (Li, Na, K, Rb and Cs). It reacts rapidly with oxygen in the atmosphere to produce a dull grey coating of Li_2O (lithium's analogue of water). To avoid such corrosion, lithium must be shipped in a container that coats it and protects it from oxidation. Lithium minerals comprise only seven parts per million of the Earth's crust. These have found uses in certain glasses, porcelain, batteries and nuclear reactors. Pure metallic lithium is usually produced by the electrolysis of a lithium chloride solution. Its name derives from lithos, Greek for "stone".

Lithium forms structural alloys with both aluminum and magnesium. Mg–Li alloys have the highest strength-to-weight ratio of all structural materials. Li is used as a degasser to scavenge oxygen in the production of steel and of copper. It is also used as the anode (positive terminal) of some batteries.

The most interesting aspect of lithium's abundance in nature is that it is almost unique among elements in being synthesized by three physically distinct processes: in the Big-Bang origin of the universe, in stars, and by cosmic-ray interactions with interstellar matter. Models of the Big Bang reveal that its heavier isotope is probably primarily a residue of that beginning, one of the three lightest elements claiming that ancient lineage. But observed variations in the abundance ratio of the two Li isotopes in both stars and in interstellar clouds present major puzzles. Lithium isotopes are therefore rich in meaning.

Almost as interesting is the role of lithium's lighter stable isotope, ^6Li, in the production of the hydrogen bomb. The crucial tritium is produced by bombarding ^6Li with neutrons: $^6Li + n \rightarrow {}^3H$ (tritium) + ^4He. The radioactive tritium (^3H) is a major fusion fuel when reacting with deuterium (^2H) in the thermonuclear bomb. Because Li effectively absorbs neutrons it is also useful for neutron-shielding devices.

Natural isotopes of lithium and their solar abundances

A	Solar percent	Solar abundance per 10^6 Si atoms
6	7.5	4.28
7	92.5	52.8

^{6}Li | Z = 3, N = 3
Spin, parity: J, π = 1+;
Mass excess: Δ = 14.086 MeV; S_n = 5.66 MeV

^{6}Li is the lightest stable nucleus to have negative-parity nuclear orbitals. These are its 3rd proton and 3rd neutron, which each have unit orbital angular momentum. Since both orbitals possess negative parity, the double combination results in positive parity.

Abundance

From the isotopic decomposition of normal lithium one finds that the mass-6 isotope, ^{6}Li, is the lesser abundant of lithium's two isotopes; 7.5% of terrestrial Li. Lithium presents some of the most interesting abundance questions in astrophysics (see also **^{7}Li**). Using the total abundance of elemental Li = 57.1 per million silicon atoms in solar-system matter, this isotope has

solar abundance of ^{6}Li = 4.28 per million silicon atoms.

Lithium is not an abundant element, only 29th in rank of all element abundances, so that ^{6}Li is one of the least abundant nuclear species lighter than iron. Only ^{9}Be is less abundant than ^{6}Li and the comparably rare ^{10}B among the light nuclei. These are such low-abundance isotopes that their processes of nucleosynthesis must be rare or very inefficient.

Relative to silicon, the total elemental lithium is even 140 times less abundant in the solar photosphere than on Earth or in meteorites, as recorded by the solar spectrum. Since the Li/Si element abundance ratio in the meteorites should also have entered the Sun when it formed, one concludes that the Sun must be destroying its initial lithium supply as it ages. This occurs at the base of the surface convection zone of the Sun. The bottom of that zone lies at a depth that is about 1/4 of the Sun's radius. Here the high temperatures (a few million degrees kelvin, MK) that solar-surface nuclei experience is hot enough to destroy lithium, especially ^{6}Li, by nuclear interactions with protons (^{6}Li + p → ^{3}He + ^{4}He). Those proton-induced nuclear reactions destroy ^{6}Li much more readily than they do ^{7}Li because the quantum probabilities of the reaction are greater than for ^{7}Li. As a result, to deplete elemental lithium by a factor of 140 in the Sun would require that almost all ^{6}Li has already been destroyed there. But this conflicts with measurements of the solar-wind Li isotope ratio as recorded by the lunar soil, into which the solar wind blows. There one finds that ^{7}Li/^{6}Li $= 31 \pm 4$, considerably greater than the meteoritic ratio ^{7}Li/^{6}Li $= 12.14$ but not nearly as great as the huge ratio expected in the solar wind if nuclear reactions serve only to deplete the Sun's lithium. This conflict is unresolved, but its discoverers postulate that some production of ^{6}Li occurs continuously in the solar surface owing to the solar energetic protons that bombard that surface. The energetic protons are accelerated in solar flares, and they create both Li isotopes when they strike carbon and oxygen nuclei.

Nucleosynthesis origin

Big Bang   Of the very underabundant 3rd, 4th and 5th elements (Li, Be, B), only ^{7}Li, not ^{6}Li, was produced in significant amounts by the hot dense epoch of the early universe that we call the Big Bang. So the origin of ^{6}Li must be sought elsewhere.

Stellar nucleosynthesis   No production of ^{6}Li seems possible in stars, other than a very small surface abundance that can be established by nuclear reactions in solar flares. Even with that small production, stars are net destroyers of ^{6}Li, so when their ejecta return to the interstellar material it is ^{6}Li-poor. So stars are not its source.

Cosmic rays   Each of the isotopes of the very underabundant 3rd, 4th and 5th elements (Li, Be, B) is observed to be very much enriched in the cosmic rays. This is understood, because the cosmic-ray particles strike interstellar atoms, and, in those collisions involving carbon and oxygen atoms, fragments of the nuclei are broken off. These are called spallation nuclear reactions. Such collisions are not frequent enough to produce abundances of the common elements, but so rare are Li, Be, and B that almost their entirety is produced by such reactions. Li, Be and B are prominent fragments of C and O atoms. In the cosmic rays arriving at Earth they are about 5 to 20% of the abundance of carbon. This is very much more abundant than their interstellar and solar abundances. The cosmic rays in their travels through the interstellar gas have encountered an accumulated thickness of 5 to 10 grams of matter per cm^2. This is enough target material for the spallation fragments to "fill in" the very underabundant Li, Be, and B to the several percent level of carbon in the cosmic rays. When these stop and are diluted by the interstellar atoms, they become very rare.

Only for the ^{7}Li isotope is cosmic-ray production not the source of most of the nuclei. The cosmic rays produce ^{7}Li/^{6}Li in a ratio near 2, with exact details depending on the energy spectrum assumed for the unseen low-energy cosmic rays; but in the solar system that ratio is 12.5. The cosmic rays are the source of ^{6}Li, which means they are the source of only about 1/6 of the ^{7}Li. The remainder is a combination of Big-Bang relic plus production of ^{7}Li in AGB stars. But ^{6}Li, which is produced neither in stars nor Big Bang is entirely the result of cosmic-ray interactions in the interstellar medium.

Astronomical observations

Lithium is detectable in many stars by virtue of its doublet of atomic lines having wavelengths near 6707.8 Å, and is a diagnostic of the chemical enrichment of the Galaxy during its ageing. The spectroscopy is usually not able to distinguish the two isotopes of Li in the weak and "noisy" signal from distant stars. The same doublet from ^{6}Li has wavelengths that are only 0.160 Å greater, so distinguishing between different relative amounts of Li isotopes is a very difficult task. But because ^{7}Li is the more abundant isotope and because ^{6}Li is more easily destroyed, most of the galactic variation of the Li elemental abundance is surely due to ^{7}Li, not ^{6}Li. This distinction is very important for belief in the Big Bang, for observation of abundant ^{6}Li in very old stars could rule out Big-Bang cosmology as we know it.

Very old stars   The abundance of lithium seems to be a fairly constant value in very old stars that formed before stellar nucleosynthesis had occurred. It is believed that this is ^{7}Li, because the Big Bang, which is believed responsible for the initial lithium, created only ^{7}Li. Therefore any observations of ^{6}Li in very old stars could rule out Big-Bang cosmology as it is now regarded. Recently the ^{6}Li isotope has been observed for the first time in an old metal-poor star, and it does contain some ^{6}Li. The observed ratio is ^{7}Li/^{6}Li = 20, with signifcant room for error. This shows that some cosmic-ray production (see above) has already occurred before this star was born to have produced the ^{6}Li in its gas, so that not all of its ^{7}Li was produced in the Big Bang. But the Big-Bang abundance must be 90% of the observed ^{7}Li abundance in this star since cosmic rays produce only ^{7}Li/^{6}Li = 2. So Big-Bang expectations look good.

Interstellar Li isotopes   Atomic abundances are also measured in interstellar clouds along the line of sight to stars. Their absorption produces a diminution of the light at the wavelength of the isotopes in question, and its magnitude reflects the abundance. Only recently has spectrograph resolution become accurate enough to distinguish the relative numbers of ^{7}Li and ^{6}Li along these lines of sight. Interestingly, the observed ratios are less than the solar system's value, ^{7}Li/^{6}Li = 12.5; that is, more ^{6}Li-rich. One cloud does possess approximately that solar ratio, but two others reveal ^{7}Li/^{6}Li = 3.6 and ^{7}Li/^{6}Li = 1.7, with uncertainties of about 20%. These are definitely less than the solar value, and seem to show that the cosmic-ray production of Li, which favors the ^{6}Li isotope, has been quite active in the 4.6 billion years since the Sun was born – even, perhaps, since the observed cloud was born. The observations also attest to an inhomogeneous interstellar medium, within which one cloud differs significantly from another. The smallest observed ratio (1.7) is essentially equal to the production ratio by cosmic rays, requiring that within that cloud the cosmic-ray-produced Li greatly exceeds any ^{7}Li-rich gas that may have initially been present. This is a definite puzzle, however, because that same interstellar cloud does not seem to contain a larger than normal element ratio Li/H, as it would if enough cosmic-ray Li had been produced to lower the ^{7}Li/^{6}Li ratio substantially. Full and important consequences of these observations remain to be drawn; but they are so important that they could force revaluation of Big-Bang concepts that require the Galaxy to have begun with abundant ^{7}Li-rich gas.

Anomalous isotopic abundance

This isotope of Li does not always occur in all natural samples in its usual proportion. It is a good candidate for isotopic variations among distinct meteoritic materials; but this has not been easy to find owing to Li's rarity.

Cosmic rays   The ^{6}Li/^{7}Li ratio near 0.6 observed in the cosmic rays is much larger than in the solar system because in cosmic rays both isotopes are spallation fragments of carbon and oxygen created by nuclear reactions between the cosmic rays and the interstellar atoms with which they collide. (See **Nucleosynthesis origin**, *Cosmic rays* above, and also see **^{7}Li**, below). Thus the Li isotope anomaly in the cosmic rays is understood in terms of the nuclear physics that alters the cosmic-ray composition from what it was originally when the cosmic rays began their high-speed journey. The ^{6}Li/^{7}Li ratio observed in the cosmic rays does not, therefore, represent the isotopic composition of Li in any bulk sample of stellar or planetary matter.

With yearly increases of analytical precision in the measurement of ^{7}Li/^{6}Li isotopic ratios in small meteoritic samples, especially in presolar grains, one hopes and expects to find interesting tale-telling variations.

^{7}Li | $Z = 3, N = 4$
Spin, parity: $J, \pi = 3/2-$;
Mass excess: $\Delta = 14.908$ MeV; $S_n = 7.25$ MeV

^{7}Li is the lightest stable nucleus to have either nuclear spin $J = 3/2$ or negative parity. This results from its 3rd proton, which must have unit orbital angular momentum and therefore negative parity.

Abundance

From the isotopic decomposition of normal lithium one finds that the mass-7 isotope, ^{7}Li, is the more abundant of lithium's two isotopes; 92.5% of terrestrial Li. The cosmic issues surrounding the lithium abundance are among the most complex and fascinating of any element. Using the total abundance of elemental Li = 57.1 per million silicon atoms in solar-system matter, this isotope has

solar abundance of ^{7}Li = 52.8 per million silicon atoms.

Lithium is not an abundant element, only 29th in rank, making ^{7}Li one of the least abundant nuclear species lighter than iron. Only ^{6}Li, ^{9}Be and ^{10,11}B are less abundant among the light nuclei. These are such low-abundance isotopes that their processes of nucleosynthesis must be rare or very inefficient. This abundance value for Li comes from the meteorites, which retain the initial Li, whereas the Sun has, in its 4.6-billion-year history, destroyed almost all the Li within its interior and has reduced the surface Li abundance to a few percent of the meteoritic value. This happens because convection circulates all envelope matter down to the bottom of the Sun's surface convection zone where the temperature is hot enough (about 2.5 million kelvin) to burn up most of the Li by proton-induced thermonuclear reactions. Because ^{6}Li is destroyed more easily than ^{7}Li, one expects that almost all lithium in the Sun is ^{7}Li. Measurements of Li isotopes in the solar wind by the NASA space mission *Genesis* will, in the next decade, confirm or refute this expectation; but past measurements of Li isotopes in the solar wind (below) have not confirmed that picture. There does remain another option; namely, that all of the initial Li in the Sun has been destroyed in its surface convection zone and that the Li seen there today has been created by nuclear interactions with high-energy solar-flare particles. In this case the measured ^{7}Li/^{6}Li ratio is expected to be near 2.

Observations of lunar soil have complicated the argument. Relative to silicon, lithium is 140 times less abundant in the solar photosphere, as recorded by the solar spectrum, than it is in the meteorites. Since the Li/Si element abundance ratio in the meteorites should also have entered the Sun when it formed, this confirms that the Sun is actually destroying its initial lithium supply as it ages. This occurs at the base of the surface convection zone of the Sun. The bottom of that zone lies at a depth that is about 1/4 of the Sun's radius. Here, the high temperatures (a few MK) that solar nuclei experience are hot enough to destroy lithium, especially ^{6}Li, by nuclear interactions with protons (^{6}Li $+ p \rightarrow$ ^{3}He $+$ ^{4}He). Those proton-induced nuclear reactions destroy ^{6}Li much more readily than they do ^{7}Li (by ^{7}Li $+ p \rightarrow$ ^{4}He $+$ ^{4}He), so that to deplete elemental lithium by a factor of 140 in the Sun would require that almost all ^{6}Li has already been destroyed there. But this conflicts with measurements of the solar-wind Li isotope ratio as recorded in the lunar soil, into which the solar wind blows at about 400 km/s. At such speed the Li atoms get buried to a depth of about 0.03 micrometers in the grains on the Moon's surface. In the Moon's soil one finds that ^{7}Li/^{6}Li $= 31 \pm 4$, considerably greater than the meteoritic ratio ^{7}Li/^{6}Li $= 12.14$ but not nearly as great as the huge ratio expected in the solar wind if nuclear reactions are indeed the source of the depletion of the Sun's lithium. This conflict is unresolved, but its discoverers postulate that production of ^{6}Li also occurs continuously in the solar surface owing to the solar energetic protons that bombard that surface. The energetic protons are accelerated in solar flares, and they create both Li isotopes when they strike carbon and oxygen nuclei.

The abundance of Li is quite different in old metal-poor stars than in young ones. This has been interpreted as a datum of great importance for the cosmologic history of matter (see **Astronomical observations** below).

Nucleosynthesis origin

Big Bang Of the very underabundant 3rd, 4th and 5th elements (Li, Be, B), only ^{7}Li is produced in significant quantities by the hot dense epoch of the early universe that we call the Big Bang. Its source is found among the last nuclear reactions that can occur between hydrogen and helium isotopes as the initial matter rapidly cools and thins out. The responsible nuclear reactions are a combination:

$$(a)\ ^4\text{He} + \ ^3\text{H} \rightarrow \ ^7\text{Li}, \qquad (b)\ ^4\text{He} + \ ^3\text{He} \rightarrow \ ^7\text{Be},$$

after which the ^{7}Be, which is radioactive, decays to ^{7}Li. This allows some ^{7}Li to survive the Big Bang. The amount is small, about Li/H $= 10^{-10}$, one in ten billion, but because ^{7}Li is such a rare nucleus, that production suffices to account for that level of Li abundance in the oldest stars (see **Astronomical observations**, Very old stars, below). Its production was but a tiny trickle at the last possible moment in the first minutes of the Big-Bang universe, but it was enough. Should the Big Bang prove to be an oversimplification of cosmic origins, it is noteworthy that the same production of ^{7}Li may occur in "little bangs," smaller events of creation spread through the universe.

Stellar nucleosynthesis The reaction, ^{4}He $+ \ ^3$He $\rightarrow \ ^7$Be $+$ photon, also happens at the center of the Sun and other stars. The ^{7}Be decay that follows is an important factor in the so-called "solar neutrino puzzle" (see **^{7}Be**). But negligible amounts of ^{7}Li can result there because the hot protons surrounding that ^{7}Li nucleus (following the ^{7}Be decay) rapidly destroy it through the reaction ^{7}Li $+$ p $\rightarrow \ ^4$He $+ \ ^4$He. So every production of ^{7}Li at the center of the Sun really is but a transient in the production of two helium atoms. ^{7}Li is a bit player in the main drama by which hydrogen is fusing to helium.

To achieve stellar nucleosynthesis of ^{7}Li requires a more finely timed scenario. Stars seem to provide such scenarios in a type of red-giant star having two pulsing shells – the asymptotic giant branch AGB stars (see **Glossary**). The ^{3}He nuclei needed as seeds for production are abundantly produced in hydrogen burning at lower temperatures, too cool for the ^{4}He $+ \ ^3$He reaction to occur. But in those doubly pulsing shells rapid structure changes and rapid convection of matter from one place to another occur repetitively. First the cool ^{3}He containing matter is compressed, heating it sufficiently that the ^{4}He $+ \ ^3$He reaction does occur; then, before the ^{7}Be can decay (its halflife on Earth is 53 days), it is swept by convection out to cooler environments such that, after it decays to ^{7}Li in cooler surroundings, that ^{7}Li is not quickly destroyed by protons as in the Sun. In the outer layers of such stars the ^{7}Li survives. Models of AGB stars are quite complex, so this facile account rests on uncertain results. The ^{7}Be is synthesized in hot-bottom burning; but once the initial supply of ^{3}He is exhausted by the ^{4}He $+$ ^{3}He $\rightarrow \ ^7$Be $+$ photon reaction, further hot-bottom burning depletes the Li that was established earlier. Thus it is necessary that the wind from AGB stars reduces their masses rapidly enough to carry away ^{7}Li-rich surface gas before ^{7}Li is destroyed by continuing hot-bottom burning. Similar fine tuning is required for ^{7}Li production in nova explosions, which occur in thin thermonuclear shells atop the white dwarfs after the AGB phase has concluded. It is this production by AGB stars and by novae and supernovae (see below) that accounts for most of the increase of the ^{7}Li abundance with time in the Galaxy. That increase is seen in the surface compositions of more recently born stars, reflecting the amount of ^{7}Li contained in the gas from which they formed. (see **Astronomical observations**, below).

When massive stars explode as supernovae of Type II, their cores emit staggering numbers of neutrinos. A total energy in neutrinos that is a thousand times greater than all of the past energy that has ever been emitted by the Sun during its 4.6 billion years of shining is emitted from those supernova cores in ten seconds! This theoretical expectation associated with the formation of a neutron star was confirmed by the detection of neutrinos from supernova 1987A when it exploded in February 1987. Although the neutrinos interact only very weakly with the overlying nuclei in the supernova, enough do interact to account for significant ^{7}Li production. Some neutrinos break a single nucleon from the ^{12}C nucleus and create ^{11}B. But those responsible for ^{7}Li interact in the helium shell of the supernova. Here the neutrinos break a single nucleon from the ^{4}He nucleus and create a ^{3}He nucleus. It then is captured by the ^{4}He to produce the ^{7}Be nucleus. The amount produced per supernova seems likely to account for about half of the ^{7}Li in the solar abundances. Thus supernovae may be a major source of ^{7}Li. Their role relative to that of AGB stars is yet to be fixed precisely.

Cosmic rays   Each of the isotopes of the very underabundant 3rd, 4th and 5th elements (Li, Be, B) is very much enriched in the cosmic rays. This is observed and understood. It occurs because the cosmic-ray particles strike interstellar atoms, and, in the collisions involving carbon and oxygen atoms, fragments of the nuclei are broken off. These are called spallation nuclear reactions. Li, Be and B are prominent fragments of C and O atoms. Such collisions are not frequent enough to produce abundances of the common elements, but so rare are Li, Be and B that, except for ^{7}Li, almost their entirety is produced by such reactions. These isotopes are observed in the cosmic rays arriving at Earth with large abundances, say 5 to 20% of the abundance of carbon, which is roughly a thousand times greater than their abundances in meteorites and in stars. The explanation is that the cosmic rays in their travels through the interstellar gas have encountered an accumulated thickness of 5 to 10 grams of matter per cm^2. This is enough to "fill in" the very underabundant Li, Be, and B to the several percent level of carbon. But because cosmic rays constitute a negligible fraction of the total matter in the galaxy, this large abundance in the cosmic rays does not weigh heavily into the total. That total is largely ancient cosmic rays that slowed down long ago and were diluted by mixing into the interstellar gas.

Only for the ^{7}Li isotope is cosmic-ray production not the source of most of the light isotopes of Li, Be and B. The cosmic rays produce ^{7}Li/^{6}Li in a ratio near 2, with exact details depending on the energy spectrum assumed for the unseen low-energy cosmic rays; but in the solar system that ratio is ^{7}Li/^{6}Li $= 12.5$. The cosmic rays are the sole source of the rare lighter isotope, ^{6}Li, which means they are the source of only about 1/6 of the ^{7}Li. The remainder is a combination of Big-Bang relic plus production of ^{7}Li in stars.

Astronomical observations

Lithium is detectable in many stars by virtue of its doublet of atomic lines having wavelengths near 6707.8 Å, and is a diagnostic of the chemical enrichment of the Galaxy during its ageing. The spectroscopy is usually not able to distinguish the two isotopes of Li in the weak and "noisy" signal from distant stars. The same doublet from ^{6}Li has wavelengths that are only 0.160 Å greater, so distinguishing between different relative amounts of Li isotopes is a very difficult task. But because ^{7}Li is the more abundant isotope and because ^{6}Li is more easily destroyed, most of the galactic variation of the Li elemental abundance is surely due to ^{7}Li, not ^{6}Li. This distinction is very important for belief in the Big Bang, for observation of abundant ^{6}Li in metal-free very old stars would rule out Big-Bang cosmology as we know it.

Very old stars The abundance of lithium seems to be a fairly constant value in very old stars that formed before stellar nucleosynthesis had occurred. The nucleosynthesis histories of the matter within stars is often summarized in the variable ratio of Fe to H in the stars. Larger Fe/H ratios reveal a history of more past stellar nucleosynthesis, which created the iron, in stars that evolved their lives before the observed star had formed. Because Fe/H increases in interstellar gas with time owing to stellar nucleosynthesis, it also increases with time in the composition of the stars formed from that interstellar gas. The astronomically observed result is that for stars having Fe/H between 0.001 and 0.1 of the present Fe/H ratio, the Li/H ratio is essentially constant. This observation suggests, and theory confirms, that the Li at those low initial levels was not a product of stellar nucleosynthesis in those stars. It is only still surviving today in their atmospheres. It was inherited from the beginning of the universe, from the Big Bang. This enables one to conclude that the universe was left with

$$\text{Li/H} = 2 \times 10^{-10} \text{ from the Big Bang,}$$

or 0.2 Li atoms per billion H atoms. This tells us something important about the Big Bang. However, the Big Bang can produce only the ^{7}Li isotope. Recently the ^{6}Li isotope has been observed for the first time in an old metal-poor star, so it does contain some ^{6}Li. The observed ratio is ^{7}Li/^{6}Li $= 20$, with signifcant room for error. This shows that some cosmic-ray production (see above) has already occurred before this star was born to have produced the ^{6}Li in its gas, so that not all of its ^{7}Li was produced in the Big Bang. But the Big-Bang abundance must be 90% of the observed ^{7}Li abundance in this star since cosmic rays produce only ^{7}Li/^{6}Li $= 2$. So Big-Bang expectations still look good.

It is of interest that Li is easily destroyed by (p, alpha) reactions at $T = 2.5$ MK or greater, so that when newly born stars arrive on the Main Sequence, they have already destroyed Li throughout most of their interiors by these nuclear interactions in the deep convection of the pre-Main-Sequence evolution. The observed Li has survived only in the outermost few percent by mass of the interior of the stars.

More recent stars   As the interstellar Fe/H ratio increased in our Galaxy, from 0.1 times its present value, up to the present value, the ratio of Li/H also increased, up to a value

$$Li/H = 2 \times 10^{-9} \text{ in the young stars,}$$

or 2 Li atoms per billion H atoms. This is tenfold more Li than in the oldest stars. Since this amount increases as the Fe abundance increases, and since the Fe increase occurs because of nucleosynthesis in stars, one usually concludes that Li is also synthesized in stars and eventually exceeds (in the interstellar gas) the amount that the Big Bang had left there from the beginning. It is virtually certain that all of this abundance data pertains to ^7Li, the more abundant of its two isotopes, because theory tells that only ^7Li is created in the Big Bang, and, moreover, that only ^7Li is created by nucleosynthesis in stars. This agrees with the theory of nucleosynthesis in stars, which provides a natural route to synthesize ^7Li during nuclear burning in AGB stars.

Another interpretation is possible. Because observed Li has survived only in the outermost few percent by mass of the interior of the stars, the higher Li abundance $Li/H = 2 \times 10^{-9}$ seen in younger stars could be interpreted as measuring the true Big-Bang lithium if one argues that old metal-poor stars had, owing to their great age, enough time to destroy 90% of their surface Li by deep circulations. This so-called "High-Li" option for Big-Bang nucleosynthesis can imply either high or low ratios of the total mass (normal matter + dark matter) that is required by the general theory of relativity to close the universe. The interpretation then depends on scenarios for the type and amount of dark matter in the universe; the low-mass option would require nonbaryonic dark matter in halos and clusters of galaxies, whereas the high-mass option requires little of either. This is one of many connections between the ultimate particle-theory of matter and the interpretations of cosmology, and is currently a frontier issue.

The lighter isotope, ^6Li, is not created by either event. Furthermore, it is more easily destroyed in stars than is ^7Li. This leaves the need for the cosmic-ray source of the small abundance of ^6Li (see **^{6}Li** and the next item).

Interstellar Li isotopes   Atomic abundances are also measured in interstellar clouds along the line of sight to stars. Their absorption produces a diminution of the light at the wavelength of the isotopes in question, and its magnitude reflects the interstellar abundance. Only recently has spectrograph resolution become accurate enough to distinguish the relative numbers of ^{7}Li and ^{6}Li along these lines of sight. Interestingly, the observed ratios are less than the solar system's value, ^{7}Li/^{6}Li = 12.5. One cloud does possess approximately that solar ratio, but two others reveal ^{7}Li/^{6}Li = 3.6 and ^{7}Li/^{6}Li = 1.7, with uncertainties of about 20%. These are definitely less than the solar value, and seem to show that the cosmic-ray production of Li, which favors the ^{6}Li isotope, has been quite active in the 4.6 billion years since the Sun was born – even, perhaps, since the observed cloud was born. The observations also attest to an inhomogeneous interstellar medium, within which one cloud differs significantly from another. The smallest observed ratio (1.7) is essentially equal to the production ratio by cosmic rays, requiring that within that cloud the cosmic-ray-produced Li greatly exceeds any ^{7}Li-rich gas that may have initially been present. This is a definite puzzle, however, because that same interstellar cloud does not seem to contain a larger than normal Li/H element ratio, as it would if enough cosmic-ray Li had been produced to lower a ratio initially near 10. Full and important consequences of these observations remain to be drawn; they are so important that they could force revaluation of the Big-Bang concepts that require the Galaxy to have begun with abudant ^{7}Li-rich gas.

F star gap   The abundance of Li in recent stars has created a very significant puzzle for the understanding of stellar structure. Although Li/H = 2 × 10^{-9} in most recently formed stars, it is about 100 times less in recent spectral type F stars. These Population I dwarf stars are about 15 to 50% more massive than the Sun and are accordingly somewhat hotter on their surfaces, T = 6000 to 7000 K. Because young stars that are either slightly cooler and slightly hotter than this F band possess the usual Li/H = 2 × 10^{-9}, it seems inescapable that the dwarf F stars have destroyed the Li within their atmospheres. Because those atmospheres are not suitable for nuclear reactions, it seems that surface material in these stars and only in these stars is cycled by mixing processes down to higher temperatures (above 2.5 million degrees) where the Li is destroyed by nuclear interactions with protons (ionized H atoms). The trouble is that the computations of stellar structure do not predict such mixing. So this has come to be thought of as an "anomalous mixing" that operates only in these stars. A very strong clue is that a similar destruction is observed for the element Be in these F stars. A factor 100 depletion in Li is accompanied by about a factor 10 depletion in Be. This argues for a slow mixing process, perhaps induced by the rotations of these stars, that drags surface material down to temperatures near 2 million K before mixing it back again to the surface. Li is easily destroyed by (p, alpha) reactions at T = few million K, so that when newly born stars arrive on the Main Sequence, they have already destroyed Li throughout most of their interiors by these nuclear interactions. The observed Li has survived only in the outermost few percent by mass of the interior of the stars, so that mixing with deeper matter, if it occurs, encounters there no Li to replenish the surface Li.

Anomalous isotopic abundance

This isotope of Li does not always occur in all natural samples in its usual proportion.

Cosmic rays   The ^{6}Li/^{7}Li ratio observed in the cosmic rays is much larger than in the solar system because both isotopes in cosmic rays are spallation fragments of carbon and oxygen created by nuclear reactions between the cosmic rays and the interstellar atoms with which they collide. (See **Nucleosynthesis origin**, *Cosmic rays* above, and **^{6}Li** above.) Thus that Li isotope anomaly is not really *an anomaly* and is understood in terms of the nuclear physics that alters the cosmic-ray composition from that existing when the cosmic rays began their high-speed journey.

Meteorites   With yearly improvements in the analytical precision of the measurement of ^{7}Li/^{6}Li isotopic ratios in small meteoritic samples, especially in presolar grains, one hopes and expects to find interesting tale-telling variations. Excess ^{7}Li from production as daughter of radioactive ^{7}Be has been found for the first time in 2002. Calcium–aluminum-rich inclusions (CAIs) in meteorites reveal that ^{7}Be existed in the early solar system with measureable abundance. This great surprise is inferred from the excess daughter ^{7}Li that exists in proportion to the Be abundance in the CAIs. It was produced when the ^{7}Be, which was incorporated into the CAI during its formation in the earliest events in the solar system, later decayed to ^{7}Li (see **^{7}Be**).

4 | Beryllium (Be)

All beryllium nuclei contain four protons and therefore +4 electronic units, so that four electrons orbit the nucleus of the neutral atom. Its electronic configuration is $1s^2 2s^2$. This can be abbreviated as an inner core of inert helium (a noble gas) plus two s-wave electrons in the second radial s state: $(He)2s^2$. This locates Be at the top of Group IIA (Mg, Ca, Sr, Ba) of the periodic table. Beryllium therefore has valence +2.

Beryllium is named for the mineral beryl, which is its major source on Earth. Beryl is a beryllium aluminum silicate ($3BeO.Al_2O_3.6SiO_2$). Emeralds are one type of beryl. Isolation of Be involves separating the beryllium oxide BeO from the beryl by chemical means, then separating Be from O by reduction processes or by electrolysis. Purified beryllium is a hard gray metal. Handling requires special care because it is a deadly poison. The sugary taste of this metal and its compounds is an irony in that regard.

Overwhelmingly the most interesting aspect of beryllium's abundance in nature is that it is not effectively synthesized within stars. This accounts for its rarity; Beryllium has but one stable isotope (mass nine). But three of its radioactive isotopes have played pivotal roles in nuclear astrophysics. Especially important is ^{8}Be, which can be formed by collisions of two helium nuclei; but it almost immediately breaks apart again into two He nuclei. It is unusual that an even-Z even-N light nucleus would not be stable; this is attributable more to the very tight binding of the two alpha particles separately than it is to unfavorable nuclear binding for ^{8}Be. Nonetheless, its fleeting existence in the dense He gas at the centers of evolved stars amounts to a measurable instantaneous concentration of ^{8}Be in that gas. Capture of a third alpha particle by it is nature's subtle path to the production of ^{12}C in stars (see ^{12}C). Without this key step, which was worked out by Edwin Salpeter, all of nucleosynthesis and the structure of red-giant stars would remain mysteries. Salpeter was awarded the 1997 Craafoord Prize in astronomy for this insight.

Natural isotopes of beryllium and their solar abundances

A	Solar percent	Solar abundance per 10^6 Si atoms
7	solar neutrino source; new extinct radioactivity in CAIs	
9	100	0.73
10	extinct radioactivity in CAIs; in cosmic rays; in solar wind	

^{7}Be $\quad$ Z = 4, N = 3 $\quad$ $t_{1/2}$ = 53.3 days
$\quad$ **Spin, parity:** J, π = 3/2−;
$\quad$ **Mass excess:** Δ = 15.768 MeV; $\quad$ S_n = 10.68 MeV

Abundance

^{7}Be is not a stable isotope despite its large neutron separation energy, because it can beta decay to the more stable *isobar* ^{7}Li (see under *Mass excess* in **Glossary**). As a result no ^{7}Be exists on Earth or in the meteorites, except by transient production by cosmic rays, and none has been seen (yet) in stars. But it is stable against breaking up into nuclear particles (as opposed to beta decay) and is an observable isotope in nature in two ways.

Solar neutrinos $\quad$ ^{7}Be emits neutrinos that come to Earth from the center of the Sun, recording the abundance of ^{7}Be at the solar center. The ^{7}Be is created when ^{3}He nuclei are captured by abundant ^{4}He nuclei at the solar center, whereas the ^{3}He is made frequently from ^{2}H by a subsquent proton capture. When ^{7}Be decays to ^{7}Li by capturing an electron, it emits an almost massless and chargeless neutrino having energy 0.861 MeV. This specific energy allows ^{7}Be to be identified as its source. The calculated flux of these neutrinos at Earth is an astonishing 4.7 billion per cm^2 per second – so many passing each second through your fingernail! Neutrino astronomers express this in terms of the number of captures per second expected in a huge mass of detector element, and call that expected capture rate "solar neutrino units" (written SNU). Of the expected 7.9 SNU for the chlorine detector, ^{7}Be was expected on the basis of computer models of the Sun to account for 1.1 of them, ranking ^{7}Be second in importance to ^{8}B, which produces 6.1 SNU in calculated solar models. They are linked even more strongly, however, because every ^{8}B production at the solar center passes through ^{7}Be via its reaction with hot protons: ^{7}Be + p → ^{8}B + photon. Because the total observed neutrino capture rate in the chlorine experiment has amounted to only about 1/3 of what has long been required by theory, ^{7}Be and ^{8}B are both essential parts of what has come to be called the "solar neutrino problem." The pioneering search for neutrino captures from the Sun is the chlorine absorber in the Homestake Gold Mine in Lead, South Dakota. All neutrino astronomy detectors are placed deep underground to avoid confusion with the cosmic rays that strike the surface of the Earth and would also make the signal searched for by neutrinos. But the cosmic rays cannot penetrate thick layers of soil. The reality of the solar neutrino problem was confirmed by the gallium solar neutrino detector in the Gran Sasso tunnel east of Rome. So the deficiency in observed neutrinos is a result of both ^{7}Be and ^{8}B, in some unknown ratio. But when the data from all solar neutrino experiments now existing are combined, it leaves very little room for any ^{7}Be neutrinos at all. Some now describe this as the "^{7}Be solar neutrino problem." In the attempt to clarify this, a new experiment sensitive only to ^{7}Be neutrinos is being planned, also for the Gran Sasso underground laboratory.

Whatever the outcome, great repercussions on science result from the "solar neutrino problem." The implications may be for our Sun, or for the nature of neutrinos, or both. Recent evidence is convincing that the problem is in the nature of the neutrino itself; namely, that during transit it spontaneously transforms (oscillates) to a neutrino of different flavor, and to which the detector is insensitive. This is possible because of the small rest mass of neutrinos.

Gamma-ray astronomy Another observable situation for ^{7}Be abundance occurs in a nova explosion. As heated surface matter on the white dwarf is explosively ejected, the ^{7}Be created in the explosion decays with its 53-day halflife. After each decay, its daughter (^{7}Li) emits a characteristic gamma ray having energy 0.478 MeV. Gamma-ray astronomy hopes to catch a nova in this act, for to do so would confirm much of the understanding of what a nova explosion is. By measuring the arriving gamma rays as having the energy 0.478 MeV, they will know that it is ^{7}Be that they are detecting – ^{7}Be that has just been created in the explosion. The nova explosion itself is visible to the eye, and optical astronomy routinely record about 40 per year in the Milky Way galaxy. One of the instruments on NASA's *Compton Gamma Ray Observatory* had attempted to do this since its launch in April 1991 until its atmospheric reentry in June 2000 caused it to break into burning pieces that fell into the Pacific Ocean, but no nova occurred close enough to the solar position in the Galaxy to enable successful detection. For this gamma-ray line to be sufficiently intense for detection, the companion star must donate matter enriched in ^{3}He, which is likely, because the ^{7}Be is created when a ^{3}He nucleus is captured by a ^{4}He nucleus in the thermonuclear explosion on the white dwarf.

Extinct radioactivity Calcium–aluminum-rich inclusions (CAIs) in meteorites reveal that ^{7}Be was alive in the early solar system with measureable abundance. This great surprise is inferred from the excess daughter ^{7}Li that exists in proportion to the Be abundance in the CAI. It was produced when the ^{7}Be, which was incorporated into the CAI during its formation in the earliest events in the solar system, later decayed there. This new discovery in France in 2002 may prove to be one of the clearest signs of intense solar-flare cosmic rays at the time when the CAIs were being produced. ^{7}Be does not live long enough (53-day halflife) to have been injected by a neighbor supernova, so supernova nucleosynthesis can not be the source of this extinct radioactivity. If remeasurements confirm this discovery, it proves that at least one extinct radioactive isotope has been produced by high-energy particles in the solar disk. The number of ^{7}Be nuclei initially in the CAI exceeds by a factor of 200 or so the number of the other extinct Be isotope, ^{10}Be, in the same inclusion. This is plausible in that estimates of their production ratio in cosmic-ray collisions are near 100. In fact, ^{7}Be is about 1/5 of the abundance of stable but rare ^{9}Be.

Nucleosynthesis origin

Stellar nucleosynthesis   The reaction, ^{4}He $+ \, ^3$He $\rightarrow \, ^7$Be $+$ photon, happens at the center of the Sun and other stars. The ^{7}Be decay that follows is an important factor in the so-called "solar neutrino puzzle." Although ^{7}Be is rather rare at the solar center, decaying to ^{7}Li with its 8-week halflife, there is rather a huge mass of ^{7}Be at any given time at the solar center because ^{7}Be is stable against breakup into smaller nuclear particles. That mass of ^{7}Be can be reliably calculated. But evidence of neutrinos from the Sun shows that the neutrino flux expected from that mass of ^{7}Be does not entirely arrive at the Earth.

In the explosion of a nova, the same ^{4}He $+ \, ^3$He $\rightarrow \, ^7$Be $+$ photon reaction is also responsible for the production of ^{7}Be there. In this explosive situation on the skin of a white-dwarf star the temperature is much higher than at the solar center. The ^{7}Be would not be able to survive except for the rapid cooling of the matter in a nova just after the thermonuclear runaway. That cooling enables ^{7}Be to survive for its full expectation based on its halflife. This might conceivably allow ^{7}Be to condense in dust grains grown in the nova ejecta, resulting today in isotopically pure ^{7}Li in presolar grains, where the ^{7}Be would have decayed.

The nuclear reaction ^{4}He $+ \, ^3$He $\rightarrow \, ^7$Be also occurs at the hot base of the hydrogen envelope of red-giant AGB stars. If convection can rapidly mix that to cooler temperatures before the ^{7}Be decays to ^{7}Li, the AGB envelope can become enriched in ^{7}Li. Such events may be responsible for much of the observed Li abundance in stars (see ^{7}**Li**).

---

^{9}Be | $Z = 4$, $N = 5$
Spin, parity: J, $\pi = 3/2-$;
Mass excess: $\Delta = 11.348$ MeV; $S_n = 1.67$ MeV

The neutron separation energy $S_n = 1.67$ MeV of ^{9}Be is the smallest of any stable nucleus, one of many reasons why ^{9}Be is so rare.

Abundance

^{9}Be is the only stable isotope of beryllium, with

> solar abundance of ^{9}Be $= 0.73$ per million silicon atoms

in solar-system matter. This makes it the least abundant of all nuclei having mass number less than iron. All elements more rare than Be are heavier than niobium ($Z = 41$). All isotopes of Li, Be and B are such low-abundance isotopes that their processes of nucleosynthesis must be rare or very inefficient.

Nucleosynthesis origin

Big Bang ^{9}Be is not produced in the Big Bang. Of the very underabundant 3rd, 4th and 5th elements (Li, Be, B), only a single isotope is produced in significant amounts by that hot dense epoch of the early universe. That isotope is ^{7}Li (see ^{7}**Li**).

Stellar nucleosynthesis   No ^{9}Be is produced in stellar interiors, so its source must be found elsewhere. Its low neutron separation energy and its low nuclear charge mean that it will be easily destroyed in stars.

Cosmic rays   Each of the isotopes of the very underabundant 3rd, 4th and 5th elements (Li, Be, B) is very much enriched in the cosmic rays. This is observed and understood. It occurs because the cosmic-ray particles strike interstellar atoms, and in the collisions involving carbon and oxygen atoms, fragments of the nuclei are broken off. These are called spallation nuclear reactions. Li, Be and B are prominent fragments of C and O atoms. Such collisions are not frequent enough to produce abundances of the common elements, but so rare are Li, Be and B that almost their entirety is produced by such reactions. By contrast they are observed in the cosmic rays arriving at Earth with very large abundances, say 5 to 20% of the abundance of carbon. The explanation is that the cosmic rays in their travels through the interstellar gas have encountered an accumulated thickness of 5 to 10 grams of matter per cm^2. This is enough to "fill in" the very underabundant Li, Be, and B to the several percent level of carbon.

   The cosmic-ray energy spectrum actually measured on Earth does not produce Be in the correct ratio to match the solar-system value ^{11}B/ Be $= 23$. They instead produce the ratio ^{11}B/ Be $= 12$. This was a problem for a very long time, but it has probably been solved in recent years in the following way. There exists in the interstellar medium a component of low-energy cosmic rays that cannot penetrate the solar-wind plasma flowing away from the Sun. The low-energy cosmic rays produce ^{11}B more efficiently than ^{9}Be. When both components of the cosmic rays are considered, the observed ratio ^{11}B/^{9}Be $= 23$ results. This is a satisfying explanation only because it seems plausible; but it needs much more study to be sure that it is really true (see also **^{11}B**).

Astronomical observations

The abundance of beryllium in the very early Galaxy seems from observations in very old stars to have increased in proportion to the increase in the iron abundance. This would necessarily be the stable ^{9}Be isotope. The very old stars formed before most of the stellar nucleosynthesis had occurred. The varying nucleosynthesis histories of the matter in differing stars are often summarized in the variable ratio of Fe to H in the stars. Larger Fe/H ratios reveal a prestar history of more stellar nucleosynthesis, which creates iron, from the stars that had evolved their lives before the observed star had formed. Because Fe/H increases in interstellar gas with time owing to stellar nucleosynthesis, it also increases with time in the composition of the stars formed from that interstellar gas. The astronomically observed result is that for stars forming from gas having Fe/H between 0.001 and 0.1 of the present Fe/H ratio, the ratio Be/H increases in proportion. Observations that Be and Fe increase in step can be expressed as maintaining the element ratio

$$Be/Fe = 1.4 \times 10^{-6},$$

or 1.4 Be atoms per million Fe atoms, as iron is synthesized in supernovae. This is remarkably close to the best ratio estimated in our solar system, 0.81 Be atoms per million Fe atoms. Nature creates 1.4 ^{9}Be atoms every time it creates a million Fe atoms. An incorrect inference would be that ^{9}Be is a product of stellar nucleosynthesis; for there exists a big problem in that explanation. The theory of stellar nucleosynthesis in stars creates no ^{9}Be at all. Nor does the Big Bang produce Be. Be nuclei are created by collisions of interstellar atoms with cosmic rays; and supernovae (which make the iron) also accelerate the cosmic rays. This explains their proportionate increase, though many interesting problems remain (see **Nucleosynthesis origin**).

The abundance of Be in more recent dwarf stars of spectral type F has created a very significant puzzle for the understanding of stellar structure. Although Be/Fe = 1.4×10^{-6} in old stars, that ratio is about 10 times smaller in the spectral type F stars of recent birth. These Population I dwarf stars are about 15 to 50% more massive than the Sun and are accordingly somewhat hotter on their surfaces, T = 6000 to 7000 K. Because young stars that are either slightly cooler and slightly hotter than this F band possess the usual Be/Fe = 1.4×10^{-6}, it seems inescapable that the dwarf F stars have destroyed the Be within their atmospheres. Because those atmospheres are not suitable for nuclear reactions, it seems that surface material in these stars and only in these stars is cycled by mixing processes down to higher temperatures where the Be is destroyed by nuclear interactions with protons (ionized H atoms). The trouble is that the computations of stellar structure do not predict such mixing. So this has come to be thought of as an "anomalous mixing" that operates only in these stars. A supporting clue is that a similar destruction is observed for the element Li in these F stars. A factor 100 depletion in Li accompanies the factor 10 depletion in Be. This argues for a slow mixing process, perhaps induced by the rotations of these stars, that drags surface material down to temperatures near 2 million K before mixing it back again to the surface. Be is easily detroyed by (p, alpha) reactions at temperatures of a few million K, so that when newly born stars arrive on the Main Sequence, they have already destroyed Be throughout most of their interiors by these nuclear interactions. The observed Be has survived only in the outermost few percent by mass of the interior of the stars.

Be is detectable in many stars by virtue of its doublet of atomic lines having wavelengths near 3131 Å, and is a diagnostic of the chemical enrichment of the Galaxy as time goes by. Its having negligible abundance in very old stars with negligible Fe is very important to belief in the Big Bang. Observation of more abundant Be in such very old iron-free stars would suggest that it was a Big-Bang product, which does not occur in the standard model of the Big Bang.

Anomalous isotopic abundance

Because beryllium has only a single stable isotope, there normally exist no isotopic ratios to measure, so that one can not speak of isotopic anomalies for Be. There is one exception however – within the cosmic rays. There is poetry in this, for it is in the cosmic rays that all Be is created. The cosmic rays arriving at the solar system have been recorded by NASA space missions. These reveal an unfolding drama over the abundance ratio ^{10}Be/ ^{9}Be within the cosmic-ray particles. Its value in the cosmic rays is 20%, with an uncertainty of about 0.4%. Although ^{10}Be is radioactive and does not exist on Earth, it does live for more than a million years.

^{10}**Be** | $Z = 4$, $N = 6$ $T_{1/2} = 1.51$ million years
Spin, parity: $J, \pi = 0+$;
Mass excess: $\Delta = 12.607$ MeV; $S_n = 6.81$ MeV

Abundance

^{10}Be is not a stable isotope. As a result none exists on Earth and none has been seen (yet) in stars. But it is an observable isotope in the cosmic rays. In just the same way that the stable isotopes of Li, Be and B are vastly overabundant in the cosmic rays because of their interactions with interstellar atoms, so too does ^{10}Be exist among them. Nor is it rare. Observations of isotopes in the cosmic rays show that the ratio ^{10}Be/ ^{9}Be is about 0.1 to 0.2. This enables many conclusions about the history of the cosmic rays. The production ratio ^{10}Be/ ^{9}Be by spallation reactions between cosmic rays and interstellar atoms is near 0.5; so almost 20 to 40% of all of the ^{10}Be nuclei ever created during the interstellar lifetime of the cosmic rays is still alive when those cosmic rays arrive at Earth. The ^{10}Be is a cosmic-ray clock, measuring how old the cosmic rays are. It does not measure how old most cosmic-ray nuclei are, but rather how long they have been cosmic rays. This ^{10}Be clock requires that the cosmic rays were accelerated in the last 10 million years. The cosmic rays are thus a "young" component of our universe. Since the high abundance level of Li, Be and B requires that the cosmic rays have traversed 6 to 9 g/cm^2 along their integrated paths (some more and some less), the average density along their path was only about 0.2 atoms per cm^3. This is actually a very good vacuum, better than any that can be created in labs on Earth; but it is typical for the interstellar medium of the Galaxy.

This radioactive nucleus is also found alive in meteorites and inferred to have been alive in the early solar system. It is also produced by solar-energetic particles in the Sun's flares and carried to Earth in the solar wind (see below).

Nucleosynthesis origin

No ^{10}Be is synthesized by nuclear reactions in stars. Thus the ^{10}Be in the cosmic rays has been produced by the cosmic rays themselves during their collisions with interstellar atoms. The fragments then move along with the cosmic rays as new nuclei of somewhat smaller energies.

Cosmic rays   The one sample that has been amenable to testing for variations in isotopic ratio is the cosmic rays arriving at the solar system and recorded by NASA space missions. These record an unfolding drama over the abundance ratio ^{10}Be/ ^{9}Be within their particles. Its value in the cosmic rays is large, about 20%, with an uncertainty of about 0.4 %. Although ^{10}Be is radioactive and does not exist on Earth, it does live more than a million years. Its halflife is 1.51 Myr. Such a large ratio means that the stable ^{9}Be in the cosmic rays was also created by collisions over the past 10 Myr or so.

Other applications

The existence of ^{10}Be in cosmic rays allows some other applications of natural interest. It exists in the cosmic rays owing to collisions between them and interstellar atoms. The same thing is true for radioactive ^{26}Al. The ^{26}Al/^{10}Be ratio in the cosmic rays enables one to determine that they have been traveling about 10 Myr in interstellar space before arriving at Earth. High-energy cosmic rays reach the ground and produce measurable levels of these radioactivities on Earth. The ^{26}Al/^{10}Be ratio in buried quartz allows careful measurement of the time of sedimentary burial. The production ratio $p(26)/p(10) = 6$ from cosmic-ray interactions, so differential decay allows dating of sediments, as in fluvial deposits in Mammoth Cave, KY. This works with a 100 000-year accuracy for deposit ages up to 5 Myr.

Extinct radioactivity   Calcium–aluminum-rich inclusions (CAIs) in meteorites reveal that ^{10}Be was alive in the early solar system with measureable abundance, about 0.001 to 0.0001 of stable ^{9}Be. This is inferred from the excess daughter ^{10}B that exists in proportion to the Be abundance in the meteorite sample. It was produced when the ^{10}Be decayed in the meteorite. An important controversy has arisen over the reason for the abundance of this extinct radioactivity; namely, whether it was created by cosmic rays accelerated by a supernova that caused the solar cloud to collapse, forming our solar system, or whether it was created in the planetary-formation disk when high-energy particles from solar flares on the very young Sun struck the forming meteoritic inclusion. But ^{10}Be can in any case not be produced by thermonuclear reactions in stars, so nucleosynthesis of that type is not the source of this extinct radioactivity.

Cosmogenic radioactivity ^{10}Be is created as a collision fragment in meteorites when cosmic rays strike the meteorite during its journey to the Earth. This radioactivity is counted in the lab, alive ^{10}Be today in the meteorite, and gives information on its history in space just before its fall. Most meteorites can be concluded to have been exposed to cosmic rays for only about a million years during travel to Earth. Prior to that time they were shielded by deep burial within an asteroid, from which they were liberated by an asteroidal collision about a million years ago.

Solar wind   The fast-moving ionized gas from the Sun is called the "solar wind." It is known to contain a tiny but measureable concentration of ^{10}Be because it is found on the dust brought back to Earth from the Moon by *Apollo 17* astronauts. This ^{10}Be is known to be deposited from the solar wind rather than being there as the result of cosmic-ray interactions because the ^{10}Be coats the surfaces of the lunar dust grains. From the observed amount one concludes that about 0.03 ^{10}Be nuclei strike each square meter every second on the lunar surface. The speed of the solar wind, about 400 km/s, implants those atoms in the grain surfaces to a depth of only about 1 micrometer, whereas the ^{10}Be made by cosmic rays is more energetic so that it is spread through their grain volumes. Incidentally, radioactive ^{14}C, which is used in artifact dating in paleontology, is also found coating the same grains and is also produced in the solar-surface gas by energetic particles accelerated by solar flares and which reimpact the surface. The accelerated particles are trapped on closed magnetic loops that cause them to reimpact the solar atmosphere, where they collide with C and O atoms, spalling off fragments and leaving ^{10}Be nuclei. The steady-state abundance expected is the abundance that can be created during one mean lifetime of ^{10}Be (about 40% longer than the halflife). It is unsurprising that solar gases contain young radioactive nuclei, because study by NASA's *Compton Gamma Ray Observatory* of gamma-rays lines emitted from the solar surface show that the surface is heavily bombarded by solar cosmic rays during solar flares.

5 | Boron (B)

All boron nuclei have charge $+5$ electronic units, so that five electrons orbit the nucleus of the neutral atom. Its electronic configuration is $1s^2 2s^2 2p$. This can be abbreviated as an inner core of inert helium (a noble gas) plus three additional electrons: (He) $2s^2 2p$, which locates it at the top of Group IIIA of the periodic table. Boron therefore has valence $+3$ as do other IIIA elements (Al, Ga, In and Tl).

Boron draws its name from the mineral *borax* ($Na_2B_4O_7$), which is its major source on Earth. It is found more concentrated than on average in the salts of the ocean and in dry beds of ancient seas. Deposits of borax, the hydrated sodium salt of boron, are found mainly in the middle east. Boric acid is found in high concentrations in volcanic vapors in Italy. Today's supply is dominated by another hydrated sodium borate from the Mojave Desert. The elemental form of B can be isolated by heating borax with carbon. It is a semimetal, resembling the metal aluminum in many ways, but also resembling the nonmetal carbon. It forms as a shiny black crystal, and is almost as hard as diamond (carbon). It does not melt until 2079 °C. Boron has outstanding lightness, stiffness, resistance to stretching and hardness. Crystalline boron is almost chemically inert, similar to crystalline carbon. Boron was isolated with the other rare light elements Li and Be in the decade after 1800.

One of its two stable isotopes, ^{10}B, is such a good absorber of neutrons that it is used in control rods in nuclear reactors. This property also makes it useful for construction of neutron detectors. Boron is used to make windows that are transparent to infrared radiation, for high-temperature semiconductors, and for electric generators of a thermoelectric type.

Boron is one of the three light elements (Li, Be, B) that are not effectively synthesized by nuclear reactions in stable stars. Its origin in nature must be sought in other astrophysical processes. These involve cosmic-ray collisions with interstellar atoms and neutrino-burst nucleosynthesis in supernova matter. Transient production at the solar center of one of its radioactive isotopes, however, produces energetic neutrinos that have been the easiest of the solar neutrinos to detect.

Natural isotopes of boron and their solar abundances

A	Solar percent	Solar abundance per 10^6 Si atoms
8	radioactive; source of neutrinos from the Sun	
10	19.9	4.22
11	80.1	17.0

^{8}**B** | Z = 5, N = 3 $t_{1/2}$ = 0.77 seconds
Spin, parity: J, π = 2+;
Mass excess: Δ = 22.920 MeV; S_n = 13.14 MeV

^{8}B is probably the shortest-lived radioactive nucleus to have a measureable abundance of physical significance in the natural world.

Abundance

^{8}B is not a stable isotope. It decays by positron + neutrino emission to ^{8}Be, which then breaks apart into two helium nuclei. As a result no ^{8}B exists on Earth or in the meteorites, and none has been seen (yet) in the light from stars. But it is an important observable isotope by virtue of emitting neutrinos that come to Earth from the center of the Sun, recording the abundance of ^{8}B at the solar center. When the ^{7}Be at the solar center captures a proton instead of capturing an electron, it produces ^{8}B. Its prompt decay at the solar center is accompanied by the emission of a neutrino. The calculated flux of these neutrinos at Earth is an astonishing 5.8 million per cm^2 per second – so many passing unfelt through one's fingernail! Because the ^{8}B neutrinos are more energetic than those from ^{7}Be, they produce more expected captures per second in chlorine than does the numerically larger flux of neutrinos emitted by ^{7}Be in the Sun. Neutrino astronomers express this in terms of the number of captures per second expected in a huge mass of detector element, and call that expected capture rate "so-lar neutrino units" (written SNU). Of the expected 7.9 SNU for the chlorine detector, ^{8}B accounts for 6.1 of them, ranking ^{8}B first in importance in calculated solar models (for the chlorine detector). These two radioactive nuclei are linked even more strongly, however, because every ^{8}B production at the solar center arises from ^{7}Be via its capture reaction with hot protons: ^{7}Be + p $\rightarrow$ ^{8}B + photon. Because the total observed neutrino capture rate in the chlorine experiment has amounted to only about 1/3 of what has long been required by theory, ^{8}B's expectations seem to bear the lion's share of the guilt; however, ^{7}Be and ^{8}B are both essential parts of what has come to be called the "solar neutrino problem." The pioneering search for neutrino captures from the Sun is the chlorine absorber in the Homestake Gold Mine in Lead, South Dakota. All neutrino astronomy detectors are placed deep underground to avoid confusion with the cosmic rays that strike the surface of the Earth and would also create the signal produced by neutrinos. But the cosmic rays cannot penetrate thick layers of soil, whereas the neutrinos can. The solar neutrino problem discovered by the chlorine absorber was confirmed by the gallium solar neutrino detector in the Gran Sasso tunnel east of Rome. But that experiment also demonstrates that the deficiency in observed neutrinos is a result of both ^{7}Be and ^{8}B, in some unknown ratio. Experiments based on the recoil of electrons when the solar neutrinos scatter from them have pinpointed the ^{8}B problem; namely, the ^{8}B flux is only about half of the number required by calculated models of the Sun. Whatever the eventual resolution of these and of several new experiments now beginning, great repercussions on science result from the "solar neutrino problem." The implications may be for our Sun, or for the nature of neutrinos, or both. Latest observations from the SNO detector in Canada point convincingly to an explanation in terms of the oscillation of neutrinos during their journey from Sun to Earth into other neutrino flavors. This great discovery requires that neutrinos not be totally massless.

Nucleosynthesis origin

Stellar nucleosynthesis   The origin of ^{8}B at the solar center requires the existence of ^{7}Be there. The reaction, ^{4}He + ^{3}He → ^{7}Be + photon, happens at the center of the Sun and other stars and creates the progenitor ^{7}Be. The ^{7}Be decay that follows is one important factor in the so-called "solar neutrino puzzle." Although ^{7}Be is rather rare at the solar center, decaying to ^{7}Li with its human-scale halflife, there is rather a huge mass of ^{7}Be at any given time at the solar center. That mass can be reliably calculated, or so one believes. The point is that about 1 out of 750 ^{7}Be nuclei captures a proton from the hot surrounding gas instead of decaying radioactively. That reaction, ^{7}Be + p → ^{8}B + photon, produces the ^{8}B neutrino emitter at the solar center. It is a weak branch of the so-called "pp chains" of thermonuclear fusion of hydrogen into helium. These chains provide the thermonuclear power for the Sun, which would otherwise have cooled long ago. The earliest attempts at reconciling the solar neutrino problem consisted of trying to rationalize a slightly cooler solar center, so that the ^{8}B branch would be extinguished entirely. But these attempts have not credibly succeeded. Modern results confirm that the neutrino is a more complicated elementary particle than had previously been thought and that some of them alter their observable stripes on the way from Sun to Earth.

^{10}B | $Z = 5$, $N = 5$
Spin, parity: J, $\pi = 3+$;
Mass excess: $\Delta = 12.051$ MeV; $S_n = 8.44$ MeV

^{10}B is the lightest stable nucleus to have nuclear spin $J = 3$. This results from its 5th neutron and 5th proton both having unit orbital angular momentum and coupling their spin angular momenta to give the maximum possible $J = 3$. The reaction ^{10}B + n → ^{7}Li + ^{4}He has such a large cross section that boron is useful in neutron detectors, which easily detect the 2.8-MeV alpha particle.

Abundance

From the isotopic decomposition of normal boron one finds that the mass-10 isotope, ^{10}B, is the lesser abundant of boron's two isotopes; 19.9% of terrestrial B. Using the total abundance of elemental B = 21.2 per million silicon atoms in solar-system matter, this isotope has

solar abundance of ^{10}B = 4.22 per million silicon atoms.

Boron is not an abundant element, ranking 34th overall, so that ^{10}B is certainly one of the least abundant nuclear species lighter than iron. Only ^{9}Be is less abundant than ^{10}B among the light nuclei. These are such low-abundance isotopes that their processes of nucleosynthesis must be rare or very inefficient.

Nucleosynthesis origin

Big Bang Of the very underabundant 3rd, 4th and 5th elements (Li, Be, B), only a single isotope, ^{7}Li, not B, is produced in significant quantities by the hot dense epoch of the early universe that we call the Big Bang. So the origin of ^{10}B must be sought elsewhere.

Stellar nucleosynthesis No ^{10}B is produced in stellar interiors, so its source must be found elsewhere.

Cosmic rays Each of the isotopes of the very underabundant 3rd, 4th and 5th elements (Li, Be, B) is very much enriched in the cosmic rays. This is observed and understood. It occurs because the cosmic-ray particles strike interstellar atoms, and in the collisions involving carbon and oxygen atoms fragments of those nuclei are broken off. These are called spallation nuclear reactions. Li, Be and B are prominent fragments of C and O atoms. Such collisions are not frequent enough to produce abundances of the more abundant elements, but so rare are Li, Be and B that almost their entirety can be produced by such reactions. They are observed in the cosmic rays arriving at Earth with large abundances, say 5 to 20% of the abundance of carbon. The explanation is that the cosmic rays in their travels through the interstellar gas have encountered an accumulated thickness of 5 to 10 grams of matter per cm^2. This is enough to "fill in" the very underabundant Li, Be, and B to the several percent level of carbon.

The cosmic-ray energy spectrum actually measured on Earth does not produce the boron isotopes in the correct ratio to match the solar-system value ^{11}B/^{10}B = 4.05. They instead produce the ratio ^{11}B/^{10}B = 2.5. For the likely resolution of this old problem, see ^{11}B, below.

Astronomical observations

The abundance of boron varies in very old stars in proportion to their iron abundance. The measurements are very difficult because the useful transitions are in the ultraviolet. The very old stars formed before most of the stellar nucleosynthesis had occurred. The varying nucleosynthesis histories of the matter in differing stars are often summarized in the variable ratio of Fe to H in the stars. Larger Fe/H ratios within a star reveal a history of more stellar nucleosynthesis prior to that star's birth. Both the iron content and the boron content had been synthesized previously by stars that lived their lives before the observed star had formed. Accordingly, Fe/H increases in interstellar gas with time owing to stellar nucleosynthesis; and so it also increases with time in the composition of the stars formed from that interstellar gas. The astronomically observed result is that for stars having Fe/H between 0.001 and 0.1 of the present Fe/H, the ratio B/H increases in proportion. This observation suggests that B is a product of stellar activity. The B and Fe increase in step, maintaining the element ratio

$$B/Fe = 2.1 \times 10^{-5},$$

or 21 B atoms per million Fe atoms. This is essentially equal to the ratio measured in our solar system, 23 B atoms per million Fe atoms. The conclusion seems to be that nature creates 23 B atoms every time it creates a million Fe atoms. The conclusion that boron is a product of stellar nucleosynthesis has a big problem, however; nucleosynthesis in stars does not produce enough boron. It creates no ^{10}B at all; and the rate of stellar production of ^{11}B involves a neutrino process that is probably not sufficiently prolific. Nor did the Big Bang produce boron. Boron is produced primarily by cosmic rays; and the correlation with Fe occurs because supernovae (which make iron) also accelerate the cosmic rays that create the B by interstellar collisions.

Isotopes in interstellar gas   With the aid of the *Hubble Space Telescope* it has been possible for the first time to measure the boron isotopic ratio within diffuse clouds of the Milky Way. The interstellar ultraviolet radiation renders B ionized (B^+) in the diffuse clouds; therefore its spectrum is similar to that of the element Be, but at shorter wavelengths. The strongest resonance line lies in the ultraviolet, visible to *Hubble* spectrometers. The smaller mass of the ^{10}B isotope shifts its line by 0.013 Å (about 0.001%) toward longer wavelengths. Very detailed analysis of the line pair has shown in several clouds that today's interstellar abundance ratio is ^{11}B/^{10}B = 3.4 ± 0.7, which is consistent with the solar ratio 4.05. For the first time one can conclude that the solar ratio is not an abnormal one, but is shared by interstellar gas at a value larger than the ratio 2.5 that is produced by cosmic-ray collisions in the interstellar gas. Another source of ^{11}B is needed.

Anomalous isotopic abundance

The isotopes of boron do not always occur in their usual proportions. In particular, certain meteoritic samples reveal both deficiencies and excesses of ^{10}B. Comparisons with differing samples of very primitive meteorites, ones that have revealed independent evidence of carrying unequilibrated presolar materials, have been shown to also have ^{11}B/^{10}B isotopic ratios varying between 3.84 and 4.25, large variations for an isotopic ratio. Samples having the largest chemical ratios of B/Si are those having the largest

¹¹B/¹⁰B ratios. That is, the B abundance in the meteoritic sample correlates with extra ¹¹B. So although it is not possible for elements having but two isotopes to reveal which of the two is anomalous, it is sensible owing to the correlation with B abundance to think of varying admixtures of a boron component that is enriched in ¹¹B. This has been interpreted as a component of boron produced by low-energy cosmic-ray interactions, perhaps even in the presolar cloud or even in the early solar system itself. (See ¹¹**B** for more on this.)

Extinct radioactivity   Calcium–aluminum-rich inclusions (CAIs) in meteorites reveal that ¹⁰Be was alive in the early solar system with measureable abundance, about 0.001 to 0.0001 of stable ⁹Be. This is inferred from the excess daughter ¹⁰B that exists in proportion to the Be abundance in the meteorite sample. That is, the ratio ¹⁰B/¹¹B is measured in CAIs to be greater in the portions having larger Be/B element abundance ratio. This implicates ¹⁰Be as the source of the extra ¹⁰B. It was produced when the ¹⁰Be decayed in the meteorite. An important controversy has arisen over the reason for the abundance of this *extinct radioactivity* (see **Glossary**); namely, whether it was created by cosmic rays accelerated by a supernova that caused the solar cloud to collapse, forming our solar system, or whether it was created in the planetary-formation disk when high-energy particles from solar flares on the very young Sun struck the forming meteoritic inclusion. But ¹⁰Be can in any case not be produced by thermonuclear reactions in stars, so nucleosynthesis in stars is not the source of this extinct radioactivity in the early solar system.

> ¹¹**B** | $Z = 5, N = 6$
> **Spin, parity:** $J, \pi = 3/2-$;
> **Mass excess:** $\Delta = 8.668$ MeV; $S_n = 11.46$ MeV
>
> *¹¹B is a very tightly bound nucleus. Its neutron separation energy exceeds that of any lighter stable nucleus save ⁴He, and its alpha separation energy exceeds that of all lighter stable nuclei. Its proton separation energy, $S_p = 11.23$ MeV, is also large.*

Abundance

From the isotopic decomposition of normal boron one finds that the mass-11 isotope, ¹¹B, is the more abundant of boron's two isotopes; 80.1% of terrestrial B. Using the total abundance of elemental B = 21.2 per million silicon atoms in solar-system matter, this isotope has

solar abundance of ¹¹B = 17.0 per million silicon atoms.

Boron is not an abundant element, ranking 34th overall, so that ¹¹B is certainly one of the least abundant nuclear species lighter than iron. One must go up the chart to element 33, arsenic, to find an element of less abundance than ^{11}B. Only ^{9}Be, ^{6}Li and ^{10}B are less abundant than ^{11}B among the light nuclei. These are such low-abundance isotopes that their processes of nucleosynthesis must be rare or very inefficient.

Nucleosynthesis origin

Big Bang   Of the very underabundant 3rd, 4th and 5th elements (Li, Be, B), only a single isotope, ^{7}Li, not ^{11}B, is produced in significant amounts by the hot dense epoch of the early universe that we call the Big Bang. So the origin of ^{11}B must be sought elsewhere.

Stellar nucleosynthesis   No ^{11}B is produced by thermonuclear reactions in stellar interiors. It is instead destroyed if heated above several million degrees. This means that the only boron that exists in normal stars is that inherited when the star formed from the interstellar gas, and, moreover, that stellar boron is limited to the outer layers of the atmospheres of the stars. So its natural source must be found elsewhere.

When massive stars explode as supernovae of Type II, their cores emit staggering numbers of neutrinos. A total energy in neutrinos that is a thousand times greater than all of the past energy that has ever been emitted by the Sun during its 4.6 billion years of shining is emitted from those supernova cores in ten seconds! This theoretical expectation associated with the formation of a neutron star was confirmed by the detection of neutrinos from supernova 1987A when it exploded in February 1987. Although the neutrinos interact only very weakly with the overlying nuclei in the supernova, enough do interact to account for significant ^{11}B production. Some neutrinos break a single nucleon from the ^{12}C nucleus to create ^{11}B. The amount needed per supernova to account for the ^{11}B abundance is a few ten-millionths of solar mass of ^{11}B; but this agrees pretty well with calculations of the ^{11}B mass created in this way. Thus supernovae may be the major source of ^{11}B.

Cosmic rays   Each of the isotopes of the very underabundant 3rd, 4th and 5th elements (Li, Be, B) is very much enriched in the cosmic rays. This is observed and understood. It occurs because the cosmic-ray particles strike interstellar atoms; and during the collisions involving carbon and oxygen atoms fragments of those nuclei are broken off. These are called spallation nuclear reactions. Such collisions are not frequent enough to produce abundances of most elements, but so rare are Li, Be and B that almost their entirety may be produced by such reactions. They are observed in the cosmic rays arriving at Earth with large abundances, say 5 to 20% of the abundance of carbon. The explanation is that the cosmic rays in their travels through the interstellar gas have encountered an accumulated thickness of 5 to 10 grams of matter per cm^2. This is enough to "fill in" the very underabundant Li, Be, and B to the several percent level of carbon.

The cosmic-ray energy spectrum actually measured on Earth does not produce the boron isotopes in the correct ratio to match the solar system value ^{11}B$/$ ^{10}B $= 4.05$. They instead produce the ratio ^{11}B$/$ ^{10}B $= 2.5$. This was a problem for a very long time, but it has probably been solved in recent years in the following two ways. Firstly, extra ^{11}B is created in supernovae by the neutrino interactions (see above). This could augment the ratio created by cosmic-ray interactions (^{11}B$/$ ^{10}B $= 2.5$) to the observed ratio (4.05). Secondly, a low-energy part of the cosmic rays may augment ^{11}B. There exists in the interstellar medium a component of low-energy cosmic rays that cannot penetrate the solar-wind plasma flowing away from the Sun. Erroneous evidence for that component was reported by NASA's *Compton Gamma Ray Observatory*, which seemed to record gamma rays coming from the Orion interstellar clouds. Analysis showed that these gamma rays required low-energy ^{12}C cosmic rays in Orion, and these also produce ^{11}B but not ^{10}B, which needs high-energy cosmic rays. When both components of the cosmic rays are considered, the observed ratio ^{11}B$/$ ^{10}B $= 4.05$ results. This is a satisfying explanation only because it seems plausible; it needs much more study to be sure that it is really true, especially since years of analysis of growing data from the *Compton Gamma Ray Observatory* revealed that the early evidence of the long-sought low-energy cosmic rays in Orion was spurious. But one still expects that a gamma-ray telescope of higher sensitivity will eventually detect those cosmic rays. And then to be settled will be how much of the extra ^{11}B comes from supernova neutrinos and how much from the low-energy cosmic rays.

Astronomical observations

The abundance of boron seems to increase in value in very old stars in proportion to the increase in the iron abundance. See ^{10}B for this account, which implicates supernova explosions in the production of boron.

Isotopes in interstellar gas   With the aid of the *Hubble Space Telescope* it has been possible for the first time to measure the boron isotopic ratio within diffuse clouds of the Milky Way. The interstellar ultraviolet radiation renders B ionized (B$^+$) in the diffuse clouds; therefore its spectrum is similar to that of the element Be, but at shorter wavelengths. The strongest resonance line lies in the ultraviolet, visible to *Hubble* spectrometers. The smaller mass of the ^{10}B isotope shifts its line by 0.013 Å (about 0.001%) toward longer wavelengths. Very detailed analysis of the line pair has shown in several clouds that today's interstellar abundance ratio is ^{11}B$/$ ^{10}B $= 3.4 \pm 0.7$, which is consistent with the solar ratio 4.05. For the first time one can conclude that the solar ratio is not an abnormal one, but is shared by interstellar gas at a value larger than the ratio 2.5 that is produced by cosmic-ray collisions in the interstellar gas. Another source of ^{11}B is needed.

Anomalous isotopic abundance

Meteorites   This isotope of boron does not always occur in all natural samples in its usual proportion. In particular, certain meteoritic samples reveal both deficiencies and excesses of this isotope. Comparisons with differing samples of very primitive meteorites, ones that have revealed independent evidence of carrying unequilibrated presolar materials, have been shown to also have $^{11}B/^{10}B$ isotopic ratios varying between 3.84 and 4.25. This 10% spread in values far exceeds the 1% spread found in rocks from the mantle of the Earth. Meteoritic samples having the largest chemical ratios of B/Si are also those having the largest $^{11}B/^{10}B$ ratios. That is, the B abundance in the meteorite sample correlates with extra ^{11}B. So although it is not possible for elements having but two isotopes to assert which of the two is anomalous, it is sensible owing to the correlation with B abundance to think of varying admixtures of a boron component that is enriched in ^{11}B. This has been interpreted as a component of boron produced by low-energy cosmic-ray interactions, perhaps even in the presolar cloud or even in the early solar system itself. As described above in cosmic-ray nucleosynthesis of these isotopes, a low-energy component of energetic particles (meaning 10–30 MeV per nucleon) favors production of ^{11}B over ^{10}B. This exciting data has been brought into question, however, by subsequent analysis by another research group who do not confirm in their samples these large variations in the $^{11}B/^{10}B$ ratios. More work is needed to resolve this conflict.

Cosmic rays   The observed cosmic rays are anomalous in boron isotopic ratio in the sense that the cosmic-ray ratio is not the average ratio measured in the solar system. But this is expected, as the cosmic-ray B consists entirely of spallation fragments of carbon and oxygen, and only of the normal high-energy cosmic rays at that. The low-energy component does not penetrate the solar wind's magnetic field. Cosmic rays are measured to be either deficient in ^{11}B or have an excess in ^{10}B when compared with the solar ratio (see above). This is now understood as an excess of ^{10}B in the normal high-energy cosmic rays, which is expected from both B isotopes being spallation fragments in the ratio $^{11}B/^{10}B = 2.5$. Conversely, there exists an excess of ^{11}B in the low-energy cosmic-ray component. This is then anomalous only in the sense of not being equal to the bulk B istope ratio in the solar system.

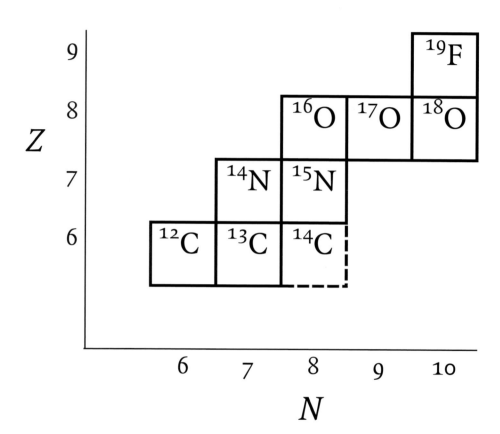

6 | Carbon (C)

All carbon nuclei have charge $+6$ electronic units, so that six electrons orbit the nucleus of the neutral atom. Its electronic configuration is $1s^2 2s^2 2p^2$. This can be abbreviated as an inner core of inert helium (a noble gas) plus four more electrons: $(He)2s^2 2p^2$, which locates it at the top of Group IV of the periodic table. Carbon therefore has valence -4 or $+4$, depending on whether it takes on four more electrons or whether it gives up four in order to complete the electronic shell. Carbon becomes very reactive with oxygen at high temperatures, making either the CO molecule or, if enough O is available, the CO_2 molecule. The CO_2 molecule is an abundant constituent of the Earth's atmosphere (about 0.03% and increasing owing to human activity), as is methane, CH_4. Carbon is not as abundant on Earth as it is in the Sun because C in the presolar gas was largely in volatile forms that were not incorporated into planets. Although carbon is the fourth most abundant element in the Sun and the universe, and is the most abundant of all solid-building elements, its abundance is not even in the top ten of elements found on Earth. The loss of so much carbon when planets were assembled must have been chemically significant.

Carbon is so oxygen hungry that it can take O away from metal oxides, thereby reducing them to free metal. Free iron, for example, is obtained by the reduction of iron oxides in the presence of hot carbon, called *coke*, in blast furnaces. In simplified form the nature of that reduction reaction can be represented $C + 2FeO \rightarrow 2Fe + CO_2$. Such furnaces are important to the world economy in the production of steel from iron.

Life on Earth is based on carbon chemistry. Every living cell, plant or animal, is built from molecules that contain carbon and hydrogen. The famed DNA molecule, which carries the genetic code of life forms, is a double-helix hydrocarbon molecule of grand size and information content. Carbon atoms on Earth cycle repeatedly between carbon-dioxide gas in atmosphere and oceans, and cellulose in plants. Green plants trap the C atoms with the aid of sunlight (photosynthesis); many plants are eaten by animals in which carbon is converted to many hydrocarbon structures; and finally plants and animals die and decay, amounting to combustion and returning the carbon to CO_2.

Because of its ability to make strong bonds with other C atoms, carbon also is able, by itself, to make three different well-known solid forms, called *allotropes*. These are amorphous solid carbon, graphite, and diamonds. These carbon solids differ in their crystalline structure, which is the spatial relationships of their atoms. Graphite has stacked sheets of graphene, which consist of a pattern of hexagons. In addition C is able to construct the very tightly bound large molecule C_{60}, called *Buckminsterfullerene*, in which the sixty C atoms arrange themselves as the sixty vertices of the common soccer ball, an approximate sphere composed of hexagons with five added pentagons, which induce the graphene sheet to curl into a sphere. (Appropriately, one of the codiscoverers of Buckminsterfullerene was Robert Curl.) Other closed surfaces of C atoms create a suite of similar large molecules called fullerenes, of which Buckminsterfullerene is the most stable member. Also important to astronomy is carbon's ability to make long-chain molecules of itself. Each allotrope reflects carbon's extraordinary ability to bond strongly with itself. These nucleate the condensation of solid carbon from hot hydrogen-free and oxygen-rich supernova gases. Carbon's allotropic forms are so constructed that they do not easily transform into each other with rising temperature, as do allotropes of some elements.

Small solid grains of these allotropes exist as *presolar grains* (see **Glossary**) found in meteorites. These presolar grains are older than the solar system and have very unusual ratios of C isotopes. The were formed when stars dying before the Sun was born ejected gas containing newly synthesized nuclei of carbon and other elements. They can be recovered today in the meteorites because these carbon allotropes do not dissolve in acid solutions, whereas the remainder of the meteorite is dissolved. Presolar carbon allotropes have introduced a new frontier to nucleosynthesis and associated insights into the synthesis of the separate isotopes of the elements and of the cosmic history of matter (*Cosmic chemical memory*, see **Glossary**).

Carbon monoxide (CO) is a colorless and odorless gas that is highly poisonous to humans because it reacts with blood hemoglobin, which normally carries oxygen to needy tissues, and thereby renders the hemoglobin incapable of oxygen transport. The CO gas is also of great importance to the astrophysics of the isotopes. By detecting its radio emissions radio astronomers measure the rotational structure of the Galaxy and also the relative numbers of $^{12}C^{16}O$, $^{12}C^{17}O$, $^{12}C^{18}O$, $^{13}C^{16}O$, $^{13}C^{17}O$, and $^{13}C^{18}O$ molecules. Furthermore, CO is such a tightly bound molecule (binding energy 11.1 eV) that cooling stellar gases normally form that molecule prior to any other chemistry among the hot atoms. Since oxygen is normally more abundant than carbon in normal stars, the consequence is that the gas cannot (upon further cooling) condense carbon solid dust grains because all of the carbon is already bound in CO, which is relatively inert at the low densities common in astrophysics. The exception to this rule occurs in supernova explosions, where the radioactivity produced causes radiations that break the CO molecule apart again, thereby restoring free carbon for the condensation of carbon grains in supernovae.

The nuclear properties of carbon furnish one radioactive isotope, ^{14}C, that lives sufficiently long to participate in the carbon cycle of life. With its 5730-yr halflife,

^{14}C would not still exist on Earth except for its steady creation by collisions of cosmic rays with oxygen and nitrogen in the Earth's atmosphere. When living things die, they contain a predictable fraction of this radioactive ^{14}C isotope. Because its numbers decay after death within the fossil, the remaining abundance of ^{14}C gives the age of the fossil. This technique has been very helpful in establishing chronology for the recent (between 500 and 50 000 years) history of life.

The abundance of carbon in each star (relative to hydrogen) is a key indicator of the amount of past nucleosynthesis prior to each star's formation; in some cases it also indicates nucleosynthesis or destruction of carbon within that very star. Ordinarily, oxygen-rich stars can become carbon-rich if carbon synthesized within them is brought convectively to the surface, in which case they are called "carbon stars" (see **Glossary**, AGB stars.) These stars are the sites of condensation of the presolar mainstream SiC grains found in meteorites. The evolution of the carbon isotope ratio owing to nuclear reactions in stars is one of the main diagnostics of the relationship of presolar grains to the stars in which they condensed. Rich new astronomical studies of carbon stars is made possible by them.

The rich chemistry of carbon also makes it a constituent of a large number of interstellar molecules in addition to CO. Radioastronomical studies of these interstellar molecules reveal variations in the abundance ratio of the two stable C isotopes. Such studies also illuminate the field of interstellar chemistry.

The importance of carbon to the chemistry of the universe, including life, is immense. From this simple fact one of the legendary tales of astrophysics occurred, in which the nuclear properties of carbon governing its nucleosynthesis in stars were predicted in advance by Fred Hoyle because, Hoyle argued, were it not so, life would not exist! Although the ^{12}C nucleus has been shown to have the right properties for creating the composition of the universe as we know it, details of its nuclear reaction with helium remain uncertain. The ^{12}C(alpha, gamma)^{16}O nuclear reaction has been called the greatest uncertainty in the fabric of nucleosynthesis because it governs the relative amounts of C and O that survive the burning of helium in stars. That C/O ratio then determines many subsequent details of the evolution of massive stars and the nucleosynthesis that they produce as their matter is redistributed to the interstellar medium.

Carbon also plays a key role in generating thermonuclear power within the stars. A cycle of thermonuclear reactions called the "carbon–nitrogen (CN) cycle" was discovered by Hans Bethe in 1939. By a series of four captures of protons and two beta decays, hydrogen is fused into helium, providing the main energy source for starlight in the universe (see ^{13}C, **Nucleosynthesis origin**).

Carbon has the distinction of being chosen as the basis for the unit of atomic mass. The atomic mass unit (amu) is defined as 1/12th of the mass of the ^{12}C atom.

Natural isotopes of carbon and their solar abundances

A	Solar percent	Solar abundance per 10^6 Si atoms
12	98.9	9.99×10^6
13	1.10	1.10×10^5
14	radioactive in Earth's biosphere; used for fossil dating	

^{12}C | $Z = 6$, $N = 6$
Spin, parity: $J, \pi = 0+$;
Mass excess: $\Delta = 0$ MeV; $S_n = 18.72$ MeV

The atomic mass excess (see **Glossary**) of ^{12}C is set to zero by definition. Doing so defines the atomic mass unit as being $1/12$ of the mass of the ^{12}C atom. But because the mass excesses of its constituent six protons and six neutrons are all greater than zero, we see that ^{12}C is considerably less massive than the sum of its twelve building blocks. The disappearing mass is caused by the negative energy of the binding of those twelve nucleons together. Einstein's $E = mc^2$ explains the amount exactly!

Abundance

Carbon is the fourth most abundant element in the universe. Its abundance in the Sun is about one-half that of oxygen, but reveals differing ratios to oxygen in other stars and in nebulae. The most abundant isotope of carbon, ^{12}C, is the fourth most abundant nucleus in the universe. The two most abundant, ^{1}H and ^{4}He, are remnants of the Big Bang, whereas ^{16}O, the third most abundant, and ^{12}C are created during the evolution of stars. Carbon ranks therefore as one of the great successes of stellar nucleosynthesis. The evolution of stars makes evident why this is so. From the isotopic decomposition of normal carbon one finds that the mass-12 isotope, ^{12}C, is 98.9% of all C isotopes.

Using the total abundance of elemental C $= 10.1$ million per million silicon atoms (i.e. 10.1 times more abundant than Si) in solar-system matter, this isotope has

solar abundance of ^{12}C $= 9.99 \times 10^6$ per million silicon atoms.

This abundance must not be thought of as universal. The interstellar matter today seems to contain 225 ± 50 atoms of C per million H atoms, whereas the solar value given above corresponds to 355 ± 40 atoms per million H. A leading carbon puzzle is how the Sun, which formed 4.56×10^9 years ago, contains more C atoms than the interstellar medium today. In the subsequent 4.56×10^9 years the C abundance should have increased, making today's ISM richer in C than is the Sun. This inconsistency might be explained if the Sun formed from material that was exceptionally self-enriched by recent stellar nucleosynthesis. This would more likely have been a supernova that exploded near the presolar cloud than a nearby AGB star. Both make new carbon; but the supernova makes 10–100 times more new carbon than a single AGB star. If either local source rendered the presolar gas carbon-rich, it may also have made the solar isotopic ratio different than the interstellar average. This puzzle is unresolved, impacting many areas of astrophysics with uncertainties. About half of the interstellar carbon exists in the form of atoms in the gas; the other half is contained within molecules and carbonaceous particles.

Nucleosynthesis origin

The ^{12}C atom enjoys the distinction of being the second most abundant nucleus formed by nucleosynthesis in the stars. This happens because it is the initial natural product of the fusion of helium, or *helium burning* (see **Glossary**) as it is called in astronomy. The basic nuclear fusion reaction is $3\ ^4\mathrm{He} \to\ ^{12}\mathrm{C}$, which creates ^{12}C from helium gas by bringing together three helium nuclei. So important is this reaction for stellar evolution and nucleosynthesis that it is known everywhere by a nickname, "the triple-alpha reaction," the name also recognizing that the nuclei of He atoms are called "alpha particles" for historical reasons. This reaction's significance is reflected too in a great deal of interesting historical lore about the rate of this reaction in the helium gas at the center of a red-giant star. The chance coming together of three alpha particles within a gas of alpha particles is so rare that the first calculation yielded an inadequate rate for this reaction. Then it was recognized that the reaction could happen in two stages: firstly, two colliding alpha particles stick together in an unbound combination (unstable ^{8}Be) for a sufficiently long duration prior to the disintegration of the ^{8}Be into its original two alpha particles such that, secondly, a third alpha particle has a chance to interact with it. This was the key to overcoming the obstacle that ^{8}Be (the combination of two ^{4}He) is not a stable nucleus. The lack of any stable nucleus at mass number 8 was long recognized to be an obstacle to nucleosynthesis. Schematically, the triple-alpha reaction could then be thought of in the following way:

$$^4\mathrm{He} +\ ^4\mathrm{He} \to\ ^8\mathrm{Be}\ (\text{for a very short time}) \to\ ^4\mathrm{He} +\ ^4\mathrm{He};\ \text{and}$$

$$^4\mathrm{He} +\ ^8\mathrm{Be} \to\ ^{12}\mathrm{C}\ (\text{using the tiny fraction of}\ ^8\mathrm{Be}\ \text{nuclei that has not decomposed}).$$

The next obstacle was that the ^{12}C nucleus so created is in an excited-energy configuration initially, one that unfortunately also breaks apart again (in a very short time) into three alpha particles, undoing all that has been accomplished. A leak from this cycle into the stable ^{12}C ground state was found via another quantum fact: namely, in a small fraction of the occurrences (roughly ten times in a million), the excited ^{12}C nucleus formed in this way emits a gamma ray rather than undergoing decomposition into alpha particles. After emitting a gamma ray, the ^{12}C nucleus is stable and can eventually find its way into the chemistry of life. This nuclear sequence, painstakingly demonstrated in nuclear physics laboratories, can be thought of as something that nature attempts exceedingly often, almost always with failure, but that succeeds in a few rare sequences. Although this view is correct, it still was not adequate. The problem is that it does not occur often enough to keep the helium cores of stars from cooling. The final chapter was foreseen by the famed English cosmologist Fred Hoyle, known on other grounds as the father of the theory of nucleosynthesis in stars. Hoyle argued that a very special configuration of three alpha particles at just the correct energy exists naturally in every ^{12}C nucleus, so that the alpha-particle gas, when putting the three together, strikes a resonant chord with the internal structure of the ^{12}C nucleus itself. This may be likened to chaotic noise in an organ pipe finding a resonance at the note of A. This prediction of an excited resonance in the ^{12}C nucleus was made on the grounds that it must be so in order for carbon to be as abundant as it undeniably is – a radical argument at that time. In a dramatic sequence of events that would change the future of nuclear astrophysics, laboratory experiments found that the ^{12}C nucleus does indeed have such a resonance at exactly the predicted energy. With success at last, nuclear astrophysicists could appreciate that when helium becomes hot enough to fuse, its main products will be ^{12}C and ^{16}O.

About half of the carbon abundance is created in massive stars that eventually become Type II supernovae, wherein the carbon exists primarily in two separate spherical shells at the time of explosion; but the other half occurs in intermediate-mass (say 1.5–6 solar masses) AGB stars (see **Glossary**). This occurs in the red AGB stars, which are creating new ^{12}C by the triple-alpha process in a helium-burning shell beneath their H-containing outer envelope. These stars undergo periodic dredgeup episodes that convect this newly created ^{12}C to the surface, eventually turning the AGB star into a carbon star. The outer envelope and its new ^{12}C are eventually lost to the interstellar gas when the core of C and O arrives at the end of its nuclear evolution and becomes a white-dwarf star. This is *primary* ^{12}C because it can be produced in stars intially composed of only H and He. Some of this may be converted to *primary* ^{13}C (see ^{13}C, **Nucleosynthesis origin**, Primary nucleosynthesis). The action of these distinct star classes as sources of carbon is well documented by astronomers (see *Astronomical measurements*, below).

When He burns in the hydrogen-exhausted core at the center of a star, it liberates more nuclear power by making ^{16}O than by making ^{12}C, as the smaller atomic mass excess for ^{16}O reveals. The relative amounts of ^{12}C and ^{16}O made during He burning depend on the rate of the subsequent nuclear reaction, $^{12}C + ^4He \rightarrow ^{16}O$. The quantum probability for this reaction to occur has also been very hard to pin down. Its precise value is very important, not only by virtue of determining the resultant abundance ratio $^{12}C/^{16}O$ in nature, but also because that final ratio when all of the He has been burned sets the conditions for the subsequent thermonuclear evolution of the core of ^{12}C and ^{16}O that remains. For example, the $^{12}C/^{16}O$ ratio determines how much C can itself burn when it ignites at the core of the larger $^{12}C/^{16}O$ central sphere of the star, and since carbon fusion is the source of magnesium, it directly influences the Mg/O ratio in nature. Many other such consequences attend the rate of the $^{12}C + {}^{4}He \rightarrow$ ^{16}O reaction. Even after 30 years of research, nuclear physicists still try to improve its precision. Both of these celebrated nuclear reactions are excellent examples of "Murphy's Law" in astrophysics, in this case that the most important nuclear rates in the universe prove to be the hardest to measure.

Astronomical measurements

The abundances of carbon relative to other elements have been measured in stars and in the interstellar clouds and ionized nebulae. These derive from high-resolution spectroscopy of stars, from the strength of forbidden line emissions and of ultraviolet lines from ions in nebulae, and from millimeter radio spectroscopy in interstellar clouds. These show that relative to the hydrogen and helium abundance, those of C, O, Si, Fe and some others have grown more or less in step with one another as the Galaxy aged. They have painted a rather clear picture of rapid growth of carbon abundance after the first stars have begun to form, helping understanding that carbon is indeed produced in the explosions of massive stars, called Type II supernovae. Carbon is also involved in the demonstration that not all elements grow together in their galactic abundance. The best measure of this has been the ratio of carbon to oxygen. Both are very abundant in stars. Both are primary products of nucleosynthesis. But observations of old stars show convincingly that the O/C ratio was larger in the gas from which those oldest stars formed than in that gas giving birth to later forming stars. The earliest values of O/C are about three times greater than those in the Galaxy today. The reason is that only massive core-collapse supernovae can be born and evolve fast, and in their ejecta the O/C ratio is several times the ratio in solar abundances. But after 100 million years or so, the intermediate-mass red giants have evolved to their carbon-star phase (after which they become white-dwarf stars), and they eject large quantities of carbon into so-called planetary nebulae (which have nothing to do with planets!). When mixed into the interstellar gas these cause the C/O ratio to increase to a value that seems to hold steady thereafter.

Carbon abundances are measured in planetary nebulae by observing from space the ultraviolet lines of doubly ionized C^{++}. The observed C/O abundance ratio is large and variable, indicating C production in the intermediate-mass stars that are the immediate precursors to the planetary nebulae. Evidence of different nuclei synthesized in different types of stars is established by such data. They also reveal the nucleosynthesis of carbon to be more complex than that of oxygen.

Atomic carbon plays an important role in both heating and cooling of interstellar clouds. Because its ionization potential (11.3 eV) is less than the Lyman limit of H, carbon is largely ionized in the interstellar medium, unless shielded within dark molecular clouds. In regions exposed to ultraviolet starlight, the 157.7-micrometer infrared emission of C^+ is an important coolant and is used to map the Galaxy. Deeper within clouds the neutral C emission at 609 micrometers plays similar roles. Atomic C would not exist freely within molecular clouds but for the photodissociation of the CO molecule, which is abundant there. C^{+++} is detected from lower-density ISM by its ultraviolet emissions at 155 nanometers. The absorption lines of atomic and ionized carbon lie in the ultraviolet and have been observed by space telescopes, most notably the *International Ultraviolet Explorer* and the *Hubble Space Telescope*.

Molecular astronomy of carbon molecules is very rich. Of about 120 known interstellar molecules more than three-quarters contain carbon atoms; diatomic molecules include CO, CN, C_2, CH, CH^+, CN^+, and CO^+; polyatomic include CH_2, CH_4, C_2H_2, CH_3OH, CH_3CH_2OH, H_2CO and HNC; large complex unsaturated radicals and polycyclic aromatic hydrocarbons are also detected. These all play a role in the thermochemistry of interstellar clouds. The 2.6-millimeter line of CO diagnoses density and temperature in molecular clouds, as do other molecules.

Carbon solids are a significant part, maybe the dominant part, of the opacity of the interstellar gas. Their known sizes range from nanometers to micrometers. Their most prominent spectral signature is the 217.5-nanometer ultraviolet absorption feature. Most of the mass of these solid grains is probably acquired by condensation of cold gas onto the mantles of interstellar dust grains in molecular clouds; but a significant component of the total grain mass that has been recovered comes from the stardust grains that form in stellar matter as it departs and cools from the stars (see *Presolar grains*, in **Glossary**).

Isotopes in stars Relative numbers of the two isotopes of carbon have been measured in a large number of stars by the spectroscopy of the 5-micrometer vibration-rotation band of the CO molecule, which differs for each carbon isotope owing to the different mass on one end of the CO dumbbell. These show that in red-giant stars the isotope ^{13}C is often variable with respect to the abundance of ^{12}C. These variations are understood as an increase in the ^{13}C abundance in these stars owing to the dredging up to the surface of carbon that has been exposed to very hot protons (temperature of order 20×10^6 K) deeper within the star. This has caused the nuclear transmutation of a fraction of the ^{12}C into ^{13}C by the capture of a proton by ^{12}C. This is an example of a star changing its own surface composition. It occurs first as stars evolve up the red-giant branch after their Main-Sequence lifetimes. The surface convection zone deepens as the luminosity rises and the star reddens, and an initial ratio near $^{12}C/^{13}C = 90$ falls to values near 40 when the first dredgeup of fresh ^{13}C is complete; and it would remain at that value if no further dredgeup of matter occurs. But observations show that further dredgeup does occur, because at the higher luminosities the surface $^{12}C/^{13}C$ ratio continues to fall to values near 10, with intermediate values also observed. This evidence is taken to indicate that slow mixing induced by rotation and mass loss continues to convert ^{12}C to ^{13}C, accounting for the increasing ^{13}C richness of the atmosphere. Evidence of this deep slow mixing in red giants is also found in the burning up of ^7Li and of ^3He. In some carbon stars the ratio $^{13}C/^{12}C$ is as large as 1/4, which is the largest value that can be achieved during the CN cycle in stars. This understanding is only schematic, however, for many problems with the degree of mixing required for these stars remain to be solved. These same isotopic patterns of alteration by nuclear reactions are confirmed by the isotopes in presolar carbon-carrying grains (see **Anomalous isotopic abundance**, below), dust grains that formed in red-giant stars and in supernova explosions.

Isotopes in interstellar clouds   The carbon isotopic abundance ratio is the best measured of any element in the Galaxy. Millimeter-wave radio spectroscopy has been able to measure the relative isotopic abundances in carbon-bearing molecules in interstellar clouds. Of about 120 known interstellar molecules more than three-quarters contain carbon atoms. Most prominent of these is the CO molecule. Because the chemistry of making these molecules depends only weakly on the mass of the isotope involved, the relative strengths of the radio lines from the molecules ^{13}CO and ^{12}CO, which differ significantly in the wavelengths of the emitted radiation, can measure the relative abundances of the carbon isotopes in the interstellar clouds. H_2CO, CS and HC_3N are the other molecules that have proven most useful. Too many $^{12}C^{16}O$ molecules along the line of sight render the clouds thick to this isotopomer's radio-transition at 110 GHz, however; therefore the line is measured in two different less abundant isotopomers of CO, viz: $^{13}C^{16}O$ and $^{12}C^{18}O$. This means that the oxygen ratio $^{18}O/^{16}O$ must be employed to extract a carbon ratio $^{12}C/^{13}C$. From the CO data it is found that $^{12}C/^{13}C = (7.8 \pm 1.4) R_{GC}$, where R_{GC} is the distance of the interstellar gas from the galactic center (in kiloparsecs, kpc). The molecule H_2CO gives the same dependence on distance from the center, but the $^{12}C/^{13}C$ ratio tends to be slightly larger, $^{12}C/^{13}C = (7.8 \pm 2.3) R_{GC} + 20$. This difference is understood; the chemistry of H_2CO has a slight systematic preference for ^{12}C relative to ^{13}C in comparison with CO, so these two molecules retain slightly different proportions of the gaseous ratio. This phenomenon is called isotopic fractionation, and is well studied on Earth. The observed $^{12}C/^{13}C$ ratio increases with distance from the galactic center and has the best-value ratio $^{12}C/^{13}C = 65$ in gas near the Sun today in comparison with a solar isotopic value of 89 (4.6 Gyr ago). The difference between the value of the local gas and that in the Sun may be understood, considering that this ratio should decline in time if ^{13}C is secondary and ^{12}C is primary. The increase of the ratio with distance R_{GC} is also consistent with that belief. But that may not be the whole story; perhaps the solar cloud was enriched by a nearby supernova, altering its $^{12}C/^{13}C$ ratio.

Chemical mechanisms for C isotopic fractionation in clouds It was discovered in the lab and in molecular clouds that purely chemical processes exist that can alter the ratio $^{13}C/^{12}C$. This is observed in the abundance of CO isotopomers in the clouds. Ultraviolet light capable of disrupting the CO molecule penetrates much deeper into CO-containing cold clouds at frequencies for ^{13}CO than for ^{12}CO excitation, because the much higher abundance of ^{12}CO makes the cloud more opaque to its excitation. The consequence is the alteration of the $^{13}C/^{12}C$ ratio in the CO molecules that remain and, by implication, of the phases formed by the liberated CO atoms. This also causes oxygen fractionation.

Anomalous isotopic abundance

The two stable isotopes of carbon do not always occur in all natural samples in their usual ratio. In addition to the vaying ratios detected in interstellar molecules, certain meteoritic samples reveal a deficiency and some an excess of the $^{13}C/^{12}C$ ratio in comparison with the solar value. Because carbon possesses only two stable isotopes, it is a matter of semantics to assert which isotope is varying in samples that have different isotope ratios. Saying "this sample is ^{12}C-rich" says no more or less than saying "this sample is ^{13}C-poor." Only elements having more isotopes than two can support more specific identities for the varying isotope.

Because the allotropes of carbon (amorphous carbon, graphite, diamond, fullerenes) and closely related silicon-carbide solids will not dissolve in strong acids, they are left as a solid residue following the dissolution of a chunk of meteorite in acid. For this reason these were the first presolar grains recovered. This technique for finding them has been called "burning down the haystack to find the needle." These micron-sized solids contain a million million C atoms, more or less, enough to accurately count the relative numbers of its two isotopes. This isotopic counting reveals the following highlights.

Mainstream SiC grains $^{12}C/^{13}C = 40$–70, smaller than its value of 89 on Earth, but in good agreement with astronomical observations of carbon stars, where these grains condensed. Even the distribution of the measured ratios within the grains resembles well the distribution of carbon-star isotopic ratios. When these carbon isotope data are combined with measurements of Si isotopes in the same grains, a strong case is made that these grains condensed in the carbon-star atmosphere during its loss from the star as a slow dense wind.

Type X SiC and low-density graphite grains $^{12}C/^{13}C$ ratios are mostly greater than 89 (Earth), in good agreement with expectations in supernovae, where these grains condensed. Large variations of $^{12}C/^{13}C$ are found, however, ranging from about 3 to 5000. The very large isotopic ratio within many is a sure sign of supernova interior gas condensing. How this can occur without oxidation by the more abundant oxygen was long a mystery, now recently solved. The radioactivity within supernovae causes the CO molecules (the oxidized C) to be broken apart, restoring free C atoms to the gas. In this way the carbon is not allowed by the ^{56}Co radioactivity to combust permanently. This allows carbon to condense within the cooling gases even when oxygen is more abundant. Another puzzle still to be solved is the observation of a wide spread in the ^{12}C/^{13}C ratios within the supernova grains. This is related in ways not yet understood to the radial distribution of ^{12}C/^{13}C ratios that is established in supernovae. The ratio ^{12}C/^{13}C < solar (called ^{13}C-rich) is established in the H-burning outer shell; but this quickly changes to almost pure ^{12}C more inward within the He-burning shell and within the (C, O) core that has exhausted its He. Almost all C is exhausted within the carbon-burning shell; but the supernovae each create smaller abundances of C in their central regions wherein the ratio ^{12}C/^{13}C is again ^{13}C-rich. As that lively carbon profile expands and cools, it creates a suite of carbon grains that have ^{12}C/^{13}C ratios varying by almost a factor of 1000 in measured presolar supernova grains. This is one of the greatest discoveries of modern science, testifying to the processes that created the chemical elements.

Diamonds ^{12}C/^{13}C = 93, slightly more than 89 on Earth. The meaning of this nearly solar value is not known. Individual diamonds cannot be measured, so existing measurements represent an average in a collection of millions of diamonds. As a collection of diamonds is combusted in controlled steps, the isotopic ratio being lost by the collection becomes gradually larger in ^{12}C/^{13}C ratio (but only by about 1%). It has been speculated that this is a significant indicator of changing gas composition during growth in the fast-moving gas of a supernova.

So much new information has been obtained that the study of presolar grains has opened up as new a field in astronomy.

^{13}C Z $=$ 6, N $=$ 7
Spin, parity: J, $\pi = 1/2-$;
Mass excess: $\Delta = 3.125$ MeV; $S_n = 4.95$ MeV

Because ^{13}C is more massive than 13 atomic mass units, one concludes that the extra neutron in ^{13}C is less tightly bound (has more energy, and therefore more mass) than are the average nucleons in ^{12}C. Addition of a neutron (having $\Delta = 8.071$ MeV) to a ^{12}C nucleus (having $\Delta = 0$) produces a ^{13}C nucleus having $\Delta = 3.125$ MeV, 4.95 MeV less than the mass of the added neutron. This sets the binding energy of that last neutron in ^{13}C as 4.95 MeV. The setting free of that bound ^{13}C neutron in the helium-burning inner shells of AGB stars seems to provide the major source of neutrons for the s process of nucleosynthesis.

Abundance

Carbon is the fourth most abundant element in the universe. From the isotopic decomposition of normal carbon one finds that the mass-13 isotope, ^{13}C, is 1.11% of all C isotopes. Using the total abundance of elemental C = 10.1 million per million silicon atoms (i.e. 10.1 times more abundant than Si) in solar-system matter, this isotope has

solar abundance of ^{13}C = 1.12×10^5 per million silicon atoms.

This is still a high abundance, exceeding those of the elements Na and Al, for example. These solar abundances yield an isotopic ratio $^{12}C/^{13}C = 89$. Interestingly different values are found in other samples of matter in the universe. (See ^{12}C, *Anomalous isotopic abundance*).

Nucleosynthesis origin

The ^{13}C isotope is a *secondary nucleus* because it is not efficiently created in the first stars formed after the Big Bang. Those stars are made of H and He, and the thermonuclear evolution of H + He does not produce ^{13}C. The latter is created from ^{12}C that is present (in later forming stars) in the initial gas of the stars. Its nucleosynthesis occurs primarily in stars of insufficient mass to become supernovae, especially in *AGB stars* (see **Glossary**). When that gas is heated enough in stellar interiors, ^{13}C is produced by the capture of a proton by the ^{12}C nuclei: $^{12}C + {}^1H \rightarrow {}^{13}N$, and ^{13}N, which is radioactive, undergoes beta decay to ^{13}C. This is in fact the first reaction of the so-called CN cycle in stars. The CN cycle was discovered by Hans Bethe during his search for reactions that could fuse hydrogen into helium in stars. A series of four proton captures, of which the above is the first, produces one ^4He nucleus when the last proton reaction results in the ejection of an alpha particle. When summed they total $4\,{}^1H \rightarrow {}^4He + 2$ positrons. The carbon isotopes are but catalysts to this CN cycle, which provides the power to keep most of the bright stars in the universe hot. As a byproduct, however, they set the $^{12}C/^{13}C$ abundance ratio in that matter to approximately 4, much less than the solar isotopic ratio 89. Even in those stellar regions where the entire CN cycle does not operate, the first reaction, $^{12}C + {}^1H \rightarrow {}^{13}N$, enriches the ^{13}C abundance. This happens at the bottom of the convectively mixed atmospheres of red-giant stars. It creates ^{13}C at the bottom regions, which are hotter, and the convection mixes it to the surface, thereby lowering the $^{12}C/^{13}C$ abundance ratio. This newly created ^{13}C remains in the gas when it is lost from the star, a phenomenon seen by astronomers as a planetary nebula. It enriches the interstellar gas in ^{13}C. This secondary production, from ^{12}C, is the source of the ^{13}C in the universe.

Primary nucleosynthesis Some astronomical evidence suggests that ^{13}C nucleosynthesis has a primary component; that is, a production rate in stars that does not rely on the prior presence of carbon in the interstellar gas from which the stars formed (see below). This is attested to by nitrogen also having a primary component (see ^{14}N). This can occur in a red AGB star, which is creating new ^{12}C by the triple-alpha process in a helium-burning shell beneath its H-containing envelope. Such stars undergo periodic dredgeup episodes that convect this new ^{12}C to the surface, eventually turning the AGB star into a carbon star. That surface envelope has its own convective properties, bringing the surface atoms repeatedly to the high temperatures at the bottom of the convection zone. If that base temperature is high enough (40×10^6 K), the new (primary) ^{12}C can be in part converted there to ^{13}C and ^{14}N by the first two reactions of the CN cycle, as above. This results in *primary* ^{13}C and ^{13}N since it is effected in stars that need not have initial C abundance. The chemical evolution of the ^{12}C/^{13}C abundance ratio during the ageing of the Galaxy has provided one of the richest of element studies in all of astrophysics.

The populations of nuclei depend upon their destruction rates as well as on their creation rates. The destruction of ^{13}C after H burning ends occurs by one of the important reactions in stellar nucleosynthesis: ^{13}C + ^{4}He → ^{16}O + n. Measurements in the laboratory of this reaction at the low kinetic energies of stellar temperatures have been made. Its importance derives primarily from this reaction being one of the two major sources of free neutrons for the *s process* (see **Glossary**). During helium burning in massive stars, the cross section for this reaction determines how much ^{13}C is destroyed, and thereby how many neutrons are liberated, by the time the helium is totally consumed. In this setting ^{13}C does not survive until the end; it is all burned. This determines the neutron fluence for the s process in massive stars at that moment, which in turn determines the abundance of s-process nuclei created. In the He-burning shell of an AGB star, however, the situation is quite different. The ^{13}C is first created from the convective cycling which mixes protons into the hot ^{12}C, which was itself the result of the He-burning flashes in the shell. The He is not consumed in each thermonuclear pulse, but the cross section for this reaction determines how many neutrons are liberated from ^{13}C during and preceding each pulse. This sequence appears to be the major site for the s process of the heaviest elements, and the ^{13}C isotope is its neutron source.

Astronomical measurements

The abundances of carbon relative to the other elements has been measured in stars and in the interstellar clouds and ionized nebulae. These measurements derive from high-resolution spectroscopy of spectra from stars, from the strength of forbidden line emissions and of ultraviolet lines from ions in nebulae, and from millimeter radio spectroscopy in interstellar clouds (see ^{12}C, *Astronomical measurements* for a summary).

Isotopes in stars Relative numbers of the two isotopes of carbon have been measured in a large number of stars by the spectroscopy of the 5-micrometer vibration-rotation band of the CO molecule, which differs for each carbon isotope owing to the different mass on one end of the CO dumbbell. These show that in red-giant stars the isotope ^{13}C often varies with respect to the abundance of ^{12}C. These variations are understood as an increase in the ^{13}C abundance in these stars owing to the dredging up to the surface of carbon that has been exposed to very hot protons (temperature of order 20×10^6 K) deeper within the star. This has caused the nuclear transmutation of a fraction of the ^{12}C into ^{13}C by the capture of a proton by ^{12}C (see ^{12}C, **Astronomical measurements**, *Isotopes in stars*, for details).

Isotopes in interstellar clouds   Millimeter-wave radio spectroscopy has been able to measure the relative isotopic abundances in molecules in interstellar clouds. Most prominent of these is the CO molecule. Because the chemistry of making these molecules depends only weakly on the mass of the isotope involved, the relative strengths of the radio lines from the molecules ^{13}CO and ^{12}CO, which differ significantly in the wavelengths of the emitted radiation, can measure the relative abundances of the carbon isotopes in the interstellar clouds. Other molecules are also used (see ^{12}C, **Astronomical measurements**, for details). These show a gradient with respect to distance from the center of the Milky Way galaxy of the value of the isotopic ratios. The central regions are more rich in ^{13}C relative to ^{12}C than are increasingly outer regions. The ratio ^{12}C/^{13}C increases from values near 30 near the center of the Milky Way to values near 100 in gas more distant from the galactic center than is the Sun (see ^{12}C). The solar ratio is 89. This gradient matches a theoretical expectation that central regions should be both more C-rich and more ^{13}C-rich in comparison with ^{12}C. That expectation derives from the production of the ^{13}C nucleus being a *secondary process* in the theory of nucleosynthesis. Its production in stars depends upon the stars having already substantial initial abundance of ^{12}C, which is built first in the Galaxy's history. Abundance evolution has progressed further in central regions, as confirmed by larger ratios of element abundance C/H; therefore the ^{13}C also is relatively more abundant in central regions.

Anomalous isotopic abundance

Both interstellar molecules and certain meteoritic samples reveal sometimes a deficiency and sometimes an excess of the ^{13}C/^{12}C ratio. Because carbon possesses only two stable isotopes, it is a matter of semantics to assert which isotope is varying in samples that have different isotope ratios. Saying "this sample is ^{12}C-rich" says no more or less than saying "this sample is ^{13}C-poor." Only knowledge of the site of origin for these presolar grains can give clear meaning to such statements. For that reason this topic is discussed under ^{12}C. But carbon is a prominent building block of the best known presolar grains. (See ^{12}C, **Anomalous isotopic abundance** for the outline of the observed isotope ratios in the prominent presolar carbon grains – SiC, diamonds and graphite.)

7 | Nitrogen (N)

All nitrogen nuclei have charge $+7$ electronic units resulting from their seven protons, so that seven electrons orbit the nucleus of the neutral atom. Its electronic configuration is $1s^2 2s^2 2p^3$. This can be abbreviated as an inner core of inert helium (a noble gas) plus five more electrons: $(He)2s^2 2p^3$, which locates it at the top of Group VA of the periodic table, above phosphorus, arsenic, antimony and bismith. Nitrogen is almost inert in the form of its tightly bound molecule N_2; but when the N atoms are separated, nitrogen becomes highly reactive. Its electrons are able to interact with the electrons of other atoms in a wide number of different configurations. Almost no other element has such a wide range of oxidation states (or valences) as nitrogen. Its various positive oxidation states give rise to an impressive number of oxides: N_2O, NO, N_2O_3, NO_2, N_2O_5, and NO_3. And it has negative oxidation states -1 and -2, allowing it to make nitrides of metals.

Many commonly think of the Earth's atmosphere as oxygen, considering that humans require it; but in fact the atmosphere's molecules are 78% N_2. Nitrous oxide, N_2O, known also as "laughing gas" is also present at 0.5%, perhaps accounting for humankind's jolly nature. But in the solid Earth, N is not abundant because it is not incorporated heavily into rocks. In the universe, however, or in the gaseous Sun, one finds the true abundance of nitrogen, which is very large. It is the sixth most abundant element, about 1/3 as abundant as carbon and about three times more abundant than silicon. Understanding this large abundance was one of the earliest triumphs of the theory of nucleosynthesis in stars.

Nitrogen is an important industrial element. It is obtained in bulk by liquefying air, from which N_2 boils away at $-150\,°C$. This low boiling point makes liquid nitrogen one of the most common coolants for low-temperature scientific and technological research, and great fuming canisters of it are commonly seen in the technological world. Many space experiments are cooled by liquid nitrogen. The most common industrial application is the preparation of ammonia, NH_3, much of which is used in fertilizers. Nitric acid is a common element in explosives; and much compressed nitrogen is used in enhanced oil recovery, because it can be forced into liquid oil deposits without chemically reacting.

Small solid grains containing measurable amounts of nitrogen exist as *presolar grains* (see **Glossary**) found in meteorites. These presolar grains are older than the solar system and have very unusual ratios of N isotopes. They were formed when stars dying before the Sun was born ejected gas containing newly synthesized nuclei of nitrogen and other elements. The abundance ratio of the two isotopes of nitrogen has become a sorting diagnostic for the families of presolar grains. Especially N-rich are the solid grains of Si_3N_4 found in the meteorites, and which appear to have originated within supernova gas after the explosions had begun to cool.

The abundance of nitrogen in stars (relative to hydrogen) is a key indicator of the amount of past nucleosynthesis prior to the star's formation; and in many red-giant stars it also indicates nucleosynthesis of nitrogen within that star itself. In many red giants in globular clusters (which are old low-mass stars) astronomers notice that as the N abundance is larger, the C abundance is smaller; this agrees with the conversion of C to N in subsurface convective mixing within those stars. Lines from the CN and NH molecules that can form in those red-giant surfaces also reveal this conversion within them. In HII regions of interstellar gas nitrogen atoms are ionized, and bright emission lines of the ion identify it and measure the abundance of nitrogen within the interstellar gas. The radial decline of the N/H ratio with increasing distance from the center of the Galaxy testifies to a radial gradient in nucleosynthesis products. This is a key to the dynamic chemical evolution of the Galaxy.

The rich chemistry of nitrogen also makes it a constituent of a large number of interstellar molecules. Radioastronomical studies of these interstellar molecules reveal variations in the abundance ratio of the two stable N isotopes. Such studies also illuminate the field of interstellar chemistry.

Nitrogen plays a key role in generating the thermonuclear power within the stars. A cycle of thermonuclear reactions called the "carbon–nitrogen cycle" was discovered by Hans Bethe in 1939. By a series of four nuclear captures of protons by carbon and by nitrogen isotopes, with two intervening beta decays, hydrogen is fused into helium within stars, providing the main energy source for most of the starlight in the universe. The reaction rate of ^{14}N with protons is the controlling factor for that power.

Natural isotopes of nitrogen and their solar abundances

A	Solar percent	Solar abundance per 10^6 Si atoms
14	99.63	3.12×10^6
15	0.366	1.15×10^4

¹⁴N | $Z = 7$, $N = 7$
| **Spin, parity:** J, $\pi = 1+$;
| **Mass excess:** $\Delta = 2.863$ MeV; $S_n = 10.55$ MeV

The unit spin of this isotope gave early evidence that the atomic nucleus does not consist of protons and electrons. For if it did, ¹⁴N would consist of 14 protons and 7 electrons, making 21 Fermi–Dirac particles in all. Since each has half-integral spin, such a nucleus would necessarily have half-integral spin as well owing to the odd number (21) of fermions. The integral spin (J = 1) showed that the nucleus consisted of an even number of fermions, 7 protons and 7 neutrons, requiring the neutron to be an independent particle of half-integral spin rather than a composite of protons and electrons.

¹⁴N is also noteworthy as an "odd–odd nucleus" that is not radioactive. Among those nuclei whose natural abundance is primarily synthesized in stars, all other odd-Z and odd-N nuclei are radioactive, although a few live so long that they are effectively stable. The other stable odd–odd nuclei are residues of the Big Bang (²H) or of cosmic-ray interactions (⁶Li and ¹⁰B).

Abundance

Nitrogen is the sixth most abundant element in the universe. Its abundance in the Sun is about one-eighth that of oxygen, but reveals differing ratios to oxygen in other stars and in nebulae. The most abundant isotope of nitrogen, ¹⁴N, is also the sixth most abundant isotope in the universe. The two most abundant, ¹H and ⁴He, are remnants of the Big Bang, whereas ¹⁶O, ¹²C, and ²⁰Ne are created from H and He during the evolution of stars. Nitrogen provides one of the great successes of stellar nucleosynthesis. The evolution of stars makes evident why this is so.

From the isotopic decomposition of terrestrial nitrogen one finds that the mass-14 isotope, ¹⁴N, is 99.63% of all N isotopes, essentially the entire N abundance. (See **¹⁵N** for evidence from Jupiter for an even smaller ¹⁵N fraction in the solar system.) Using the total abundance of elemental N = 3.13 million per million silicon atoms (*i.e.* 3.13 times more abundant than Si) in solar-system matter,

solar abundance of ^{14}N $= 3.12 \times 10^6$ per million silicon atoms.

Nucleosynthesis origin

The ¹⁴N atom is the fourth most abundant nucleus that is produced by nucleosynthesis in the stars (after ¹⁶O, ¹²C and ²⁰Ne). Its nucleosynthesis occurs primarily in stars of insufficient mass to become supernovae, especially in *AGB stars* (see **Glossary**). It is the natural byproduct of the fusion of hydrogen into helium in the carbon–nitrogen cycle. The ¹⁴N is created by proton-capture reactions transmuting the initial C and O within the star to ¹⁴N during the CN cycle. Production from initial C and O makes ¹⁴N a *secondary* nucleus, since it can be made only in proportion to the amount of C and O initially there. ^{14}N is not efficiently created in the first stars formed after the Big Bang. The first stars are made of H and He, and the thermonuclear evolution of H + He does not produce ^{14}N. During the CN cycle of hydrogen burning, when carbon isotopes are already present, ^{14}N is produced by the capture of a proton by ^{13}C nuclei: ^{13}C + ^{1}H → ^{14}N. The ^{13}C is in turn replenished from proton capture by ^{12}C (see ^{13}C, **Nucleosynthesis origin**). The ^{14}N builds to large abundance because it is itself destroyed very slowly. The CN cycle was discovered by Hans Bethe during his search for reactions that could fuse hydrogen into helium in stars. The conversion of carbon isotopes to ^{14}N catalyzes the fusion of H into He, which provides the power to keep most of the bright stars in the universe hot. As a byproduct, however, those same nuclear reactions set the ^{14}N/^{12}C abundance ratio in that matter to approximately 50 to 100 (depending on the stellar temperature), much greater than the solar ratio 0.31. This happens at the bottom of the convectively mixed atmospheres of red-giant stars and in all of the matter interior to it where H has been consumed. It creates ^{14}N at the bottom of the convection region, and convection carries it to the surface, thereby increasing the ^{14}N/^{12}C abundance ratio at the surface. The ^{13}C/^{12}C abundance ratio increases at the same time. This newly created ^{14}N remains in the gas when it is lost from the star, a phenomenon seen by astronomers as winds from red-giant stars and as planetary nebulae, which are observed to be very rich in nitrogen. This return of mass from stars enriches the interstellar gas in ^{14}N and is the major source of ^{14}N in the universe.

When heated helium begins to fuse in stars following the exhaustion of hydrogen, the first highly abundant nucleus to react is the ^{14}N that was left there while the CNO cycle was converting carbon and oxygen to ^{14}N during the H exhaustion. That *helium burning reaction*, ^{14}N + ^{4}He → ^{18}F + gamma, is very significant for nucleosynthesis. The ^{18}F is radioactive and decays to ^{18}O, becoming the source of ^{18}O in nature and, even more significantly, creating the first neutron-rich nucleus of large abundance. The two extra neutrons in the ^{18}O nucleus become the fountainhead for many heavier neutron-rich isotopes that the star will create using those extra neutrons, which are taken away from ^{18}O (for example, see ^{25}Mg, **Nucleosynthesis origin**).

Primary nucleosynthesis Some astronomical evidence suggests that ^{14}N nucleosynthesis has a primary component; that is, a production rate in stars that does not rely on the prior presence of carbon in the interstellar gas from which the stars formed (see below). This can occur in a red AGB star, which is creating new ^{12}C by the triple-alpha process in a helium-burning shell beneath its H-containing envelope. Such stars undergo periodic dredgeup episodes that convect this new ^{12}C to the surface, eventually turning the AGB star into a carbon star. That surface envelope has its own convective properties, bringing the surface atoms repeatedly to the high temperatures at the bottom of the convection zone. If that base temperature is high enough (40×10^6 K), the new (primary) ^{12}C can be in part converted there to ^{14}N by the first two reactions of the CN cycle, as above. This results in *primary* ^{14}N since it is effected in stars that need not have an initial C abundance. The increase in the $^{14}N/^{15}N$ ratio with distance from the galactic center (see below) supports the existence of a primary source of ^{14}N. The action of these distinct star classes as sources of nitrogen is well documented by astronomers (see **Astronomical measurements**, below). Primary nitrogen may also be synthesized in rapidly rotating massive stars, where the rotation may induce mixing of new ^{12}C into H-burning envelopes where it is rapidly converted to nitrogen. The huge overabundance of N with respect to Fe in a recently observed metal-poor red-giant star suggests that prior massive-star synthesis of primary N had occurred in the early Galaxy.

Astronomical measurements

The abundances of nitrogen relative to other elements have been measured in stars and in the interstellar clouds and ionized nebulae. These measurements derive from high-resolution spectra of stars, from the strength of forbidden line emissions and of ultraviolet lines from ions in nebulae, and from millimeter radio spectroscopy in interstellar clouds. These show that relative to the hydrogen and helium abundance, the nitrogen abundance in the early Galaxy initially increased more slowly than those of C, O, and Fe. The latter elements are *primary* because they can be produced from initial H and He in the explosions of massive stars, called Type II supernovae. By contrast, nitrogen is mostly *secondary* because it is made in stars by transmutation of C and O into N during the CNO cycles of thermonuclear power. This requires C and O to have already existed in the initial matter of the stars. Nitrogen-rich stars are also seen by astronomers. Noteworthy are the red-giant stars that have convected and mixed deep material to their surfaces, greatly increasing the nitrogen content of the surface gas. And planetary nebulae reveal shining circumstellar gas that had earlier been lost by a red-giant star, and which may be seen to have had more N than C owing to the conversion of C to N in the earlier stellar life of that matter.

Graphing the abundance ratio N/C, as observed in interstellar gas and in the surfaces of Main-Sequence stars, against the abundance of C/H in the same objects can reveal both the *secondary* and the *primary* parts of nitrogen nucleosynthesis. The linear increase of N/C with C abundance reflects the dominant secondary nature of nitrogen, *viz*: nitrogen production is larger in stars having more initial carbon. But at small carbon abundance the ratio N/C seems more constant, less dependent on the C abundance, showing nitrogen production that does not depend on initial carbon. One very metal-poor red giant, having Fe/H ratio 6000 times smaller than in the Sun, reveals an astounding overabundance of nitrogen; namely, N/Fe ratio 500 times greater than solar. The initial metals of this red giant must surely have been synthesized by the first massive Type II supernovae, because those stars evolved fastest after star formation began. In that case the N synthesis must be *primary*. Such primary nitrogen can also result more slowly from the AGB stars creating first ^{12}C and then, while dredging up those new nuclei to the surface, converting them in part to ^{14}N within the same surface convective burning, as described above.

Isotopes in stars   Relative numbers of the two isotopes of nitrogen have been measured in a large number of carbon AGB stars by the spectroscopy of the vibration-rotation band of the HCN molecule. The wavelength differs for each nitrogen isotope owing to the different mass on one end of the HCN dumbbell. These show that in these evolved red-giant stars the isotope abundance ratio ^{15}N/^{14}N is variable and less than in the Sun. These variations are understood as a decrease in the ^{15}N abundance in these stars owing to the dredging up to the surface of nitrogen that has been exposed to very hot protons (of order 30×10^6 K) deeper within the star, where ^{15}N is destroyed more rapidly than ^{14}N. This is an example of a star changing its own surface composition. In AGB red-giant carbon stars ^{14}N/^{15}N > 500 is mostly seen, being a ratio as huge as 5300 in IRC+10216, one such well-studied carbon star. These values are greater than the solar ratio ^{14}N/^{15}N $= 272$.

Isotopes in dark interstellar clouds   Millimeter-wave radio spectroscopy can measure the relative isotopic abundances in nitride molecules in interstellar clouds; however, the low abundance of ^{15}N makes this very difficult. Most prominent of the N-bearing molecules is the CN molecule. Other diatomic nitrides observed include NH, NO, NS, PN, and SiN. Simple polyatomics include HCN, HNC, HNO, MgCN, MgNC, N_2H^+, N_2O, NaCN, and NH_2. The most abundant observed nitrogen-bearing molecule is probably ammonia, NH_3. Nitrogen is ubiquitous in larger interstellar molecules up to $HC_{11}N$, the largest N-bearing molecule yet observed. Of these only the molecules HCN and NH_3 have usefully yielded ^{15}N/^{14}N isotopic data. Too many HNC molecules along the line of sight render the clouds thick to this radiotransition at 88 GHz, however; therefore, the emission line is measured in two different less abundant isotopomers of HCN, *viz* $H^{13}C^{14}N$ and $H^{12}C^{15}N$. This means that the better measured carbon ratio ^{13}C/^{12}C must be employed to extract a nitrogen ratio ^{15}N/^{14}N. From such data it is found that ^{14}N/^{15}N $= (290 \pm 65) + 20R_{GC}$, where R_{GC} is the distance of the interstellar gas from the galactic center (in kiloparsecs, kpc). This ratio increases with distance from the galactic center and today has the ratio ^{14}N/^{15}N $= 460$ in gas near the Sun in comparison with a solar isotopic ratio of 272 (4.6 Gyr ago). The difference between the local gas and the Sun is difficult to understand (see also ^{12}C, *Astronomical measurements, Isotopes in interstellar clouds*), because it had been felt that this ratio should decline in time if ^{15}N is secondary and ^{14}N is partly primary. The increase of the ratio with distance R_{GC} supports that belief. Perhaps the solar cloud was enriched by a nearby supernova, giving it ^{15}N richness.

Isotope-dependent chemistry in these cold clouds can fractionate the ^{15}N/^{14}N ratio. It appears that ^{15}N-richness in N_2 molecules and in solid ammonia will characterize the results of specific ion–molecule cycles operating efficiently there. Atomic N is expected to be ^{14}N-rich. The many varied ^{15}N/^{14}N ratios in solar materials may reflect a memory of that molecular cloud state prior to solar birth.

Isotopes in ultraviolet diffuse interstellar clouds   Ultraviolet radiation can fractionate N isotopes in thin, or diffuse, molecular clouds. Ultraviolet starlight can penetrate such clouds. The ^{15}N can be separated (fractionated) from the ^{14}N isotopes by selective dissociation by ultraviolet of N_2 molecules. Because ^{14}N is so much more abundant than ^{15}N, the degree of self absorption of the uv light that dissociates the ^{14}N^{14}N molecule is much greater than the degree of self absorption for the nearby uv lines of the ^{15}N^{14}N isotopomer. Consequently the ^{15}N^{14}N molecules are dissociated throughout the diffuse cloud whereas the ^{14}N^{14}N molecule is dissociated only in the skin depth of the cloud. As for the similar cases of C and O isotopomers, the atoms that react freely with solids in the cloud cores are thereby enriched in the rarer and heavier isotopes ^{15}N, ^{13}C and ^{17,18}O (see **8 Oxygen (O)**). The chemical memory of analogous processes in the solar accretion disk probably has generated meteorite parent bodies of differing N isotopic compositions.

Anomalous isotopic abundance

Nitrogen is the champion of isotopic anomalies in meteorites. The ^{15}N/^{14}N abundance ratio of its two isotopes differs among bulk meteorites from −10% to 200% terrestrial. This is considerably larger than the range for C and O and has never been explained. It is probably a chemical memory effect, because it is implausible to imagine nuclear variations taken up by bulk meteorites. Because nitrogen possesses only two stable isotopes, it is a matter of semantics to assert which isotope is varying in samples that have different isotope ratios. Only elements having more isotopes than two can support more specific identitification of the varying isotope. In addition, it is uncertain whether the solar isotopic ratio for N should be taken to be the terrestrial ratio or the ratio within Jupiter. Measurements of Jupiter's atmosphere by instruments on NASA's *Galileo* spacecraft reveal a Jovian ratio ^{15}N/^{14}N $= 0.0023$, only 63% of the terrestrial value. The discoverers argue that it is the terrestrial value that is misleading owing to isotopic differences between the N_2 molecule and other more abundant N-bearing molecules in interstellar space. This issue is of importance for interpreting isotopically anomalous N in presolar grains. For those grains the location of a point on a plot of ^{15}N/^{14}N versus ^{13}C/^{12}C is instrumental in identifying families of presolar grains. If the terrestrial ratio ^{14}N/^{15}N $= 272$ has no special significance in the early solar system, the cluster of ^{14}N/^{15}N ratios near 272 within presolar graphite particles would seem to be terrestrial contamination rather than presolar N.

Because the allotropes of carbon (graphite, diamond, fullerenes) and closely related silicon-carbide solids will not dissolve in strong acids, they are left as a solid residue following the dissolution of a chunk of meteorite in acid. For this reason these were the first presolar grains recovered. This technique for finding them has been called "burning down the haystack to find the needle." These micron-sized solids contain 10 billion N atoms, more or less, enough to accurately count the relative numbers of its two isotopes. This isotopic counting reveals the following classes of presolar grains.

Mainstream SiC grains Nitrogen isotopic ratios vary between 50 and 19 000, but in more than 70% of the grains ^{14}N/^{15}N = 500–5000, in good agreement with astronomical observations of ^{14}N-richness in carbon stars, where these grains apparently condensed. Their surface convection zones consume ^{15}N and may dredge up newly synthesized ^{14}N along with the carbon that changes ordinary giants into carbon stars.

Type X SiC and low-density graphite grains In these presolar grains ^{14}N/^{15}N = 13–250, much more ^{15}N-rich than the Earth and Sun, in rough agreement with expectations in supernovae, where these grains condensed. The ^{15}N-richness is one defining characteristic of the supernova class of SiC grains, commonly called "X grains." The supernova ^{15}N is created by neutrino inelastic scattering from the hugely abundant ^{16}O nucleus, followed by ejection of one nucleon to leave a mass-15 isotope. One problem is that these supernova calculations produce N that is even more ^{15}N-rich than are these X grains. No good explanation exists for the N isotope ratio in X grains lying between the values calculated in supernovae and the more ^{14}N-rich gas found elsewhere.

Diamonds In bulk collections of large numbers of diamonds, ^{14}N/^{15}N near 400 is typical, confirming their presolar nature by the difference from the terrestrial ratio 272 (but see comments re. Jovian value, above and below). This isotopic ratio (as well as that of C) changes during stepwise combustion of the diamond collection, being much nearer to terrestrial ^{14}N/^{15}N = 272 in the earliest N released. The explanation is not known.

^{15}N | $Z = 7, N = 8$
Spin, parity: $J, \pi = 1/2-$;
Mass excess: $\Delta = 0.101$ MeV; $S_n = 10.83$ MeV

> ^{15}N is just barely more massive than 15 atomic mass units, by $\Delta = 0.101$ MeV, making its mass closer to an integral multiple of the atomic mass unit than any other stable isotope (save ^{12}C, which is defined as 12 amu). This means that the binding energy per nucleon of the entire ^{15}N nucleus is almost identical to that of ^{12}C.

Abundance

Nitrogen is the sixth most abundant element in the universe, but its mass-15 isotope is but the 30th most abundant isotope. From the isotopic decomposition of atmospheric nitrogen one finds that the mass-15 isotope, ^{15}N, is but 0.366% of the N isotopes. Using the total abundance of elemental N = 3.13 million per million silicon atoms (i.e. 3.13 times more abundant than Si) in solar-system matter, this isotope has

$$\text{solar abundance of } ^{15}\text{N} = 1.15 \times 10^4 \text{ per million silicon atoms.}$$

This is still a moderately high abundance, about equal to that of the element phosphorus. The terrestrial isotopic ratio ^{15}N/^{14}N = 0.00367, or ^{14}N/^{15}N = 272, is often taken to represent solar abundances; but in fact the solar ratio is not known. Interestingly different values are found in other samples of matter in the universe. In particular, analyses of Jupiter's atmosphere by instruments on NASA's *Galileo* spacecraft reveal a Jovian ratio ^{15}N/^{14}N = 0.0023, only 63% of the terrestrial value. The discoverers argue that the terrestrial value is misleading owing to isotopic differences between the N_2 molecule and other more abundant N-bearing molecules in interstellar space. This issue is of importance for interpreting isotopically anomalous N in presolar grains.

Nucleosynthesis origin

The ^{15}N isotope is only partly a *secondary nucleus* because it is not efficiently created in the CN cycle in the first stars formed after the Big Bang. In the CN cycle ^{15}N and ^{14}N are both created from ^{12}C that is present (in later forming stars) in the initial gas of the stars. ^{15}N is produced by the capture of a proton by the ^{14}N nuclei: ^{14}N + ^{1}H → ^{15}O, and the ^{15}O, which is radioactive, undergoes beta decay to ^{15}N. This is in fact the slowest reaction of the so-called CN cycle in stars; therefore it determines the thermonuclear power rate for the brilliant Main-Sequence stars. The CN cycle was discovered by Hans Bethe during his search for reactions that could fuse hydrogen into helium in stars. A series of four proton captures, of which the above is the slowest, produces one ^{4}He nucleus when the last proton reaction results in the ejection of an alpha particle. When summed they total 4 ^{1}H → ^{4}He + 2 positrons. The C and N isotopes are but catalysts to this CN cycle, which provides the power to keep most of the bright stars in the universe hot. As a byproduct, however, they set the ^{15}N/^{14}N abundance ratio in that matter to approximately 1/30 000, much less than the solar ratio 1/272. This happens at the bottom of the convectively mixed atmospheres of red-giant stars. It creates ^{15}N and ^{14}N at the bottom regions, which are hotter, and the convection mixes it to the surface, thereby increasing the N abundance there but lowering the ^{15}N/^{14}N abundance ratio. This newly created nitrogen remains in the gas when it is lost from the star, a phenomenon seen by astronomers as planetary nebulae. It somewhat enriches the interstellar gas in ^{15}N.

This secondary production by the CN cycle from ^{12}C is not the major source of ^{15}N in the universe; most of it cannot arise from the intermediate-mass AGB stars, because their production ratio ^{15}N/^{14}N is too small to make most of nature's ^{15}N even though it does account for most ^{14}N. Most of the ^{15}N production occurs in explosive stars, novae and supernovae. In novae, the time is short, so when abundant ^{14}N captures a proton in the "hot CN cycle," leaving radioactive ^{15}O, that ^{15}O cannot capture another proton. The gas quickly cools, during which time the ^{15}O decays to ^{15}N in greater abundance than in the normal CN cycle. Thus nova ejecta carry large ^{15}N/^{14}N ratios. But their contribution to bulk ^{15}N is still not accurately fixed; an informed guess is that half of ^{15}N comes from this source. Within the interior shells of supernovae, on the other hand, an even more exotic phenomenon produces a portion of the ^{15}N supply. The intense barrage of high-energy neutrinos leaving the newly formed neutron star at the collapsing center causes significant neutrino absorption by overlying ^{16}O, which is very abundant, again creating ^{15}O by ejecting a neutron, whereafter ^{15}O decays to ^{15}N during the cooling expansion. This is *primary* ^{15}N. Both types of explosion constitute nature's ^{15}N sources.

Astronomical measurements

In general, molecular spectroscopy of the N-bearing molecules in interstellar matter determines the ^{15}N/^{14}N abundance ratio (see **^{14}N**).

Anomalous isotopic abundance

Because nitrogen possesses only two stable isotopes, it is a matter of semantics to assert which isotope is varying in meteoritic presolar-grain samples that have different isotope ratios (see **^{14}N** for that data in presolar grains). Identifying the correct solar abundance ratio for N, whether terrestrial or Jovian (see **Abundance**, above) will facilitate interpretation of yet other ratios found in presolar grains.

8 | Oxygen (O)

All oxygen nuclei have charge $+8$ electronic units, so that eight electrons orbit the nucleus of the neutral atom. Its electronic configuration is $1s^2 2s^2 2p^4$. This can be abbreviated as an inner core of inert helium (a noble gas) plus six more electrons: $(He)2s^2 2p^4$, which locates it at the top of Group VI of the periodic table. Oxygen therefore has valence -2 and combines readily to make oxides with almost all other elements. Its readiness to combine chemically does not prevent it from being found as pure (free) oxygen in the diatomic molecular form O_2 in the Earth's atmosphere. But free atoms are not found on Earth because O is too reactive. It can be condensed to liquid at $-183°C$ and tens of billions of kilograms of liquid oxygen are used annually in industrial and space efforts.

Oxygen is vital for sustaining life on Earth. Respiration enables it to be available to the blood for the oxidation of fuels derived from food. It was shown to be an element at the time of the American revolutionary war. This discovery showed that air itself was not a pure element, but a more complex mixture of gases. A decomposition of air into two gases was first argued by Leonardo da Vinci, but the 1775 discovery of oxygen is usually attributed to Priestly and Lavoisier. Historically chemists noticed that metals heated in air gain weight whereas those heated in vacuum do not, a phenomenon caused by the oxidation of the metal. Today we know of another naturally occurring form of oxygen, namely O_3, or ozone. That molecule is an air pollutant at ground level but a protector from the solar ultraviolet at high altitude. It also reacts with nonmetals to make several of the other prominent molecules of the Earth's air; *viz*, CO_2, CO (rare and toxic), N_2O, and SO_2.

When many substances combine with oxygen, heat and light are evolved. This is known as combustion (as in the burning of a fuel). Water is the combustion product of hydrogen, the oxide of hydrogen. Many combinations with oxygen proceed more slowly and are known as simply oxidation (the rusting of iron, for example). Oxygen itself is noncombustible, although it actively supports the combustion of other elements. All substances which burn in air burn with increased brilliancy in pure oxygen. The temperature of an acetylene flame is much higher burning pure oxygen than it is burning air. Many substances which will not burn in air will burn in pure oxygen. These properties give to oxygen many valuable commercial applications.

The body heat of all animals is furnished by oxidation. This heat requires that animals have cooling systems. Slow oxidation of organic materials, many that could otherwise cause disease, occurs in nature. Waters and streams are automatically purified by absorbing oxygen from the air, providing the key idea for the treatment and discharge of sewage. Some oxidations are also detrimental to human purposes, as in unwanted spontaneous combustions.

Oxygen is also one of the most important elements for nucleosynthesis. It possesses three stable isotopes ($A = 16, 17, 18$), each of which has a unique and interesting consequence for understanding the origin of the abundances. Oxygen participates in the most common nuclear fusion process by which stars are prevented from cooling, the so-called CNO cycle. Nucleosynthesis of oxygen in massive stars has been demonstrated by the existence of oxygen-rich gas in the ejecta of exploding stars. These supernovae have gradually enriched the interstellar gases in oxygen, primarily the mass-16 isotope, which is by far the most abundant. This gradual enrichment is noticed in the chemical evolution of galaxies. Later-forming stars possess higher intial abundance ratios O/H than those that formed long ago.

The discovery in 1973 that refractory minerals found in meteorites are enriched in the ^{16}O isotope launched a new era of cosmic chemical memory of nucleosynthesis. Though only 5%, the excess showed that not all rocks in the solar system were assembled from well mixed gas of the solar system. It suggested that presolar materials also can be found. The study of isotopes in natural history was accelerated by this pivotal discovery.

Natural isotopes of oxygen and their solar abundances

A	Solar percent	Solar abundance per 10^6 Si atoms
16	99.76	1.52×10^7
17	0.038	5.78×10^3
18	0.200	3.04×10^4

^{16}O | $Z = 8, N = 8$
Spin, parity: $J, \pi = 0+$;
Mass excess: $\Delta = -4.737$ MeV; $\qquad S_n = 15.67$ MeV

^{16}O is the lightest nucleus to have a negative mass excess Δ. This means that its nucleons are on average more tightly bound than those of ^{12}C, the standard ($\Delta = 0$).

Abundance

Oxygen is the third most abundant element in the universe. Its most abundant isotope, ^{16}O, is the third most abundant isotope in the universe. The most abundant two, ^1H and ^4He, are remnants of the Big Bang, whereas oxygen is created during the evolution of stars. In that sense ^{16}O is the greatest success of stellar nucleosynthesis.

It is 1.5 times more abundant than carbon, the next most abundant element by a wide margin. From the isotopic decomposition of normal oxygen one finds that the mass-16 isotope, ^{16}O, is 99.76% of all O isotopes. Using the total abundance of elemental O = 15.2 million per million silicon atoms (i.e. 15.2 times more abundant than Si) in solar-system matter, this isotope has

$$\text{solar abundance of } ^{16}O = 15.2 \times 10^6 \text{ per million silicon atoms.}$$

For decades the abundance of O relative to H in the interstellar medium has been believed to be inexplicably smaller than in the Sun. Astronomers puzzled "Why is the Sun oxygen-rich?" But in the last few years the solar abundance has been lowered by more careful analyses to the abundance O/H = 500–550 per million H atoms which has caused the conflict to disappear. In the solar neighborhood the abundance ratio O/H = 500 ± 50 per million H atoms. Recent measurements by NASA's FUSE ultraviolet telescope have confirmed the lower O abundance in the gas surrounding the Sun by showing that in the local bubble the ratio D/O= 0.038 (see ^2H). Observations further suggest that in the local ISM about 60–65% of O is in gas atoms, and about 30–35% resides in grains, much as ice. For that reason the actual observed value for interstellar O atoms in the gas is near 330 per million H atoms. The lowering of the O abundance has been an important development of the past decade, because it modifies several important arguments about nucleosynthesis that were based on a larger O abundance or a larger O/C ratio in the Sun.

Nucleosynthesis origin

The ^{16}O atom enjoys its distinction as the most abundant nucleus that is synthesized in the stars because it is the natural end product of the fusion of helium, or *He burning* (see **Glossary**) as it is called in astronomy. Those basic nuclear fusion reactions are: $3\,^4He \rightarrow\,^{12}C$; $^{12}C + \,^4He \rightarrow\,^{16}O$. The bulk of this occurs in stars more massive than ten solar masses. They become Type II *supernovae* (see **Glossary**), each of which ejects a few solar masses of ^{16}O. From its negative mass excess, one sees that the 16 nucleons in ^{16}O are bound by 4.74 MeV more tightly than those of the ^{12}C nucleus, the reference standard for atomic mass excess and itself also one of the most abundant nuclei in the universe. The ^{16}O nucleons are on average 0.30 MeV more tightly bound than the average nucleon in ^{12}C. This has important consequences for stellar evolution and nucleosynthesis. Although ^{16}O exists and is created in minor amounts in other epochs of nuclear burning in stars, it exists primarily owing to this He-burning process of stellar nucleosynthesis, in which ^{16}O and ^{12}C are the two most abundant products.

When He fuses in the hydrogen-exhausted core at the center of a star, it liberates more nuclear power by making ^{16}O than by making ^{12}C, as the smaller atomic mass excess for ^{16}O reveals. The relative amounts of ^{12}C and ^{16}O made during He burning depend on the rate of the nuclear reaction $^{12}C + \,^4He \rightarrow\,^{16}O$. The quantum probability for this reaction to occur has been very hard to pin down. Its precise value is very important, not only by virtue of determining the resultant abundance ratio ^{12}C/^{16}O in nature, but also because that final ratio when all of the He has been burned sets the conditions for the subsequent thermonuclear evolution of the core of ^{12}C and ^{16}O that remains. For example, the ^{12}C/^{16}O ratio determines how much C can itself burn when it ignites at the core of the larger ^{12}C/^{16}O central sphere of the star, and since carbon fusion is the source of magnesium, it directly influences the Mg/O ratio in nature. Many other such consequences attend the rate of the ^{12}C $+$ ^{4}He $\rightarrow$ ^{16}O reaction. Even after 30 years of research, nuclear physicists still try to improve its precision. It is an excellent example of "Murphy's Law" in astrophysics, in this case that the most important nuclear rates prove to be the hardest to measure.

Astronomical measurements

The abundance of oxygen relative to other elements has been measured in stars and in the interstellar clouds and ionized nebulae. These have largely relied on three types of measurements: (1) A "forbidden" O atom line at 6300 Å; (2) ultraviolet and infrared lines of the OH molecule; (3) an allowed O atom triplet near 7773 Å. These element abundance measurements are effectively of the ^{16}O abundance because it is so much greater than those of ^{17}O and ^{18}O. Measurements derive from high-resolution optical and infrared spectra from stars, from the strength of forbidden line emissions from ions in hot diffuse nebulae, and from millimeter radio spectroscopy in interstellar clouds. These show that relative to the hydrogen and helium abundance, those of O, Mg, Si, S, Ca, and some others have grown more or less in step with one another. They have painted a rather clear picture of a rapid increase in O abundance after the first stars had begun to form, helping our understanding that O is indeed produced in the explosions of massive stars, called Type II supernovae. Oxygen is also involved in the demonstration that not all elements grow together in their galactic abundance. The best measure of this has been the ratio of oxygen to iron. Both are very abundant in stars. Both are primary products of nucleosynthesis. But observations of old stars show convincingly that the O/Fe ratio was larger in the gas from which those old stars formed than in that giving birth to later-forming stars. The earliest values of O/Fe are about three times greater than those in the middle-aged Galaxy. The reason is that only massive core-collapse supernovae can be born and evolve fast, and in their ejecta the O/Fe ratio is several times the ratio in solar abundances. But after 100 million years or so, the white-dwarf population had been established, so that their later explosions as Type I supernovae can eject large quantities of iron. These cause the O/Fe ratio to decline to a value that seems to hold steady thereafter. This evidence of different nuclei synthesized in different types of supernovae is established by these and similar data.

Isotopes in stars   Relative numbers of the three isotopes of oxygen have been measured in a large number of stars by the spectroscopy of the 5-micrometer vibration-rotation band of the CO molecule, which differs for each of the three O isotopes owing to the different mass on one end of the CO dumbbell. These show that in red-giant stars the isotopes ^{17}O and ^{18}O are often variable with respect to the abundance of ^{16}O. These are understood as alterations of the initial ^{17}O and ^{18}O abundances within stars, because stars do not easily change their surface ^{16}O abundance. (See ^{17}**O** and ^{18}**O**.)

Isotopes in dark interstellar clouds   Millimeter-wave radio spectroscopy has been able to measure the relative isotopic abundances in oxide molecules in dark interstellar clouds, where molecules dominate. Most prominent of these is the CO molecule. Because the chemistry of making these molecules depends only weakly on the mass of the O isotope involved, the relative strengths of the radio lines from the molecules $C^{16}O$, $C^{17}O$ and $C^{18}O$ can measure the relative abundances of each O isotope in the interstellar clouds. Other molecules are also used. The ground-state line of OH at 18 cm and formaldehyde (H_2CO) lines have been useful, especially for the ^{16}O/^{18}O ratio. Too many H_2CO molecules along the line of sight render the clouds thick to this radio-wavelength transition, however; therefore the line is measured in two different less abundant isotopomers of H_2CO; *viz.* $H_2^{12}C^{18}O$ and $H_2^{13}C^{16}O$. This means that the better measured carbon ratio ^{13}C/^{12}C must be employed to extract an oxygen ratio ^{16}O/^{18}O. From the data it is found that ^{16}O/^{18}O $= (59 \pm 12)R_{GC} + (37 \pm 83)$ where R_{GC} is the distance of the interstellar gas from the galactic center (in kiloparsecs, kpc). This ratio increases with distance from the galactic center and today has the value ^{16}O/^{18}O $= 540$ in gas near the Sun in comparison with 500 in the Sun (4.6 Gyr ago). These data show a gradient of the isotopic ratios with respect to distance from the center of the Milky Way galaxy. The central regions are richer in ^{17}O and ^{18}O relative to ^{16}O than are increasingly outer regions. This matches the theoretical expectation that derives from the nuclei ^{17}O and ^{18}O being *secondary nuclei* in the theory of nucleosynthesis. Their production in stars depends upon the stars having already a substantial abundance of ^{16}O, which is built first. Abundance evolution has progressed further in central regions, as confirmed by larger ratios of element abundance O/H; therefore the heavy O isotopes are relatively more abundant. Less satisfactory is the measured value for the ratio ^{18}O/^{17}O, which holds constant at 3.2 in interstellar clouds but is about 60% greater in the solar system. The reason for this discrepancy is not understood. It could be a problem in interpretation of the millimeter spectroscopy, a problem with the chemistry of making the observed molecules, or a special circumstance of the birth of the Sun that endowed it with more ^{18}O. Resolving this question is an important challenge to astrophysics, with big consequences in all areas of stellar astronomy. The NASA space probe *Genesis*, launched in 2001, will resolve this.

Isotopes in diffuse interstellar clouds   Ultraviolet radiation can fractionate O isotopes in thin, or diffuse, molecular clouds. Ultraviolet starlight can penetrate such clouds. The ^{16}O can be separated (fractionated) from the ^{17}O and ^{18}O isotopes by selective dissociation by uv of CO molecules. Because ^{16}O is so much more abundant than ^{17}O and ^{18}O, the degree of self absorption of the uv light that dissociates the C^{16}O molecule is much greater than the degree of self absorption for the nearby uv lines of the C^{17}O and C^{18}O isotopomers. Consequently, the C^{17}O and C^{18}O molecules are dissociated throughout the diffuse cloud whereas the C^{16}O molecule is dissociated only in the skin depth of the cloud. Ultraviolet astronomers using the *Hubble Space Telescope* have measured the effect of this in the form of fivefold greater isotopic ratios for both C^{16}O/C^{18}O and for C^{16}O/C^{17}O in such clouds. This same strong fractionation effect has been suggested as a possible explanation for the ^{16}O-rich solids found in meteorites (see *Chemical mechanisms for ^{16}O enrichment*, below).

Anomalous meteoritic isotopic abundance

The isotopes of O do not always occur in all natural samples in their usual proportions. Of special historic interest, certain meteoritic samples reveal a deficiency and some an excess of the ^{16}O isotope.

Bulk oxygen in meteorites, Earth and Moon   When the abundances of all three O isotopes are measured in any sample, it is possible to locate that composition as a point on a graph whose y axis is the ratio ^{17}O/^{16}O and whose x axis is the other ratio ^{18}O/^{16}O. Such a graph has been named the *three-isotope plot* (see **Glossary**), because three isotopes make only two independent ratios, the two coordinates of the composition. Each bulk sample is represented by its location point on that graph. In all bulk samples of Earth and Moon, the values obtained lie very close to one another in this plot. The ^{17}O/^{16}O ratio varies from its average value by amounts that range from -20 parts per thousand deficit to $+15$ parts per thousand excess, and, significantly, in these same samples the ^{18}O/^{16}O ratio deviates from average in the same direction but by exactly twice as much. For example, if ^{17}O/^{16}O $= -12$ parts per thousand, then ^{18}O/^{16}O $= -24$ parts per thousand (when measured from an appropriate average value). This correlated behavior, namely deviations twice as large for ^{18}O/^{16}O as for ^{17}O/^{16}O is a well-understood consequence of thermal chemistry. These chemical effects are those that depend to some degree on the mass of the isotope that is reacting. The mass-18 isotope differs in mass from the mass-16 isotope by twice the amount that the mass-17 isotope differs from it. Temperature and velocity effects are thus twice as large for the ^{18}O/^{16}O as for ^{17}O/^{16}O, so that such chemical factors can "fractionate" one isotope from another. One says that the samples lie on a correlation line having slope $1/2$. This is called "mass-dependent chemical fractionation", and measurements of it can be used as a tool to study the history of terrestrial samples. But the effect is one of mass-dependent chemistry rather than of initial isotopic variations.

Meteorites fall into different classes, and the distinct classes have distinct oxygen isotopic compositions. The slope $1/2$ mass-dependent chemical fractionation line defines the set of all compositions that can be generated by mass-dependent chemical processes from a single starting composition. Should a measured sample fall far off that line, one safely concludes that an additional physical effect is influencing the abundance of at least one of the three isotopes. Within the carbonaceous chondrite class of meteorites, the oxygen isotopic composition from one type of chondrite, say CM, lies off the slope 1/2 line defined by the oxygen samples within a separate type, say CR, and lines are displaced from that of the Earth. It follows that the two classes have genuinely different oxygen isotopic compositions, and can not be generated one from the other by chemical processes. Only the enstatite chondrite class of meteorites has oxygen isotopic compositions lying on the Earth's fractionation line, as does the Moon. These bodies may have had a common history (as far as oxygen is concerned); but the chondrite classes contain fundamental isotopic differences in their oxygen, making them distinct in their physical parentage. It is common to measure this difference by the amount of extra ^{17}O ($\Delta^{17}O$) in one class with respect to another. This is done more as a quantitative measure of their difference than as a belief that they differ primarily in ^{17}O, however.

Calcium–aluminum-rich inclusions   It was the historic discovery that certain gem-like minerals that are found within Ca–Al-rich inclusions (CAIs) within meteorites contain a 5% excess of ^{16}O that spawned scientific excitement over anomalous isotopic abundances in general within meteorites. The FUN (Fractional and Unknown Nuclear) subgroup of the CAIs also carries large mass-dependent fractionation of O isotopes in their suite of minerals, enriching progressively the two heavier O isotopes. A 5% excess of ^{16}O, with the other two isotopes retaining their usual ratio between them, may not sound like much; but its magnitude was considered huge in 1973 when it was first discovered. This means that on the three-isotope plot the $^{17}O/^{16}O$ and $^{18}O/^{16}O$ ratios are both smaller than normal by 50 parts per thousand. And this is interpreted most simply if those rocks contain an extra 5% of ^{16}O. Except for the presolar grains, the normal variations of isotope ratios in meteorites are typically but a few parts per thousand, and these small variations are all that the modest mass differences among isotopes can generate during the normal course of mass-dependent chemistry on the Earth. A 5% excess, 50 parts per thousand excess, of ^{16}O was far larger than had ever been known in any isotope that is part of the common geochemistry of the Earth. It created a puzzle that is not totally resolved to the present day. The energetic pursuit of this puzzle by isotopic chemists soon revealed others, and the whole process has not stopped short of discovering small presolar grains in meteorites that are older than the Earth and that inherited their isotope ratios from other stars. In this sense, ^{16}O plays a pivotal role in the history of isotopic anomalies. As far as the gem-like CAI minerals are concerned, it appears that those small rocks, the CAIs that are extracted from meteorites, were assembled from progenitor grains that were, in bulk, rich in the ^{16}O isotope. Those initial grains may themselves be totally gone, erased by internal chemistry; but the excess ^{16}O remains unless the dust assembly exchanged oxygen atoms with the gas. If dust were initially ^{16}O-rich, the initial gas of the planetary system would necessarily have been slightly deficient in ^{16}O. The subsequent chemical exchange of atoms between dust asssembly and gas has moved the CAI mix toward the standard abundance ratios. This attributes to the CAIs a memory of earlier progenitor solids that were ^{16}O-rich. Because many mineral assemblages in the CAIs appear to have been condensed from gas, a vaporization of presolar ^{16}O-rich dust to make presolar ^{16}O-rich gas in a local region having an enhanced ratio of dust to gas may be an indicated intermediate step within the total chemical memory. A generic theory of isotopic anomalies of this type has been called *cosmic chemical memory* (see **Glossary**). The theory of cosmic chemical memory for isotopic anomalies was advanced by this writer during the 1970s, when alternative views held sway. Especially favored in the literature of the 1970s was an altogether different idea of a neighbor exploding star mixing different isotopes into the solar cloud, and that some solid samples reveal a heightened admixture of atoms from that supernova. This theory, sometimes called the "supernova trigger for formation of the solar system", has not been accepted for the phenomenon of excess ^{16}O, although it is still retained as a possibility for inserting short-lived radioactivity, the *extinct radioactivity* (see **Glossary**), into the solar gases. A great many such issues and experimental discoveries were stimulated by that 1973 discovery by Robert Clayton and his colleagues of excess ^{16}O in the CAI minerals. Recent measurements showing that ^{16}O-richness correlates to some degree with ^{25}Mg-richness among the magnesium isotopes provides data in favor of the cosmic chemical memory theory.

The first explanation for the enrichment of the ^{16}O isotope utilized nucleosynthesis theory to locate ^{16}O-rich gas. The discoverers of the ^{16}O excess in CAIs had attributed that excess to formation of carrier grains "near supernovae"; but it was soon recognized that this was not specific enough – condensation deep within the expanding and cooling supernova interior was the only way to manufacture progenitor grains having sufficiently large ^{16}O excess to account for the data through cosmic chemical memory. Present debate centers on whether this nucleosynthesis cause is correct or whether chemical mechanisms can have operated to produce the same result.

Chemical mechanisms for ^{16}O enrichment   It was discovered in the lab, in molecular clouds and in the Earth's atmosphere that purely chemical processes exist that can enrich only ^{16}O, leaving the ratio ^{17}O/^{18}O unchanged. This is observed in the formation of ozone. It is understood by quantum mechanical symmetries, in which the ozone ^{16}O–^{16}O–^{16}O is more symmetrical than either ^{17}O–^{16}O–^{16}O or ^{18}O–^{16}O–^{16}O and therefore has different numbers of quantum transition states.

A different chemical mechanism has been found in diffuse molecular clouds (see *Astronomical measurements*, above) by ultraviolet astronomy. Ultraviolet light capable of disrupting the CO molecule penetrates much deeper into CO-containing cold clouds at frequencies for $C^{17}O$ and $C^{18}O$ excitation than for $C^{16}O$ excitation, because the much higher abundance of $C^{16}O$ makes the cloud opaque to its excitation. The consequence may be the breakup of only the $C^{17}O$ and $C^{18}O$ molecules in the shielded regions, so that ^{17}O and ^{18}O more easily enter the chemistry of solid formation than does ^{16}O. Existence of these chemical mechanisms has created uncertainty about whether the ^{16}O excess in CAIs is a manifestation of initial nuclear isotopic abundances or of such chemical effects. Resolution of this uncertainty is one of the front-line questions in the cosmic history of isotopes.

Presolar oxide grains   Meteorites contain far more oxide grains than carbon grains; in this they are similar to the Earth. But the searches for presolar grains of carbon were easier than those for presolar oxide grains. The presolar oxide grains are but a small number within a much larger population of oxide grains made by chemistry during the early solar system. Those made from gas in the solar system all lie on the same chemical fractionation line having slope $1/2$ in the three-isotope plot . The challenge was to identify presolar oxides in the face of that large background of this normal material. The only sure way to identify them is by unusual isotopic ratios, by their lying off the fractionation line in the the three-isotope plot. These mark a particle's origin as being in another time and place, when the isotopes had different relative abundances. These sites are the stars themselves, or more exactly the gaseous matter that they lose late in their lives. That outflowing matter cools toward the temperatures when dust can form, like soot, but having the isotopic and chemical compositon of that star's matter. Painstakingly, oxide grains have been isolated from meteorites and placed in a sputtering beam of ions that can be followed by automated isotopic counting of the sputtered ions. When these are anomalous, those grains have been selected for more careful study.

Approximately 120 presolar Al_2O_3 grains have been found to date. Because an abnormal isotopic ratio does not identify which isotope is abnormal (the numerator or the denominator), it has been necessary to identify patterns or families within these 120 particles and to identify the abnormal on the basis of all three isotopic ratios. The most populous group is characterized by excesses of ^{17}O and modest depletions in ^{18}O (relative to ^{16}O). This pattern is characteristic of normal ratios that have been modified by nuclear reactions with hot protons. Accordingly, these grains are suspected of condensing within matter that was previously participating in hydrogen burning, but that has cooled upon ejection from those stars. The ^{17}O nucleus is produced in hydrogen burning by the reaction $^{16}O + {}^1H \rightarrow {}^{17}O$, whereas the ^{18}O is consumed by the reaction $^{18}O + {}^1H \rightarrow {}^{15}N + {}^4He$. The change of the ^{16}O abundance is small during this process, so that it is reasonable to think of it as the isotope to compare with. The stars in question are most likely common red giants or AGB stars. Another class of oxide grains has much larger depletions of the ^{18}O, as if the depletion owing to hydrogen reactions was greater, and it has been suggested that those grains originate in stars that mixed surface oxygen down to hotter depths than are predicted by the theory of convection in stars. If this be so, it is an example of the presolar grains revealing unsuspected facts about the structure of stars.

Presolar ^{16}O-rich oxide grains from supernovae were expected. They had been predicted in a series of papers by the writer, so it was somewhat surprising that the overwhelming majority of oxide grains discovered seem to have originated in red-giant stars. The great paucity of ^{16}O-rich supernova grains is still not understood. Perhaps they are too tiny, smaller than those that have been visible for study; or perhaps the intense radioactivity within the supernova, which breaks apart its molecules when they form, interferes with the condensation process for oxides more than it does for graphite. However, at least one such grain has been found, having negligible ^{17}O, as expected for supernovae, and excess ^{16}O and ^{18}O. Perhaps more will be found as this quest continues.

^{17}O	$Z = 8, N = 9$
	Spin, parity: $J, \pi = 5/2+$;
	Mass excess: $\Delta = -0.809$ MeV; $\qquad S_n = 4.14$ MeV

Although its neutron separation energy, S_n, is small, the proton binding energy of ^{17}O is greater than in any heavier stable odd-A nucleus. It is the lightest stable ground state to require a spin 5/2+, as is required in the nuclear shell model by the exclusion principle applied to neutrons.

Abundance

From the isotopic decomposition of normal oxygen one finds that the mass-17 isotope, ^{17}O, is 0.038% of all O on Earth. It is the least abundant, by a wide margin, of the three stable O isotopes, 2630 times less abundant than the most abundant of the O isotopes, namely ^{16}O. This much smaller abundance reflects the history of the Galaxy in an important way; namely, it shows that ^{17}O cannot be synthesized in stars directly from H and He, as ^{16}O can be. Using the total abundance of elemental O = 15.2 million per million silicon atoms in solar-system matter, this isotope has

solar abundance of $^{17}O = 5780$ per million silicon atoms.

This abundance is by no means small, just small in comparison with ^{16}O. The ^{17}O abundance is comparable to those of the elements phosphorus, chlorine, titanium, or manganese. It is the 35th most abundant isotope.

Nucleosynthesis origin

In the mix of interstellar atoms from which the solar system formed, ^{17}O exists primarily owing to the hydrogen-burning process of stellar nucleosynthesis. ^{17}O is made by converting a fraction of the ^{16}O to ^{17}O by the nuclear reaction

$^{16}O + {}^{1}H \rightarrow {}^{17}F$, which quickly undergoes beta decay to ^{17}O. The primary site for this is low-mass stars, either red giants or the explosions on the surfaces of white dwarfs that are called novae. The created $^{17}O/^{16}O$ ratio is about 0.01 at locations in and inside of the H-burning shell in red giants. This is 25 times larger than the solar ratio and is a good source of ^{17}O. Although ^{17}O is also formed this way in supernovae, its high nuclear rate of destruction there suggests that less than half of its abundance owes its origin to supernovae. The ^{16}O exists during the fusion of H into He by virtue of having been present in the initial composition of the stars. Because of this, stars that form without initial ^{16}O cannot produce ^{17}O. This is very different from the case of ^{16}O, which is a natural product of nuclear burning in stars that form with only H and He in their initial compositions. As such stars died, their ejected matter slowly increased the abundance of ^{16}O in the interstellar gas, and this slowly increased the capability for making some ^{17}O from the initial ^{16}O that was inherited by later-forming stars. A so-called *primary nucleus* (e.g. ^{16}O) can be synthesized in stars from the abundant H and He that all stars inherited from the Big Bang. A *secondary nucleus* on the other hand (e.g. ^{17}O) can not be synthesized so directly; it must instead utilize the smaller abundances that stellar nucleosynthesis itself creates, and it must do so in a later-generation star that ingested the ^{16}O created by an earlier-generation star. The production of ^{17}O can utilize only the ^{16}O created by an earlier-generation star because the ^{16}O created in stars never comes in contact with hydrogen in that same star, because H was exhausted before its new ^{16}O is created. This gives ^{17}O an inferior status, a secondary nucleus, and explains why it is so much less abundant than ^{16}O. (See **^{18}O** for another example of a secondary nucleus.)

After ^{17}O has been created during the fusion of H, it can be injected into the interstellar gas along with the hydrogen-burned envelope. The ^{17}O does not last through more advanced burning stages of the stars, being consumed by $^{17}O + {}^{4}He \rightarrow {}^{20}Ne + n$ whenever the He gets hot enough to fuse. This makes ^{17}O one of the sources of free neutrons during early stellar evolution; but its abundance is so low that it is not a very significant neutron source. The low separation energy for its extra neutron, $S_n = 4.14$ MeV, ensures that in thermally advanced burning ^{17}O will be absent, because the extra neutron that it holds will have found its way to much greater binding in some other nucleus, so fierce is the competition for extra neutrons.

Astronomical measurements

The abundance of oxygen relative to other elements has been measured in stars and in the interstellar clouds and ionized nebulae. These measurements derive from high-resolution spectroscopy of spectra from stars, from the strength of forbidden line emissions from ions in nebulae, and from millimeter radio spectroscopy in interstellar clouds.

Isotopes in stars   Relative abundances of the three isotopes of oxygen have been measured in a large number of stars by the spectroscopy of the 5-micrometer vibration-rotation band of the CO molecule, which differs for each of the three O isotopes owing to the different mass on one end of the CO dumbbell. These show that in red-giant stars the isotopes ^{17}O and ^{18}O are often variable with respect to the abundance of ^{16}O. These variations are understood as an increase in the ^{17}O abundance in these stars owing to the dredging up to the surface of oxygen that has been exposed to very hot protons (of order 25×10^6 K) deeper within the star. This has caused the nuclear transmutation of a portion of the ^{16}O into ^{17}O by the capture of a proton by ^{16}O. These stars also often reveal a deficit of ^{18}O, which is also understood as the result of destruction of ^{18}O by the same hot protons. In carbon AGB stars, for example, ^{16}O/^{17}O = 550–4100, mostly less than the solar ratio 2600, indicating ^{17}O formation in the AGB stars. This understanding is only schematic, however, for many problems with the degree of mixing required for these stars have not been solved. These same isotopic patterns of alteration by nuclear reactions are confirmed by the isotopes in presolar oxide grains (see below), dust grains that formed in red-giant stars.

Isotopes in dark interstellar clouds   Millimeter-wave radio spectroscopy has been able to measure the ^{17}O/^{16}O isotopic ratio only with difficulty owing to the rarity of ^{17}O. The relative strengths of the radio lines from the molecules ^{12}C^{17}O and ^{12}C^{18}O have been measured in 14 dark interstellar clouds. These show a gradient in the ^{17}O/^{16}O ratio with respect to distance from the center of the Milky Way. The central regions are more rich in ^{17}O and ^{18}O relative to ^{16}O than are increasingly outer regions. This matches the theoretical expectation that derives from the nuclei ^{17}O and ^{18}O being *secondary nuclei* in the theory of nucleosynthesis. Their production in stars depends upon the stars having already substantial abundance of ^{16}O, which is built first. Abundance evolution has progressed further in central regions, as confirmed by larger ratios of element abundance O/H; therefore the heavy O isotopes are relatively more abundant (see ^{16}O).

Less satisfactory is the measured value for the ratio ^{18}O/^{17}O, which holds constant at 4.1 in interstellar clouds, but is about 60% greater in the solar system. The reason for this discrepancy is not understood. It could be a problem in interpretation of the millimeter spectroscopy, a problem with the chemistry of making the observed molecules, or a special circumstance of the birth of the Sun that endowed it with more ^{18}O.

Isotopes in diffuse interstellar clouds   Ultraviolet radiation is observed to fractionate O isotopes in diffuse molecular clouds (see ^{16}O).

Anomalous meteoritic isotopic abundance
This isotope of O does not always occur in all natural samples in its usual proportion. In particular, aluminum oxides in meteorites usually reveal an excess of the ^{16}O isotope. In such refractory minerals, the ^{18}O/^{17}O ratio stays almost unchanged, and for that

reason it is preferable to think that ^{16}O is enriched (see ^{16}O for that story). Because of mass-dependent chemical fractionation it is not possible to assign an anomaly in just one isotope from measurement of a single anomalous isotope ratio. Use of the *three-isotope plot* (see **Glossary**) has revealed that the different classes of chondritic meteorites differ slightly in their bulk ratio $^{18}O/^{17}O$.

FUN CAIs   The FUN (Fractionated and Unknown Nuclear) subgroup of the CAIs carries a large mass-dependent fractional enrichment of ^{17}O that is half of the fractional enrichment of ^{18}O. Some unknown process progressively enriched these two heavier isotopes.

Presolar oxide grains   Meteorites contain far more oxide grains than carbon grains; in this they are similar to the Earth. But the searches for presolar grains of carbon were easier than those for presolar oxide grains. The presolar oxide grains are a small number within a much larger population of oxide grains made by chemistry during the early solar system. Approximately 120 presolar Al_2O_3 grains have been found to date. Because an abnormal isotopic ratio does not identify which isotope is abnormal (the numerator or the denominator), it has been necessary to identify patterns or familes within these 120 particles and to identify the abnormal on the basis of all three isotopic ratios. The most populous group is characterized by excesses of ^{17}O and modest depletions in ^{18}O (relative to ^{16}O). This pattern is characteristic of normal ratios that have been modified by nuclear reactions with hot protons. Accordingly, these grains are suspected of condensing within matter that was previously exposed to hydrogen-burning temperatures, but that has cooled upon ejection from such stars. The ^{17}O nucleus is produced in hydrogen burning by the reaction $^{16}O + {}^{1}H \rightarrow {}^{17}O$, whereas the ^{18}O is consumed by the reaction $^{18}O + {}^{1}H \rightarrow {}^{15}N + {}^{4}He$. The change of the ^{16}O abundance is small during this process, so that it is reasonable to think of it as the isotope to compare with. The stars in question are most likely common red giants or AGB stars. One class of grains has much larger depletions of the ^{18}O, as if the depletion arising from hydrogen reactions was greater, and it has been suggested that those grains originate in stars that mixed surface oxygen down to hotter depths than are predicted by the theory of mixing in stars. If this be so, it is another example of the presolar grains revealing unsuspected facts about the structure of stars.

^{18}O | $Z = 8$, $N = 10$
Spin, parity: J, $\pi = 0+$;
Mass excess: $\Delta = -0.782$ MeV; $S_n = 8.05$ MeV

^{18}O is the lightest stable nucleus to have two excess neutrons. This gives it unique importance for the evolution of neutron-richness in stars. Its proton binding energy is greater than in any heavier stable nucleus.

Abundance

From the isotopic decomposition of normal oxygen one finds that the mass-18 isotope, ^{18}O, is 0.20% of all O isotopes, making it the second most abundant isotope. But it is far behind ^{16}O, being 499 times less abundant than the most abundant O isotope. This much smaller abundance reflects the history of the Galaxy in an important way; namely, it shows that ^{18}O cannot be synthesized in stars directly from H and He, as ^{16}O can be. Using the total abundance of elemental O $= 15.2$ million per million silicon atoms in solar-system matter, this isotope has

solar abundance of ^{18}O $= 30\,400$ per million silicon atoms.

Although 499 times less abundant than ^{16}O, ^{18}O is nonetheless one of the more abundant isotopes in the universe. Its abundance is comparable to that of the elements sodium and calcium. It is the 25th most abundant isotope.

Nucleosynthesis origin

In the mix of interstellar atoms from which the solar system formed, ^{18}O exists primarily owing to the helium-burning process of stellar nucleosynthesis. But ^{18}O is not made directly by that process; instead, it is made by converting a fraction of the ^{14}N to ^{18}O by the nuclear reaction ^{14}N $+$ ^{4}He $\rightarrow$ ^{18}F. Radioactive ^{18}F then beta decays to stable ^{18}O. The primary site for this is the helium-burning shells of massive stars that become Type II *supernovae* (see **Glossary**). The ^{14}N was in turn created from initial ^{12}C in the same star during its earlier hydrogen-burning phase. This is a common aspect of the CNO cycle of H burning (see **^{14}N**). When hydrogen has been exhausted and the carbon converted to ^{14}N in the process, that stellar matter can be compressed to the point of igniting helium burning, and it is therein that the ^{18}O is synthesized by the nuclear reaction above. The ^{18}O is at that time very abundant within the He-burning matter. But ^{18}O synthesis is *secondary nucleosynthesis*. Stars that form without initial ^{12}C can not produce ^{18}O. This is very different from the case of ^{16}O, which is a natural product of nuclear burning in stars that form with only H and He in their initial compositions. As such stars died, their ejected matter slowly increased the abundance of ^{12}C and ^{16}O in the interstellar gas, and this slowly increased the capability for making some ^{18}O from their initial abundances in later-forming stars. A *primary nucleus* (e.g. ^{16}O) can be synthesized in stars from the abundant H and He that all stars inherited from the Big

Bang. A *secondary nucleus* on the other hand (*e.g.* ^{18}O) can not be synthesized so directly; it must instead utilize the smaller abundances that stellar nucleosynthesis itself creates, and it must do so in a later-generation star that ingested the ^{12}C and ^{16}O created by an earlier-generation star. Its production can utilize only the ^{12}C and ^{16}O created by an earlier-generation star because the ^{12}C and ^{16}O created by the same star never comes in contact with hydrogen, which was exhausted before its new ^{12}C and ^{16}O can be created. This gives ^{18}O an inferior status, a secondary nucleus, and explains why it is so much less abundant than ^{16}O. (See ^{17}O for another example of a secondary nucleus.)

During helium burning, ^{18}O is being both synthesized from ^{14}N and consumed by the nuclear reaction ^{18}O $+ ^4$He $\rightarrow ^{22}$Ne. The rate of the latter reaction therefore plays an essential role in determining how many ^{18}O nuclei survive the exhaustion of helium. This reaction is also the primary source of ^{22}Ne in nature. Although ^{18}O surviving from He burning is the major source of ^{22}Ne, that He zone must be ejected from the star before the ^{18}O can be destroyed by free alphas during later burning. ^{18}O will be absent during more advanced burning phases of the stars. So ^{18}O is injected into the interstellar gas when the He-burning shells of massive stars are ejected during the end of their stellar lives. Some of nature's ^{18}O is also ejected from AGB stars in the wind that carries their surfaces away. It is produced in exactly the same way during the He-burning flashes in AGB red giants, then dredged up to the stellar surfaces. But that ^{18}O may be consumed by convective mixing of those stellar envelopes downward toward hot hydrogen: ^{18}O $+ ^1$H $\rightarrow ^{15}$N $+ ^4$He. Perhaps 20% of the ^{18}O abundance is attributable to production by this sequence in AGB stars.

Astronomical measurements

The abundance of oxygen relative to other elements has been measured in stars and in the interstellar clouds and ionized nebulae. These measurements derive from high-resolution spectroscopy of spectra from stars, from the strength of forbidden line emissions from ions in nebulae, and from millimeter radio spectroscopy in interstellar clouds.

Isotopes in stars Relative numbers of the three isotopes of oxygen have been measured in a large number of stars by the spectroscopy of the 5-micrometer vibration-rotation band of the CO molecule, which differs for each of the three O isotopes owing to the different mass on one end of the CO dumbbell. These show that in red-giant stars the isotopes ^{17}O and ^{18}O are often variable with respect to the abundance of ^{16}O. These variations are understood as an increase in the ^{17}O abundance in these stars owing to the dredging up to the surface of oxygen that has been exposed to very hot protons (of order 25×10^6 K) deeper within the star. This has caused the nuclear transmutation of a portion of the ^{16}O into ^{17}O by the capture of a proton by ^{16}O. These stars also often reveal a deficit of ^{18}O, which is also understood as the result of destruction of ^{18}O by the same hot protons. In carbon AGB stars, for example, ^{16}O/^{18}O $= 700$–2400, much greater than the solar value 500, owing to ^{18}O destruction in the AGB stars. This understanding is only schematic, however, for many problems with the degree of mixing required for these stars have not been solved. These same isotopic patterns of alteration by nuclear reactions are confirmed by the isotopes in presolar oxide grains (see below), dust grains that formed in red-giant stars.

Isotopes in dark interstellar clouds   The ground-state line of OH at 18 cm and the H_2CO lines have been useful, especially for the $^{16}O/^{18}O$ ratio. Too many H_2CO molecules along the line of sight render the clouds thick to this radio-wavelength transition, however; therefore the line is measured in two different less abundant iso-topomers of H_2CO; viz: $H_2{}^{12}C^{18}O$ and $H_2{}^{13}C^{16}O$. This means that the better measured carbon ratio $^{13}C/^{12}C$ must be employed to extract an oxygen ratio $^{16}O/^{18}O$. From the data it is found that $^{16}O/^{18}O = (59 \pm 12)R_{GC} + (37 \pm 83)$ where R_{GC} is the distance of the interstellar gas from the galactic center (in kiloparsecs, kpc). This ratio increases with distance from the galactic center and today has the ratio $^{16}O/^{18}O = 540$ in in-terstellar gas near the Sun in comparison with a solar isotopic ratio of 500 (4.6 Gyr ago). The central galactic regions are more rich in ^{17}O and ^{18}O relative to ^{16}O than are increasingly outer regions. This matches the theoretical expectation that derives from the nuclei ^{17}O and ^{18}O being *secondary nuclei* in the theory of nucleosynthesis. Their production in stars depends upon the stars having already a substantial abundance of ^{16}O, which is built first. Abundance evolution has progressed further in central re-gions, as confirmed by larger ratios of element abundance O/H; therefore the heavier O isotopes are relatively more abundant. Less satisfactory is the measured value for the ratio $^{18}O/^{17}O$, which holds constant at 3.2 in interstellar clouds but is about 60% greater in the solar system. The reason for this discrepancy is not understood.

Isotopes in diffuse interstellar clouds   Ultraviolet radiation is observed to frac-tionate O isotopes in diffuse molecular clouds (see **^{16}O**).

Anomalous isotopic abundance

This isotope of oxygen does not always occur in all natural samples in its usual propor-tion. Because of mass-dependent chemical fractionation it is not possible to distinguish an anomaly in this one isotope from measurement of a single isotope ratio. One must use the the *three-isotope plot* (see **Glossary**). Using it one concludes that the different classes of chondritic meteorites differ in their ratio $^{18}O/^{17}O$.

FUN CAIs   The FUN (Fractionated and Unknown Nuclear) subgroup of the CAIs carries a large mass-dependent fractional enrichment of ^{18}O that is twice as large as the fractional enrichment of ^{17}O. Some unknown process progressively enriched those two heavier isotopes.

Presolar oxide grains   Meteorites contain far more oxide grains than carbon grains; in this they are similar to the Earth. But the searches for presolar grains of carbon were easier than those for presolar oxide grains. The presolar oxide grains are a small number within a much larger population of oxide grains made by chemistry during the early solar system. Approximately 120 presolar Al_2O_3 grains have been found to date. Because an abnormal isotopic ratio does not identify which isotope is abnormal (the numerator or the denominator), it has been necessary to identify patterns or families within these 120 particles and to identify the abnormal on the basis of all three isotopic ratios that can be formed from three isotopes. The most populous group is characterized by excesses of ^{17}O and modest depletions in ^{18}O (relative to ^{16}O). This pattern is characteristic of normal ratios that have been modified by nuclear reactions with hot protons. Accordingly, these grains are suspected of condensing within matter that was previously participating in hydrogen burning, but that has cooled upon ejection from such stars. The ^{17}O nucleus is produced in hydrogen burning by the reaction $^{16}O + {}^{1}H \rightarrow {}^{17}O$, whereas the ^{18}O is consumed by the reaction $^{18}O + {}^{1}H \rightarrow {}^{15}N + {}^{4}He$. The change of the ^{16}O abundance is small during this process, so that it is reasonable to think of it as the isotope to compare with. The stars in question are most likely common red giants or AGB stars. One class of grains has much larger depletions of the ^{18}O, as if the depletion attributable to hydrogen reactions was greater, and it has been suggested that those grains originate in stars that mixed surface oxygen down to hotter depths than are predicted by the theory of convection in stars. If this be so, it is another example of the presolar grains revealing unsuspected facts about the structure of stars.

Oxide grains from supernovae were the first presolar oxide grains predicted, but the overwhelming majority that have been found originated instead in red-giant stars. However, at least one such supernova grain has been found, having negligible ^{17}O and excess ^{16}O and ^{18}O. The author predicted these presolar particles almost three decades ago; but their scarcity suggests that most presolar oxides from supernovae are simply much smaller than those from red-giant stars and, for that reason, either do not survive in the ISM or have simply not been detectable to date.

9 | Fluorine (F)

Each fluorine nucleus has charge +9 electronic units resulting from its nine protons, so that nine electrons orbit the nucleus of the neutral atom. Its electronic configuration is $1s^2 2s^2 2p^5$. Chemically it behaves as if it wants one more electron to fill the six states of the 2p shell. Its configuration can be abbreviated as an inner core of inert helium (a noble gas) plus seven more electrons: (He) $2s^2 2p^5$, which locates it at the top of Group VIIA of the periodic table. The elements in this group are known collectively as *halogens*, and include chlorine, bromine and iodine. Fluorine is a greenish-yellow gas (of F_2 molecules), resembling the better-known chlorine in that regard. Its oxidation state is -1 because one more electron will close the 2p shell. To chemists, fluorine is famous as the most powerful oxidizer (electron grabber) of any element. It is dangerously reactive, but for the same reason is never found except in molecules, which will not willingly release it. The principle minerals bearing it are *fluorite* (CaF_2) and *cryolite* (Na_3AlF_6). Placing a little F in drinking water causes reactions with teeth that make them less prone to attack by cavity-producing acids. Sodium fluoride (NaF), stannous fluoride (SnF_2) and sodium monofluorophosphate (Na_2PO_3F) are each active ingredients of toothpastes.

The pure element can be obtained by electrolysis. Its reaction with hydrogen gas is explosive in producing HF, a strong acid used in etching glass. Indeed, when George Gore isolated F from hydrofluoric acid in 1869 his electrolysis apparatus exploded because the H and F gases mixed. F. F. H. Moissan did an improved experiment in 1886 that isolated F_2 for the first time, and was awarded the 1906 Nobel prize in chemistry.

The interesting thing about fluorine from the point of view of abundances is its extreme rarity in comparison with its neighbors. Examination of the abundances of elements 6 through 14 (C, N, O, F, Ne, Na, Mg, Al, Si) reveals that fluorine is many thousands of times less abundant than any of the others except Na, which is about 80 times more abundant than F. Rare species must either have low birth rate or high destruction rate or both. Fluorine has both. Primarily it lies in a logistic void in the chart of the nuclides, a combination of charge and mass that is bypassed by the main thermonuclear burning cycles that occur in stars. It is the forgotten element. Whereas those all about it abundantly partake of the main lines of stellar reactions, fluorine was not invited to the buffet. The advantage is that rare species teach unique things, and fluorine, overlooked by stellar nucleosynthesis, has something to say about rare processes.

Fluorine possesses one radioactive isotope that is observable in the explosions of novae. This ^{18}F has a halflife of 110 minutes, and the positrons that its decay ejects annihilate with electrons. The annihilations produce gamma-ray lines of 511 keV. These can be sought in explosions common on the surface of a white-dwarf star. Such explosions are called *novae* (new stars). The Galaxy produces 50 or so of these annually, and they have long been favorite photographic objects of astronomers. Detection of the gamma-ray lines would prove that the nova model based on thermonuclear explosions is correct.

^{19}F	$Z = 9$, $N = 10$
	Spin, parity: $J, \pi = 1/2+$;
	Mass excess: $\Delta = -1.487$ MeV; $S_n = 10.43$ MeV

All odd-Z elements lighter than fluorine have a stable isotope having equal numbers of neutrons and protons (N = Z). But no odd-Z elements heavier than fluorine have a stable isotope having N = Z. Fluorine lies on this curious border, whose explanation involves the repulsive electric force. The energy of repulsion of Z protons to themselves is proportional to Z^2, and for Z > 9 that repulsive energy is so great that the odd-Z nucleus is radioactively unstable to beta decay to the even-Z element of the same mass number. The nine-neutron isotope of F, ^{18}F, decays to ^{18}O, for example.

Abundance

Fluorine is the 24th most abundant element in the universe, just after zinc and before copper. Because F has but a single stable isotope, stellar abundance observations measure ^{19}F, the 46th most abundant isotope, which has

solar abundance of ^{19}F = 843 per million silicon atoms.

This abundance is much less than those of its immediate neighbors.

Nucleosynthesis origin

The reason for the relatively small abundance of F is that it does not participate in the main nuclear burning phases of the stars. The best chance, that of bringing ^{18}O into contact with very hot protons, occurs in the He-burning shells of massive stars during supernova explosions or in nova explosions. When heated helium occurs following the exhaustion of hydrogen, the first highly abundant nucleus to react is the ^{14}N that was left there while the CNO cycle was converting carbon and oxygen to ^{14}N during hydrogen burning. That helium-burning reaction, ^{14}N + ^{4}He $\rightarrow$ ^{18}F + gamma, is very significant for nucleosynthesis. The ^{18}F is radioactive and decays to ^{18}O, becoming the source of ^{18}O in nature and, even more significantly, creating the first neutron-rich nucleus of large abundance. Hope exists of one day observing through the 511-keV gamma-ray line the annihilation of the positrons that are emitted by the ^{18}F decay in nova atmospheres. In the He-burning shells of massive stars, the nuclear sequence includes $^{18}O + {}^{1}H \rightarrow {}^{19}F +$ gamma from the modest abundance of hot protons liberated. This is the source of perhaps 15–30% of the F abundance in nature. In some novae the same reaction may occur and produce F as well; but it is difficult to judge the fraction of F produced by novae.

The lion's share of fluorine is produced by the intense burst of neutrinos that occurs when the Type II supernova core collapses. Although neutrinos interact only infrequently with matter, a tiny fraction of their intense flux during a 10-second burst drives a proton or neutron from the ^{20}Ne nucleus, in either case resulting in ^{19}F. This occurs where both ^{20}Ne and the neutrino flux are most abundant, near the core of the exploding massive star. Much of this ^{19}F is subsequently destroyed by nuclear reactions in the heated gas when the shock wave passes, but enough survives to account for the $^{19}F/{}^{20}$Ne abundance ratio in the Sun.

Primary nucleosynthesis   The neutrino-burst nucleosynthesis makes ^{19}F a primary nucleus, because it is made even in the first stars born of only H and He.

Astronomical measurements

The abundance of fluorine relative to other elements has not been easily measured in stars owing to its low abundance and unfavorable emission lines. Most measurements that have been made derive from high-resolution infrared spectroscopy of spectra from stars, from the HF molecule. These show somewhat variable F/O ratios encompassing the solar ratio. Much work remains to be done before astronomical science can use the F abundance to draw strong conclusions about the rate of F nucleosynthesis relative to those of the other elements.

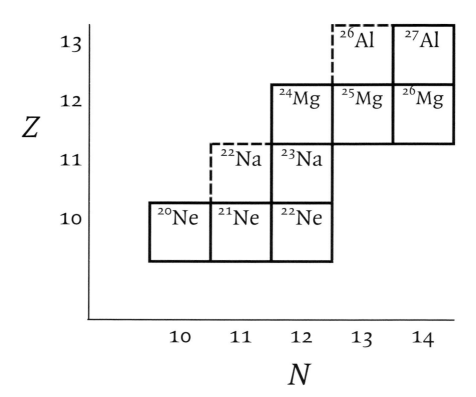

10 | Neon (Ne)

All neon nuclei have charge +10 electronic units, so that ten electrons orbit the nucleus of the neutral atom. Its electronic configuration is $1s^2 2s^2 2p^6$. Since two s electrons and six p electrons are the maximum possible populations for single $l = 0$ and $l = 1$ electronic shells, neon's configuration fills the $n = 2$ electronic shell. This gives neon's electronic structure exceptional stability, so much so that neon's electrons are unwilling (in anthropocentric terms) to rearrange into chemical bonds with other atoms. Such bonds do not produce lower energy than the separate atoms. For this reason neon is chemically *inert*. It is called a noble gas because, firstly, it is gaseous at room temperature, a gas of single atoms inasmuch as neon also does not form a molecule with itself, and, secondly, because the inertness of the closed-shell gases traditionally bears the attribute *noble*. These elements are located in Group 0 of the periodic table, with Ne standing second, between He and Ar.

Neon is the fifth most abundant gas in the Earth's atmosphere, after N_2, O_2, Ar, and CO_2 and just before He. But it is only 18 parts per million in air. And because there is precious little Ne within the Earth, it is even more rare on Earth in comparison with rock-forming elements. The interstellar gas tells a quite different story, however. Coincidentally, Ne is also the fifth most abundant element in the universe, following H, He, O, and C, and ranking just before N. It is a very abundant element, much more so than terrestrial scientists suspected. This is because the most abundant of its three isotopes is created directly by one of the most common power-generating nuclear reactions within stars.

Neon's best known use is in "neon lights", the gas-filled fluorescent tubes within which accelerated electrons cause the gases to emit their characteristic spectral colors. For neon that color is red, very red. The color can be balanced by mixing in other gases. Such tubes have the advantage of using much less power than incandescent lights (for a fixed amount of light). The element Ne is gaseous unless cooled to below $-246\ °C$, its boiling point at normal pressure. The element was first isolated in 1898 by Sir William Ramsay by the fractional distillation of liquified air. Krypton and xenon were found at the same time. Commercially the same process is used today. It takes advantage of the fact that the other molecular constituents in the atmosphere, except He, liquify at higher temperatures than does Ne, so that Ne and He remain concentrated in the gas after the liquid air is drained off.

The nucleosynthesis of neon is particularly interesting. Each of its three isotopes has a distinct origin and associated ramifications. It is unfortunate that Ne is so hard to observe in stellar spectra and incapable of forming molecules that could enable astronomical measurement of isotopic ratios. For these reasons we know little observationally about neon isotopes in astronomy. On the other hand, a rich isotopic display of neon varieties is found in the meteorites and in planetary atmospheres.

Natural isotopes of neon and their solar abundances

A	Solar percent	Solar abundance per 10^6 Si atoms
20	93.0	3.20×10^6
21	0.226	7770
22	6.79	2.34×10^5

^{20}Ne $\quad Z = 10$, $N = 10$
Spin, parity: J, $\pi = 0+$;
Mass excess: $\Delta = -7.042$ MeV; $\qquad S_n = 16.86$ MeV

Only ^{4}He, ^{12}C and ^{28}Si have greater neutron separation energy than does ^{20}Ne. But ^{20}Ne has a low alpha-particle separation energy (4.73 MeV), a curiosity that introduces into stars a nuclear burning epoch called neon burning that is initiated by a photon-induced alpha ejection from ^{20}Ne.

Abundance

Neon is the fifth most abundant element in the universe. Its most abundant isotope, ^{20}Ne, is the fifth most abundant isotope in the universe. Among those isotopes whose natural abundances are primarily the result of nucleosynthesis in stars, only ^{12}C and ^{16}O are more abundant. ^{20}Ne is about equal in abundance to ^{14}N. From the isotopic decomposition of terrestrial neon one finds that the mass-20 isotope, ^{20}Ne, is 92.99% of all Ne isotopes. The abundance of neon is not an easy matter; but good agreement between the amount observed astronomically in HII regions with that observed in the solar corona suggests that current estimates of its abundance are not far wrong. Using the total abundance of elemental Ne = 3.44 million per million silicon atoms (i.e 3.4 times more abundant than Si) in solar-system matter, this isotope has

solar abundance of ^{20}Ne $= 3.20 \times 10^6$ per million silicon atoms.

Nucleosynthesis origin

^{20}Ne exists primarily owing to the carbon-burning process of stellar nucleosynthesis. Put simply, during carbon burning two ^{12}C nuclei collide, causing the reaction ^{12}C + ^{12}C → ^{20}Ne + ^{4}He. Some of the ^{20}Ne so produced is in turn consumed by the reaction ^{20}Ne + ^{4}He → ^{24}Mg + gamma ray; but the amount remaining at the end of carbon burning is greater than that of any other product. Almost all of this happens in Type II supernovae, which on average eject about 100 times as much ^{20}Ne as do Type Ia supernovae and are about five times more frequent as well. Nature's numbers are awesome here, each Type II supernova ejecting a mass of ^{20}Ne comparable to 100 000 times the mass of the Earth!

The above reactions are helpful in understanding why ^{20}Ne is a "primary nucleus" in nucleosynthesis parlance. That designation is given to those nuclei that are abundantly produced within stars whose initial composition is only hydrogen plus helium, the two elements left from the Big Bang. Indeed, initial hydrogen and helium are converted to ^{12}C by the natural thermonuclear evolution of matter. So ^{12}C is a primary nucleus, and so therefore is ^{20}Ne, which is made from the ^{12}C. This distinguishes ^{20}Ne from the two heavier isotopes of neon; ^{21}Ne and ^{22}Ne are *secondary isotopes* because their nucleosynthesis requires that the star manufacturing them have not only hydrogen and helium in its initial composition, but also contain carbon and oxygen. The carbon and oxygen are converted to nuclei having the excess neutrons that the heavier Ne isotopes contain.

Astronomical measurements

The abundance of neon is not normally measured in stars owing to atomic physics complications. Its lines involve high excitations and are not easily seen. The most useful observations of neon abundance arise in ionized nebulae, either planetary nebulae or the HII regions surrounding young O stars. The latter show the composition of the interstellar medium near the O stars, and measurements agree with the accepted solar abundance for neon. Since ^{20}Ne dominates that abundance, such measurements are effectively of the abundance of ^{20}Ne. But we have no measurements of Ne abundance in old stars, which are cool and red. Because Ne is inert, moreover, it does not make compounds that might allow isotopic compositions to be inferred from molecular lines.

X-ray emission from young, hot, shocked supernova remnants, as observed with NASA's *Chandra X-Ray Observatory*, confirm that core-collapse Type II supernovae are prolific nucleosynthesis sources of neon.

Anomalous meteoritic isotopic abundance

The isotopes of Ne do not always occur in all natural samples in their usual proportions. Neon was one of the first elements in which isotopic anomalies were detected in meteorites, and one which subsequent research showed to have many distinct isotopic components. To describe this requires the *three-isotope plot* (see **Glossary**). When the abundances of all three Ne isotopes are measured in any sample, it is possible to locate that composition as a point on a graph whose y axis is the ratio ^{20}Ne/^{22}Ne and whose x axis is the other ratio ^{21}Ne/^{22}Ne. Such a graph has been named a *three-isotope plot*, because three isotopes make only two independent ratios, the two coordinates of the composition. Each bulk sample is represented by its location point on that graph. The y axis shows primarily variations in the ^{22}Ne abundance rather than in the ^{20}Ne abundance; and the x axis reveals primarily the amount of ^{21}Ne, which is produced by cosmic-ray interactions in meteorites in much larger amounts relative to the solar percentages than are the other two Ne isotopes. A second technique used for these discoveries was stepwise release, in which the meteorite sample is subjected (in an excellent vacuum system) to increasingly higher temepratures, causing the still retained neon to be partly released, either by diffusion or as its host minerals melt. The released gas as a function of temperature then defines a track in the three-isotope plot.

The observed three-isotope plot is dominated by four points, four distinct isotopic compositions of neon, although others are also inferred. One is solar-wind neon, also called Ne-B, whose similarity to today's solar wind suggests that solar wind from the early Sun was implanted in some mineral while the meteorite parent bodies were aggregating from dust. Its diagnostic ratio is near ^{20}Ne/^{22}Ne = 13. Having the smaller ratio ^{20}Ne/^{22}Ne = 8, Ne-A is identified with an adsorbed component of interstellar neon. It was one of the first isotopic structures to have been identified (in 1972) as presolar, meaning a chemical memory carried within some types of presolar dust. The most exciting component identified is called Ne-E. Lying near the origin of this three-isotope plot, near ^{20}Ne/^{22}Ne = 0 and ^{21}Ne/^{22}Ne = 0, it was identified as pure or nearly pure ^{22}Ne. The fourth component lies very far away from the other three, near ^{21}Ne/^{22}Ne = 0.9. This very ^{21}Ne-rich component was easily identified as nuclear reaction fragments from cosmic-ray collisions with Mg and Si atoms in the meteorite during its transit from parent body to Earth impact. (For more details see **^{21}Ne** and **^{22}Ne**.)

^{21}Ne | $Z = 10$, $N = 11$
Spin, parity: J, $\pi = 3/2+$;
Mass excess: $\Delta = -5.732$ MeV; $S_n = 6.76$ MeV

The proton binding energy of ^{21}Ne, 13.0 MeV, is greater than in any heavier stable odd-A nucleus.

Abundance

Neon is the fifth most abundant element in the universe. Its least abundant isotope, ^{21}Ne, is the 34th most abundant isotope in the universe, placing it between ^{55}Mn and ^{17}O in rank. Although it would not seem so from terrestrial abundances, ^{21}Ne is about nine times more abundant than the element fluorine in the universe. From the isotopic decomposition of terrestrial neon one finds that the mass-21 isotope, ^{21}Ne, is but 0.226% of all Ne isotopes. Using the total abundance of elemental Ne = 3.44 million per million silicon atoms (*i.e.* 3.4 times more abundant than Si) in solar-system matter, this isotope has

solar abundance of ^{21}Ne = 7770 per million silicon atoms.

Nucleosynthesis origin

In the mix of interstellar atoms from which the solar system formed, ^{21}Ne exists primarily owing to the carbon-burning process of stellar nucleosynthesis. Put simply, during carbon burning two ^{12}C nuclei collide, causing the reaction ^{12}C + ^{12}C → ^{20}Ne + ^{4}He. Some of this initial ^{20}Ne is in turn consumed by its reactions with neutrons, producing ^{21}Ne. Only a small amount of ^{21}Ne can survive this because ^{21}Ne is also easily destroyed. That's why ^{21}Ne is rare. The amount that can survive is increased if the initial star has more C and O in its composition, resulting later in the total gas having excess neutrons that lead to the free neutrons in carbon burning. Thus ^{21}Ne is a *secondary nucleus* (see **Glossary**). Lesser amounts of ^{21}Ne are synthesized by neutron capture in other (*s process*, see **Glossary**) shells. Almost all of this happens in Type II supernovae, which, on average, eject thousands of times as much ^{21}Ne as do Type Ia supernovae.

Anomalous meteoritic isotopic abundance

The ^{20}Ne/^{22}Ne versus ^{21}Ne/^{22}Ne three-isotope plot is dominated by four points, four distinct isotopic compositions of neon, although others are also inferred. One is solar-wind neon, also called Ne-B, whose similarity to today's solar wind suggests that solar wind from the early Sun was implanted in some mineral while the meteorite parent bodies were aggregating from dust. Its diagnostic ratio is near ^{21}Ne/^{22}Ne = 0.033. Having a somewhat smaller value, ^{21}Ne/^{22}Ne = 0.024, the composition known as Ne-A is identified with an adsorbed component of interstellar neon. It was one of the first isotopic structures to have been identified (in 1972) as presolar, meaning a chemical memory carried within some types of presolar dust. The most exciting component identified is called Ne-E. Lying near the origin of the three-isotope plot, it was identified as pure or nearly pure ^{22}Ne. The fourth component lies very far away from the other three, near ^{21}Ne/^{22}Ne = 0.9. This very ^{21}Ne-rich component was easily identified as nuclear reaction fragments from cosmic-ray collisions with Mg and Si atoms in the meteorite during its transit from parent body to Earth impact. It is called spallation neon. There is as yet no good explanation for why Ne-A, the presolar component, differs as much as it does from Ne-B, the solar component. But the very large ^{21}Ne/^{22}Ne in the fourth component is well understood because measurements of high-energy spallation collisions reproduce it well.

^{22}Ne $\quad$ $Z = 10$, $N = 12$

$\quad\quad$ **Spin, parity:** J, π = 0+;

$\quad\quad$ **Mass excess:** $\Delta = -5.154$ MeV; $\quad\quad$ $S_n = 10.37$ MeV

The ^{22}Ne proton binding energy is greater than in any heavier stable nucleus. When one considers the total numbers of excess neutrons (the number in excess of the number of protons) within the total natural abundances, ^{22}Ne contains more than any other isotopic abundance except ^{56}Fe. The total number of excess neutrons (two per atom) contained in the solar abundance of ^{22}Ne is almost half of the solar abundance of silicon.

Abundance

Neon is the fifth most abundant element in the universe. Its second most abundant isotope, ^{22}Ne, is the 13th most abundant isotope in the universe, placing it between ^{3}He and ^{26}Mg in rank. From the isotopic decomposition of terrestrial neon one finds that the mass-22 isotope, ^{22}Ne, is 6.79% of all Ne isotopes. Using the total abundance of elemental Ne = 3.44 million per million silicon atoms (*i.e.* 3.4 times more abundant than Si) in solar-system matter, this isotope has

$\quad\quad$ solar abundance of ^{22}Ne = 2.34×10^5 per million silicon atoms.

Nucleosynthesis origin

^{22}Ne exists primarily owing to the helium-burning process of stellar nucleosynthesis. This happens both in massive stars that will become supernovae and the class of red-giant stars called *AGB stars* (see **Glossary**). In the AGB stars the atmosphere can be enriched in ^{22}Ne as it is mixed upward from the He-burning shells of those red giants. But ^{22}Ne is not made directly by the He-burning process, but instead is made by first converting a fraction of the ^{14}N to ^{18}O by the nuclear reaction ^{14}N + ^{4}He $\rightarrow$ ^{18}F, followed by the beta decay of ^{18}F to ^{18}O. This is the point where the excess neutrons are created. The ^{14}N had been previously created from initial ^{12}C in the same star during its earlier hydrogen-burning phase. This is an essential aspect of the CNO cycle of H burning (see **^{14}N**). After hydrogen was exhausted and the initial carbon converted to ^{14}N in the process, that stellar matter can be compressed to the point of igniting helium burning, and it is therein that the ^{22}Ne is synthesized by the nuclear reaction ^{18}O + ^{4}He $\rightarrow$ ^{22}Ne. Because of this, stars that form without initial ^{12}C cannot produce ^{22}Ne. This is very different from the case of ^{20}Ne, which is a natural product of nuclear burning in stars that form with only H and He in their initial compositions. As stars died, their ejected matter slowly increased the abundance of ^{12}C and ^{16}O in the interstellar gas, and this likewise slowly increased the capability for making some ^{22}Ne from their initial abundances in later-forming stars. Astrophysicists have terms for these dependences. A *secondary nucleus* (e.g. ^{22}Ne) can not be synthesized directly starting with H and He; it must instead utilize the smaller seed abundances that stellar nucleosynthesis itself creates, and it must do so in a later-generation star that ingested the ^{12}C and ^{16}O created by an earlier-generation star. The ^{12}C and ^{16}O created by the same star never come in contact with hydrogen, which was exhausted before its new ^{12}C and ^{16}O can be created. This gives ^{22}Ne an inferior status, a secondary nucleus, and explains why it is so much less abundant than ^{20}Ne. But its abundance is still high because it is made so directly from C and O.

The populations of nuclei depend upon their destruction rates as well as on their creation rates. The destruction of ^{22}Ne occurs by one of the most important reactions in stellar nucleosynthesis: ^{22}Ne + ^{4}He → ^{25}Mg + n. Measurements in the laboratory of this reaction at the low kinetic energies of stellar temperatures have been quite challenging. The importance derives primarily from this reaction being one of the two major sources of free neutrons for the *s process* (see **Glossary**). During helium burning in massive stars, the cross section for this reaction determines how much ^{22}Ne is destroyed, and thereby how many neutrons are liberated by the time the helium is totally consumed. This determines the neutron fluence for the s process in those stars, which in turn determines the abundance of s-process nuclei created. In the He-burning shell of an AGB star, the He is not consumed in each thermonuclear pulse, but the cross section for this reaction determines how many neutrons are liberated from ^{22}Ne during each pulse. Another importance lies in the fact that this reaction is the source for ^{25}Mg and ^{26}Mg. They become quite overabundant relative to ^{24}Mg during ^{22}Ne destruction.

Anomalous isotopic abundance

The ^{21}Ne/^{22}Ne versus ^{20}Ne/^{22}Ne three-isotope plot is dominated by four points, four distinct isotopic compositions of neon, although others are also inferred. One is solar-wind neon, also called Ne-B, whose similarity to today's solar wind suggests that solar wind from the early Sun was implanted in some mineral while the meteorite parent bodies were aggregating from dust. Its diagnostic ratio is near ^{21}Ne/^{22}Ne = 0.033. Having a somewhat smaller value, ^{21}Ne/^{22}Ne = 0.024, the composition known as Ne-A is identified with an adsorbed component of interstellar neon. It was one of the first isotopic structures to have been identified (in 1972) as presolar, meaning a chemical memory carried within some types of presolar dust. The most exciting component identified is called Ne-E. Lying near the origin of this three-isotope plot, it was identified as pure or nearly pure ^{22}Ne.

Presolar grains   When presolar grains were first isolated in 1987, it became apparent that Ne-E was carried within carbonaceous grains that had been condensed within the outflowing matter from stars. The largest amount of this ^{22}Ne-rich gas is carried within SiC grains that chemically grew in red-giant *AGB stars* (see **Glossary**), as attested to by the isotopes of other elements in these so-called "mainstream SiC grains." They carry a very low ^{20}Ne/^{22}Ne isotopic ratio, but not zero. This seems to have arisen from the ^{22}Ne-rich gas present in AGB-star envelopes as a result of the dredgeup of material that was previously in the He-burning shells of these stars. (See ^{22}Ne, *Nucleosynthesis origin*). These mainstream grains are pieces of red-giant stars! Astounded cosmochemists measure the isotopic ratios within them to an incredible (for astronomy) accuracy of much better than 1%, far more accurately than astronomical techniques allow.

Presolar low-density graphite grains   Grains of graphite (a form of carbon), but having rather low density for graphite, are also found in carbonaceous meteorites. In these the neon is essentially pure ^{22}Ne. It has resulted from ^{22}Na decay in grains that were devoid of neon. Extinct radioactive ^{22}Na is recognized in the low-density graphite grains by the excess of daughter ^{22}Ne that those grains contain. The neon within some is almost pure ^{22}Ne, arguing strongly that the condensation of chemically reactive sodium is its reason for being there. This possibility was predicted in 1975 by the writer. The identification was not easy, however, because nature provides another natural source of isotopically enriched ^{22}Ne in the helium-burning shells of stars, where it is manufactured by two alpha captures starting with ^{14}N. (See ^{22}Ne, *Nucleosynthesis origin*). The "mainstream SiC" grains from AGB stars appear to be ^{22}Ne-rich for that reason (see above). But enrichment in AGB stars is limited to the ratio ^{22}Ne/^{20}Ne becoming only slightly greater than unity by admixture of the dredged up material, whereas that ratio can be very much larger than unity in a low-density graphite grain that initially contained ^{22}Na but little trapped neon gas. This appears to be the case in low-density graphite grains. Because the ^{22}Na halflife is but 2.6 yr, these grains must have condensed within a decade of an explosive event that created ^{22}Na. Most of these explosive events appear to be supernovae judging from the isotopes of other elements contained within these grains; but a smaller number seem possibly to have been condensed in nova explosions.

Solar high-energy particles (SEP)   Solar high-energy particles are accelerated by the Sun. The so-called "impulsive events" are apparently accelerated by solar flares. NASA's *Advanced Composition Explorer* (ACE) has found large variations in the ^{22}Ne/^{20}Ne ratios within such events, ranging from 0.1 to 1. These are thought to reflect the different masses in the ratio Q/A of charge to mass for Ne isotopes, since each should have the same charge state; but the variations are much larger than would be expected from similar variations of the ratio Fe/Mg in the SEP. The Fe/Mg ratio should vary by a factor 2 in Q/A, whereas the ^{22}Ne/^{20}Ne ratio varies only by 10% in Q/A. Making matters more mysterious, the ^{22}Ne/^{20}Ne ratio is larger in those impulsive flares yielding the highest ^{3}He/^{4}He ratios, which are augmented by up to 10 000 times! But the He results favor the lighter isotope, whereas the Ne results favor the heavier one. A similar situation is observed for the ratio ^{26}Mg/^{24}Mg in the SEP. This exciting data is not well understood, but is surely of high importance to the isotopic history of neon.

11 | Sodium (Na)

All sodium nuclei have charge $+11$ electronic units, so that 11 electrons orbit the nucleus of the neutral atom. Its electronic configuration can be abbreviated as an inner core of inert neon (a noble gas) plus one more electron: $(Ne)3s^1$, which locates Na in Group IA of the chemical periodic table, between Li and K. These are called *alkali metals*. They form strong alkaline caustic compounds with the OH^- ion, sometimes called *lye*. Sodium therefore has valence $+1$ and combines readily with oxygen atoms as Na_2O, which is the sixth most abundant compound in the Earth's crust. It cannot be found free because Na is too reactive, so reactive that in pure form it must be stored in a nonreactive liquid (kerosene). The Na metal is soft, silver-white and is lighter than water. If placed in water Na reacts violently to release hydrogen gas and make sodium hydroxide: $2Na + 2H_2O \rightarrow H_2 + 2NaOH$. Interestingly, H is at the top of Group IA, so in that sense water and Na_2O are similar compounds; but this reaction shows that Na can take O away from H. The pure metal was first isolated by Sir Humphrey Davy in 1807 by electrolysis of sodium hydroxide.

Like many elements, sodium's chemical symbol seems to bear no relation to its name. In such cases the symbol often derives from the Latin if the Roman Empire used one of its minerals, as in this case *natrium*, or "metal of soda", which was in turn coined from *natron*, a naturally occurring hydrous sodium carbonate mineral. The name sodium also came from the Latin *sodanum*, a headache remedy.

As an element, Na is the sixth most abundant in the Earth's crust, whereas it is 14th in the universe. This is because its oxide Na_2O is so easily made and so tightly bound into rocks that Na remained in the crust when many metals sank to the molten core of Earth.

Sodium compounds are well known to people. Table salt, *sodium chloride* (NaCl), is the most well known and constitutes about 80% of the material dissolved in sea water. Within a cubic crystal of salt, the ions of Na^+ and Cl^- are arranged alternately, one ion at each corner of the cubic cells of which the crystal is composed. Human blood is a saline solution, making salt important to body chemistry. *Sodium hydroxide* (NaOH) is used in household drain cleaners; but it is also one of the most used chemicals of the chemical industries. *Sodium carbonate* (Na_2CO_3) is used as a cleaning agent and bleach. *Sodium bicarbonate* ($NaHCO_3$), also known as baking soda, is a leavening in bakery products and is also used as a stomach antacid. *Sodium nitrate* ($NaNO_3$) is a common component of fertilizers and of explosives. It is an ingredient in dynamite, the development of which earned Alfred Nobel his great fortune with which he endowed the annual Nobel prizes.

Because sodium has but a single stable isotope it sheds no light on isotopic anomalies within presolar material. Its nucleosynthesis derives almost entirely from a single main-line source, the fusion of carbon in massive stars that will become Type II supernovae.

Natural isotopes of sodium and their solar abundances

A	Solar percent	Solar abundance per 10^6 Si atoms
22	gamma rays in novae; extinct radioactivity in presolar grains	
23	100	5.74×10^4

^{22}Na | $Z = 11, N = 11$ $t_{1/2} = 2.602$ years
Spin, parity: $J, \pi = 3+$;   **Mass excess:** $\Delta = -5.182$ MeV

The 2.6-yr halflife means that if one had a gram of ^{22}Na, it would be but half a gram 2.6 years later. Beta decay would have transmuted the other half-gram to ^{22}Ne, the daughter stable nucleus. But since a gram contains 2.7×10^{22} atoms, the halflife also requires that during each second 2.2×10^{14} atoms would decay. This would be a very dangerous sample to handle. Counting the decay rate in much smaller samples is a standard technique for measuring the halflife. ^{22}Na is the middle member of a sequence of three odd–odd $Z = N$ nuclei that have sufficiently long halflives to be observables for astronomy.

Extinct radioactivity

This isotope of sodium has great importance for the science of *extinct radioactivity* (see **Glossary**). Its abundance was not significant in the early solar system, however, but shows itself instead within presolar grains. ^{22}Na created in explosive events, novae or supernovae, can remain alive for the time needed to grow grains within the expanding and cooling gas of those explosions. Its halflife of 2.6 yr exceeds the time (weeks for a nova and one year for supernova) required for the dust to grow.

Extinct radioactive ^{22}Na is recognized in low-density graphite grains found within the carbonaceous meteorites by the excess of daughter ^{22}Ne that those grains contain. The neon within some is almost pure ^{22}Ne, arguing strongly that the condensation of chemically reactive sodium is its reason for being there. This possibility was predicted in 1975 by the writer. The identification was not easy, however, because nature provides another natural source of isotopically enriched ^{22}Ne in the helium-burning shells of stars, where it is manufactured by two alpha captures starting with ^{14}N. Another type of presolar grain, the SiC grains from AGB stars, appears to be

^{22}Ne-rich for this reason. But enrichment in AGB stars is limited to the ratio ^{22}Ne/^{20}Ne being only slightly greater than unity, whereas that ratio can be very large in a low-density graphite grain that initially contained ^{22}Na but little trapped neon gas. This appears to be the case in low-density graphite grains.

Cosmogenic radioactivity ^{22}Na is created as a collision fragment in meteorites when cosmic rays strike the meteorite during its journey to the Earth. Atoms of Mg and Si are split into fragments, some of ^{22}Na. When meteorites fall this radioactivity is counted in the lab, alive at the time of fall, and gives information on the meteorite's history in space.

Gamma-ray astronomy

Live ^{22}Na has the possibility of being seen by its gamma-ray line at 1.28 MeV emitted following its beta decay to ^{22}Ne. The writer predicted this possibility in 1974 in work with Fred Hoyle, and it provided additonal incentive for the construction and launching of energy-sensitive spectrometers into space. This culminated with NASA's *Compton Gamma Ray Observatory*, the second of NASA's planned sequence of Great Space Observatories. The first prediction was that the 1.28 MeV gamma-ray line would be seen from classical nova explosions. In these the entire star does not explode; rather, about 0.0001 solar masses of the "skin" on a white dwarf experiences a thermonuclear runaway, some fraction of which is ejected in the nova outburst. The ejecta should contain live ^{22}Na, especially from a subset of neon-rich novae. Neon is the parent from which ^{22}Na is created by radiative capture of protons during the explosion. All attempts to date have failed to detect this gamma-ray line, primarily because novae close to the Earth are rare, despite nature providing almost 100 per year in the entire Galaxy. Some ^{22}Na is also created in supernova explosions, in which the entire star disrupts. But no galactic supernova has occurred within our lifetimes, despite expecting one every few decades on average. Still, the diagnostic possibilities from ^{22}Na detection are great, and will be achieved when more sensitive gamma-ray spectrometers are flown, and especially when more fortunately nearby events occur.

Nucleosynthesis origin

^{22}Na is primarily the result of the explosive hydrogen burning or explosive helium burning in either novae or in supernovae. Within novae two successive captures of free protons by neon atoms partially transmute them into ^{22}Na. If helium novae exist (in which the explosive material on the white dwarf is made of He rather than H), two successive captures of free helium nuclei by ^{14}N atoms partially transmute them into ^{22}Na. This last process may also happen in supernovae whenever ^{14}N still remains within He at the time of explosion. This does not seem likely in the He-burning shells of massive stars, but it seems quite likely in He-cap detonation models for Type I supernovae. Interestingly, some part of the ^{22}Na abundance in supernovae is created by the burst of neutrinos from the core. Those neutrinos create free protons which would otherwise not be present, and which may convert ^{21}Ne to ^{22}Na by radiative capture of such a proton.

It must be appreciated that ^{22}Na production is not the major source of stable daughter ^{22}Ne in nature. Most ^{22}Ne is made in a stable form in stars (see **^{22}Ne**).

^{23}**Na** | $Z = 11$, $N = 12$
Spin, parity: J, $\pi = 3/2+$;
Mass excess: $\Delta = -9.530$ MeV; $\qquad S_n = 12.42$ MeV

^{23}Na has larger neutron binding energy, S_n, than any lighter odd-A nucleus.

Abundance

^{23}Na is the only stable isotope of sodium; therefore all elemental observations of its abundance are the abundance of ^{23}Na. This isotope has

solar abundance of ^{23}Na = 57 400 per million silicon atoms.

Na is the 20th most abundant nucleus within the universe, roughly comparable to Al or ^{40}Ca, and the 14th most abundant element, between Ca (13th) and Ni (15th).

Nucleosynthesis origin

In the mix of interstellar atoms from which the solar system formed, ^{23}Na exists primarily owing to the carbon-burning process of stellar nucleosynthesis. Put simply, during carbon burning two ^{12}C nuclei collide, causing the reaction ^{12}C + ^{12}C $\rightarrow$ ^{23}Na + ^{1}H. Much of this ^{23}Na is subsequently destroyed, but enough survives to account for its high abundance in the Sun and stars. From the negative mass excess Δ of ^{23}Na + ^{1}H one sees that this liberates energy. Essentially all of this happens in Type II supernovae, which on average eject about 100 times as much ^{23}Na as do Type Ia supernovae and are about five times more frequent. Nature's numbers are awesome here, each Type II ejecting a mass of ^{23}Na comparable to 1500 times the mass of the Earth!

About 10% of the ^{23}Na is created, instead, in the hydrogen-burning shells of evolved stars, where the ^{22}Ne + ^{1}H $\rightarrow$ ^{23}Na reaction can occur. At the time of the supernova explosion, this shell burning has already doubled the ^{23}Na abundance locally in the stellar surface. Another modest portion of the ^{23}Na abundance is created by the *s process* (see **Glossary**) whenever ^{22}Ne captures a neutron. But explosive carbon burning in supernovae is the main source.

The above reactions are helpful in understanding that ^{23}Na is partially a *primary nucleus* in nucleosynthesis parlance. That designation is given to those nuclei that are abundantly produced within stars whose initial composition is hydrogen plus helium, the two elements left from the Big Bang. Indeed, initial hydrogen and helium are converted to ^{12}C by the natural thermonuclear evolution of matter. So ^{12}C is a primary nucleus, and so therefore is ^{23}Na, which is made from the reaction ^{12}C + ^{12}C → ^{23}Na + ^{1}H. On the other hand, calculations of ^{23}Na production show significant secondary behavior in that a greater fraction of the ^{23}Na produced in carbon burning survives subsequent destruction by nuclear reactions such as ^{23}Na + ^{1}H → ^{20}Ne + ^{4}He when more excess neutrons exist within the composition. Excess neutrons cause the abundance of free protons to be smaller. Because of that, excess neutrons, which depend upon initial metallicity, do increase the fraction of ^{23}Na that survives to be ejected from the supernova. This is especially noticeable in the explosions of early very massive stars, 100 to 200 times the mass of the Sun, that may have produced the first nucleosynthesis since the Big Bang. Calculations show them to be 50 times more efficient in producing Mg than Na because of the absence of initial excess neutrons. This gives ^{23}Na a partly secondary quality.

Astronomical measurements

Sodium has been studied in stellar optical spectra. The famous yellow sodium lines at 5682 and 5688 Å have long been used; but weaker lines at 6154 and 6160 Å are in some cases preferable. The former pair of lines played a role in the history of spectroscopy. The doublet was called the "D lines" after an alphabetic naming of prominent lines in the solar spectrum and in flames. As nucleosynthesis progressed the galactic Na/H abundance ratio in newly born observed stars increased from 10^{-3} of solar in some early stars to a bit in excess of solar Na/H today. If compared instead to Mg, the ratio Na/Mg remains near the solar ratio in stars of all metallicities. This is understood as the coproduction of Na and Mg in massive Type II supernovae. A hint of a gradual increase in Na/Mg ratio as stellar metallicity increased is understood by the initial metallicity aiding Na survival after carbon-burning production (see above). A few giant stars seem to be Na-rich owing to production from ^{22}Ne in the H-burning shell (see above).

Na-richness is also found in some red giants within globular star clusters. The Na-richness seems to correlate with smaller oxygen abundance. This can be understood as the operation of hydrogen-burning shells in these stars, in which O is depleted gradually by the CNO cycle while Na is enriched from the ^{22}Ne + ^{1}H → ^{23}Na reaction. Substantial enrichment is possible because initially the ^{22}Ne is about four times more abundant than ^{23}Na. However, this seems to require an unexplained mixing of matter below the H-burning shell into the surface convection zone. Alternatively, the explanation can focus on the initial compositions of stars within the same globular cluster by assuming that the gas from which they formed had been enriched to varying degrees with the ejecta of more massive AGB stars that ejected their envelopes into the cluster gas early in its life, before the present remaining low-mass stars had formed. The observed Na–O anticorrelation is inheritable from those previous stars. The correct explanation is not known.

12 | Magnesium (Mg)

All magnesium nuclei have charge +12 electronic units, so that 12 electrons orbit the nucleus of the neutral atom. Its electronic configuration can be abbreviated as an inner core of inert neon (a noble gas) plus two more electrons: $(Ne)3s^2$, which locates Mg beneath beryllium in Group II of the periodic table. Magnesium therefore has valence +2 and combines readily with oxygen atoms as MgO.

Magnesium is a lightweight silvery metal. It constitutes about 2.3% by weight of the Earth's crust, being the seventh most abundant element on Earth. It occurs in the form of *magnesite* ($MgCO_3$) which is widespread in the United States, in massive deposits of *olivine* (($Mg,Fe)_2SiO_4$), and in hydrated silicates like *serpentine* ($H_4Mg_3Si_2O_9$). These are common mineral forms within rocks. A large number of other Mg-containing minerals are common. Magnesium is reasonably abundant in sea water, about 0.13%, which is an important industrial source of the metal by electrolytic decomposition of $MgCl_2$. In the wider universe, Mg is very abundant, coincidentally taking seventh place in the cosmic list, just above silicon, whose abundance is almost equal to that of Mg. In the meteorites, Mg exists primarily in the form of *enstatite* ($MgSiO_3$) and of olivine. *Spinel* ($MgAl_2O_4$) is an important oxide within meteorites, because it has been shown to contain isotopic anomalies related to its early formation in solar-system history and excess ^{26}Mg owing to the decay of radioactive ^{26}Al.

Magnesium was obtained as a distinct element in 1808 by English chemist Sir Humphrey Davy. It is silvery white, hard, and a good conductor. It has some industrial uses as a light metal. It is also highly reactive, and is famous for its explosive combustion in oxygen, giving off brilliant blue-white light, making it a staple of fireworks and, in the old days, photographic flash bulbs.

Magnesium is composed of three stable isotopes, masses 24, 25 and 26, and is an important element in nucleosynthesis theory. It is produced in stars primarily during their carbon-burning thermonuclear phase. Its abundance in stars of widely varying metallicities has been determined by astronomical observations of optical absorption lines in their spectra.

Natural isotopes of magnesium and their solar abundances

A	Solar percent	Solar abundance per 10^6 Si atoms
24	79.0	8.48×10^5
25	10.0	1.07×10^5
26	11.0	1.18×10^5

^{24}**Mg** $\quad Z = 12, N = 12$
Spin, parity: $J, \pi = 0+$;
Mass excess: $\Delta = -13.933$ MeV; $\qquad S_n = 16.53$ MeV

Abundance

From the isotopic decomposition of normal magnesium one finds that the mass-24 isotope, ^{24}Mg, is the most abundant of the three stable Mg isotopes. It is 79.0% of all Mg isotopes. Using the total abundance of elemental Mg $= 1.074 \times 10^6$ per million silicon atoms in solar-system matter, this isotope has

solar abundance of ^{24}Mg $= 8.48 \times 10^5$ per million silicon atoms.

This is the ninth most abundant nucleus within the universe.

Nucleosynthesis origin

In the mix of interstellar atoms from which the solar system formed, ^{24}Mg exists primarily owing to the carbon-burning process of stellar nucleosynthesis. Put simply, during carbon burning two ^{12}C nuclei collide, causing the reaction ^{12}C $+ ^{12}$C $\rightarrow$ ^{20}Ne $+$ ^{4}He. The alpha particle is in turn consumed by ^{20}Ne: ^{20}Ne $+ ^4$He $\rightarrow$ ^{24}Mg $+$ gamma ray. By these two reactions one effectively gets $2\ ^{12}$C $\rightarrow$ ^{24}Mg. From its negative mass excess Δ one sees that this liberates 13.93 MeV of heat energy for each ^{24}Mg fused from ^{12}C. The large neutron separation energy, $S_n = 16.53$ MeV, ensures that ^{24}Mg will survive high temperatures well, somewhat in contrast to the two heavier isotopes of magnesium. Indeed, the two heavier isotopes of magnesium are being destroyed during neon burning at the same time that the ultimate amount of ^{24}Mg is being created. Almost all of this happens in Type II supernovae, which on average eject about ten times as much ^{24}Mg as do Type Ia supernovae and are about five times more frequent. Nature's numbers are awesome here, each Type II ejecting a mass of ^{24}Mg comparable to 30 000 times the mass of the Earth!

Primary nucleus The above reactions are helpful in understanding that ^{24}Mg is a *primary nucleus* in nucleosynthesis parlance. That designation is given to those nuclei that are abundantly produced within stars whose initial composition is hydrogen plus helium, the two elements inherited from the Big Bang. Indeed, initial hydrogen and helium are converted to ^{12}C by the natural thermonuclear evolution of matter. So ^{12}C is a primary nucleus, and so therefore is ^{24}Mg, which is made from the ^{12}C. ^{24}Mg is distinguished from the two heavier isotopes of magnesium in this way; ^{25}Mg and ^{26}Mg are *secondary isotopes* because their nucleosynthesis requires that the star manufacturing them have not only hydrogen and helium in its initial composition, but also carbon and oxygen. The initial carbon and oxygen serve as nuclear catalysts for the excess neutrons that will be generated from them during the evolution of the star and that the heavier isotopes finally contain.

Astronomical measurements

Abundances of Mg and its isotopes have been studied in stars. Because ^{24}Mg dominates the spectra of Mg in stars, and especially because ^{24}Mg is a primary "alpha nucleus," the galactic evolution of ^{24}Mg is the same as the observed galactic evolution of Mg elemental abundance. As nucleosynthesis progressed the galactic Mg/H abundance ratio in newly born observed stars increased from 10^{-3} of solar in some stars to a bit in excess of solar Mg/H. This behavior is very similar to that of oxygen, and indeed to all alpha nuclei that can be created from fusion of He, C and O. These include the elements Ne, Mg, Si, S, Ar, Ca and Ti. The ratio Mg/Fe declines, however, from about three-times solar in old metal-poor stars to solar in young stars. This is understood as a rapid coproduction of the alpha nuclei, along with O, in massive Type II supernovae. These produce less iron; but the later onset of Type Ia supernova explosions dramatically increases the total Fe nucleosynthesis and causes the interstellar Mg/Fe ratio to decline to its solar value.

X-ray emission from young, hot, shocked supernova remnants, as observed with NASA's *Chandra X-Ray Observatory*, confirm that core-collapse Type-II supernovae are prolific nucleosynthesis sources of Mg.

Transitions in the molecule MgH have been observed in stellar spectra. These enable one to distinguish among Mg isotopes in stars. In metal-poor stars the ratios ^{25}Mg/^{24}Mg and ^{26}Mg/^{24}Mg are observed to be smaller than their solar values, by factors of 2–3. This supports theory, which holds that ^{24}Mg abundance should build up faster at the beginning of galactic *chemical evolution* than do its secondary isotopes.

Anomalous meteoritic isotopic abundance

This isotope of Mg does not always occur in all natural samples from meteorites in its usual proportion. In particular, presolar grains of several different types reveal sometimes a deficiency and sometimes an excess of this isotope relative to the two heavier isotopes. It is common to represent the isotopic abundance ratios with ^{24}Mg in the denominator (traditional, because ^{24}Mg is the most abundant isotope). The anomaly is defined by the fractional difference of the measured ^{25}Mg/^{24}Mg ratio (or the ^{26}Mg/^{24}Mg ratio) from its solar value; that is, by the ratio

$$(^{25}\text{Mg}/^{24}\text{Mg})_{\text{sample}}/(^{25}\text{Mg}/^{24}\text{Mg})_{\text{solar}} = 1 + \delta,$$

where δ is defined as the fractional deviation from the solar ratio. It may be positive or negative. Rather than giving the value of δ as a decimal fraction, however, it is more common to express it in percent, in parts per thousand, or in parts per ten thousand. All are found in the science literature. There is no way to determine whether a measured magnesium ratio having a value less than unity (negative δ) is to be interpreted as an excess in the number of ^{24}Mg atoms (the denominator) or a deficiency in the number of ^{25}Mg atoms (the numerator). Either gives a ratio less than unity. And for the most part, such a distinction is not actually meaningful, representing more a bias in one's mental picture than a physical situation. The Mg isotope ratio simply is what it is, and it is to astrophyiscs that one must turn for an interpretation of its deviation from solar values. Having said that, it is conventional to present the anomaly with ^{24}Mg in the denominator of the ratio.

Thermally fractionated Mg isotopes   When the abundances of all three Mg isotopes are measured in any sample, it is possible to locate that composition as a point on a graph whose y axis is the ratio ^{25}Mg/^{24}Mg and whose x axis is the other ratio ^{26}Mg/^{24}Mg. Such a graph has been named a *three-isotope plot* (see **Glossary**), because three isotopes make only two independent ratios, the two coordinates of the composition. Each bulk sample is represented by its location point on that plot. In all bulk samples of Earth and Moon, the different samples are measured to lie very close to one another in this plot. The ^{25}Mg/^{24}Mg ratio varies significantly from its average value in the calcium–aluminum-rich inclusions in meteorites, however. The deviations are very accurately measured and give important thermal information about the formation of the CAIs. Deviations in ^{25}Mg/^{24}Mg can be a few percent in the FUN CAIs (see **Glossary**, *calcium–aluminum-rich inclusions*), which is quite large for thermal fractionation attributable to chemical effects (more normally near 0.1%). In these same samples the ^{26}Mg/^{24}Mg ratio deviates from average in the same direction but by exactly twice as much. As an example, if $\delta(^{25}\text{Mg}/^{24}\text{Mg}) = 1\%$, then $\delta(^{26}\text{Mg}/^{24}\text{Mg}) = 2\%$ owing to this thermal fractionation. (In some aluminum-rich samples the ^{26}Mg/^{24}Mg ratio is even larger due to the decay of extinct ^{26}Al.) This correlated behavior is a well-understood consequence of thermal chemistry. These chemical effects are those that depend to some degree on the mass of the isotope that is reacting (or in this case, evaporating). The mass-26 isotope differs in mass from the mass-24 isotope by twice the amount that the mass-25 isotope differs from it. Temperature and velocity effects are thus twice as large for the ^{26}Mg/^{24}Mg ratio as for ^{25}Mg/^{24}Mg, so that such chemical factors can "fractionate" one isotope from another. One says that the samples lie on a correlation line having slope 1/2. This is called "mass-dependent thermal fractionation," and can be used as a tool to study the history of terrestrial and meteoritic samples. But it is a mass-dependent chemical effect (fractionation owing to thermal chemistry), not a difference in the abundances of the parent magnesium. Isotopic data of this kind seem to imply that the CAIs were partially molten rocks in the early solar disk, and that the enrichments occurred by evaporation of atoms from the small molten CAIs. This tells us something very important about the early solar system.

Specific presolar-grain groups   (See ^{25}Mg or ^{26}Mg for this information.) Because ^{24}Mg is used as the normalizing isotope in forming the two isotopic ratios, it is more convenient to list the isotopic anomalies in these presolar groups under the two neutron-rich isotopes of Mg. In contrast to silicon, presolar grains are not rich in the element magnesium. Some grains contain too few atoms of Mg for their isotopic counting. Partly for this reason Mg has not been as influential as Si in establishing the phenomenology of presolar grains. Because of the existence of excess ^{26}Mg in presolar grains from the decay of ^{26}Al, and because Al does condense easily in presolar grains, the primary role of Mg in presolar-grain studies has been the determination of the amount of ^{26}Al within the grains at the time of their condensation (see ^{26}**Mg**, *Excess* ^{26}*Mg from* ^{26}*Al: in presolar grains*). The unfortunate concomitant is that the isotopes of Mg have not played a rich role in locating the condensation sites. Its isotopes would be highly diagnostic if their ratios could be measured and if the decay of ^{26}Al within the grains did not overwrite their original meaning. The attitude here is to be thankful for the rich diagnostic information contained in the initial ^{26}Al concentration and to not bemoan the otherwise relative lack of usefulness of the Mg data.

^{25}**Mg** | $Z = 12$, $N = 13$
Spin, parity: J, $\pi = 5/2+$;
Mass excess: $\Delta = -13.133$MeV; $S_n = 7.33$ MeV

Abundance

From the isotopic decomposition of normal magnesium one finds that the mass-25 isotope, ^{25}Mg, is 10.0% of all Mg isotopes. This is about one-eight of the most abundant isotope, ^{24}Mg. But it is an abundant nucleus in the scheme of nucleosynthesis, the 16th most abundant nucleus in the universe. Using the total abundance of elemental Mg $= 1.074 \times 10^6$ per million silicon atoms in solar-system matter, this isotope has

solar abundance of ^{25}Mg $= 1.07 \times 10^5$ per million silicon atoms.

Nucleosynthesis origin

In the mix of interstellar atoms from which the solar system formed, ^{25}Mg exists primarily owing to the carbon-burning process of stellar nucleosynthesis. This occurs only in massive stars that will become supernovae. Put simply, during carbon burning two ^{12}C nuclei collide, causing the reaction ^{12}C $+$ ^{12}C $\rightarrow$ ^{20}Ne $+$ ^{4}He. Some of the liberated alpha particles in turn consume the ^{22}Ne that had been established in high abundance during helium burning; *viz.* ^{22}Ne $+$ ^{4}He $\rightarrow$ ^{25}Mg $+$ n. This creates abundant amounts of ^{25}Mg, and liberates free neutrons as well, making it one of the truly important nuclear reactions in astronomy. The associated capture of free neutrons by ^{25}Mg also creates comparable amounts of ^{26}Mg. The small neutron separation energy, $S_n = 7.33$ MeV, prevents ^{25}Mg from surviving higher temperatures. Indeed, the two heavier isotopes of magnesium are being destroyed during neon burning at the same time that the ultimate amount of ^{24}Mg is being created.

^{25}Mg is primarily a *secondary nucleosynthesis isotope*. This means that ^{25}Mg can not be manufactured in stars from initial hydrogen and helium in the stellar gas, but requires initial carbon and oxygen in the star. The separation energy, $S_n = 7.33$ MeV, is less than half that of the abundant primary isotope, ^{24}Mg, meaning that ^{25}Mg will not survive late high-temperature events as well as does ^{24}Mg. The ^{25}Mg is derived from ^{22}Ne, which was in turn created earlier in the star from ^{14}N during helium burning. The ^{14}N in turn traces its presence back to C and O atoms in the initial H-burning shell of that same star, wherein C and O were transformed into ^{14}N. This makes it clear that the production of ^{25}Mg utilizes the initial carbon and oxygen as seed nuclei, making ^{25}Mg a secondary isotope. Almost all of this happens in Type II supernovae, which on average eject thousands of times more ^{25}Mg than do Type Ia supernovae and are about five times more frequent. Nature's numbers are awesome even for secondary isotopes, each Type II ejecting a mass of ^{25}Mg comparable to 4000 times the mass of the Earth! However, a significant fraction (perhaps 10%) of ^{25}Mg nucleosynthesis occurs in AGB stars, where the same reaction, ^{22}Ne $+$ ^{4}He $\rightarrow$ ^{25}Mg $+$ n, enriches those stellar atmospheres in the He-burning-shell dredgeup events that also convert the O-rich AGB star into a carbon star.

Despite ^{25}Mg being a secondary nucleosynthesis isotope, the first generations of AGB stars may be able to produce enough for it to be the source of observed aluminum within globular cluster red giants (see ^{26}Al). This is possible if those stars can synthesize *primary* ^{14}N from the ^{12}C that is made in their He-burning shells. This would require that those stars be able to cause nuclear reactions between the dredged-up newly synthesized C and hot protons above the H-burning shells of those stars (what astronomers call hot-bottom burning). If these events can occur, ^{25}Mg can be synthesized from that primary ^{14}N by the same sequence described above for the secondary nucleosynthesis of ^{25}Mg. Although such events contribute negligibly to the amount of ^{25}Mg that the Galaxy has today, they are of particular interest to astronomers studying the very-metal-poor red giants in the globular clusters, the oldest star clusters in the Galaxy (see the following topic).

Anomalous stellar isotopic abundance

Transitions in the molecule MgH have been observed in stellar spectra. In moderately-metal-poor stars the ratios ^{25}Mg/^{24}Mg and ^{26}Mg/^{24}Mg are observed to be smaller than their solar values, by factors of 2–3. This supports theory, which holds that ^{25}Mg abundance should initially increase more slowly than does ^{24}Mg (percentagewise), but later should increase faster than does ^{24}Mg (percentagewise) as solar abundance is approached. Secondary isotopes of galactic *chemical evolution* (see **Glossary**) have such behavior.

A curious exception occurs in very-low-metallicity red giants in globular clusters, many of which show s-process enrichments that require that those stars formed from globular-cluster gas that had already been enriched to some degree by an earlier generation of AGB stars. In some of these the observed aluminum enrichment suggests that the prior AGB stars had made the local gas ^{25}Mg-rich, so that proton capture by ^{25}Mg, creating ^{26}Al, in a subsequently formed AGB star may be the source of the Al seen in those rare red giants rather than stable ^{27}Al (see ^{26}**Al**, *Extinct radioactivity*, *Live ^{26}Al in globular-cluster red-giant stars*). Quite a lot of astronomical attention will be required to unravel the unusual correlations seen in the globular-cluster red giants. They are of particular interest to astronomers in recoding the first nucleosynthesis to occur in the Galaxy after it began with only its coterie of Big-Bang light nuclei.

Specific observations of nearer normal-metallicity galactic-disk stars enriched in s-process nuclei show that the reaction ^{22}Ne + ^{4}He → ^{25}Mg + n is not the source of most of the neutrons liberated during He burning in AGB red giants; because, if it were, the ^{25}Mg/^{24}Mg ratio would be larger than it is observed to be. This provided an important clue that a process mixing H with ^{12}C generates the free neutrons needed for the main s process.

Anomalous meteoritic isotopic abundance

This isotope of Mg does not always occur in all natural samples from meteorites in its usual proportion. In particular, presolar grains of several different types reveal sometimes a deficiency and sometimes an excess of this isotope relative to ^{24}Mg. But it has been difficult to assign a nuclear cause to the small ^{25}Mg variations since thermal fractionation also produces variations. The deviation from normal of the other heavy isotope, ^{26}Mg, can not be used to pin that down because it so often has even larger excess atoms present owing to the decay of extinct ^{26}Al.

Thermally fractionated Mg isotopes   See ^{24}**Mg**. Isotopic deviations from thermal chemistry are known to exist in the FUN (Fractionated and Unknown Nuclear) subgroup of the CAIs owing to their enhanced (up to about 1%) ^{25}Mg/^{24}Mg ratios (see ^{24}**Mg**, *Thermally fractionated Mg isotopes*). This sounds small but is actually quite large for chemical processes whose rates are sensitive to the mass of the Mg atom (see ^{28}**Si**, *Thermally fractionated Si isotopes*). It is especially interesting that those CAIs revealing this large chemical fractionation of Mg isotopes also contain large isotopic excesses (but sometimes deficits!) of heavy neutron-rich isotopes of many elements (Ti, Ca, Cr, Fe, Ni). This is certainly an important clue to the precursor materials in the early solar system from which those CAIs were formed by hot chemistry under still poorly documented conditions. When this is fully understood it seems inescapable that an important insight into the origin of the solar system will have been made.

^{26}**Mg**	$Z = 12$, $N = 14$
	Spin, parity: $J, \pi = 0+$;
	Mass excess: $\Delta = 16.215$ MeV; $S_n = 11.09$ MeV

Abundance

From the isotopic decomposition of normal magnesium one finds that the mass-26 isotope, ^{26}Mg, is 11.0% of all Mg isotopes. This is about one-seventh of the most abundant isotope, ^{24}Mg. But it is an abundant nucleus in the scheme of nucleosynthesis, the 14th most abundant nucleus in the universe. Using the total abundance of elemental Mg $= 1.074 \times 10^6$ per million silicon atoms in solar-system matter, this isotope has

solar abundance of ^{26}Mg $= 1.18 \times 10^5$ per million silicon atoms.

Nucleosynthesis origin

In the mix of interstellar atoms from which the solar system formed, ^{26}Mg exists primarily due to the carbon-burning process of stellar nucleosynthesis. Put simply, during carbon burning two ^{12}C nuclei collide, causing the reaction ^{12}C $+ ^{12}$C $\rightarrow ^{20}$Ne $+ ^4$He. The alpha particle in turn consumes the ^{22}Ne that had been established in high abundance during helium burning; *viz.* ^{22}Ne $+ ^4$He $\rightarrow ^{25}$Mg $+$ n. This creates abundant amounts of ^{25}Mg, and liberates free neutrons as well. The associated capture of free neutrons by ^{25}Mg creates the comparable abundance of ^{26}Mg . The small neutron separation energy, $S_n = 11.09$ MeV, prevents ^{26}Mg from surviving higher temperatures. Indeed, the two heavier isotopes of magnesium are being destroyed during neon burning at the same time that the ultimate amount of ^{24}Mg is being created.

^{26}Mg is a *secondary nuclei* (see **Glossary**). This means that ^{26}Mg can not be manufactured in stars from initial hydrogen and helium in the stellar gas, but in addition requires initial carbon and oxygen in the star. The separation energy, $S_n = 11.09$ MeV, is but 2/3 that of the abundant primary isotope, ^{24}Mg, meaning that ^{26}Mg will not survive late high-temperature events as well as does ^{24}Mg. The paragraph above shows how ^{26}Mg is derived from ^{22}Ne, which was in turn created from ^{14}N during helium burning. The ^{14}N in turn traces its presence back to C and O atoms in the initial H-burning shell, wherein C and O were transformed into ^{14}N. This makes it clear that the production of ^{26}Mg utilizes the initial carbon and oxygen as seed nuclei, making ^{26}Mg a secondary isotope. Almost all of this happens in Type II supernovae, which on average eject thousands of times more ^{26}Mg than do Type Ia supernovae and are about five times more frequent. Nature's numbers are awesome even for secondary isotopes, each Type II ejecting a mass of ^{26}Mg comparable to 6000 times the mass of the Earth! However, a significant fraction (perhaps 10%) of ^{26}Mg nucleosynthesis occurs in AGB stars, where the same reaction, ^{22}Ne + ^{4}He $\rightarrow$ ^{25}Mg + n, followed by ^{25}Mg + n $\rightarrow$ ^{26}Mg, enriches those stellar atmospheres in the He-burning-shell dredgeup events that also convert the O-rich AGB star into a carbon star.

Anomalous stellar isotopic abundance

Transitions in the molecule MgH have been observed in stellar spectra. In metal-poor stars the ratios ^{25}Mg/^{24}Mg and ^{26}Mg/^{24}Mg are observed to be smaller than their solar values, by factors of 2–3. This supports theory, which holds that the ^{25}Mg abundance should first increase slower than does ^{24}Mg (percentagewise), but later should increase faster than does ^{24}Mg (percentagewise) as solar abundance is approached. The secondary isotopes of galactic *chemical evolution* (see **Glossary**) have such behavior.

Specific observations of stars enriched in s-process nuclei show that the reaction ^{22}Ne + ^{4}He $\rightarrow$ ^{25}Mg + n is not the source of most of the neutrons liberated during He burning; because, if it were, since many of those neutrons cause ^{25}Mg + n $\rightarrow$ ^{26}Mg, the ^{26}Mg/^{24}Mg ratio would be larger than it is observed to be. This provided an important clue that a process mixing H with ^{12}C generates the neutrons for the main s process.

Anomalous meteoritic isotopic abundance

This isotope of Mg does not always occur in all natural samples from meteorites in its usual proportion. In particular, presolar grains of several different types reveal an excess of ^{26}Mg relative to ^{24}Mg because of excess atoms present owing to the decay of extinct ^{26}Al within the grain (see **^{26}Al**). This good luck has the down side of making detections of true initial fluctuations of ^{26}Mg abundance almost impossible.

Thermally fractionated Mg isotopes   Isotopic deviations from thermal chemistry are known to exist in the FUN (Fractionated and Unknown Nuclear) subgroup of the CAIs owing to their enhanced (about 1%) ^{25}Mg/^{24}Mg ratios. But the ^{26}Mg/^{24}Mg ratio from that chemical fractionation usually cannot be confirmed owing to excess ^{26}Mg from ^{26}Al decay (see **^{24}Mg**, *Thermally fractionated Mg isotopes*).

Excess ^{26}Mg from ^{26}Al: in solar-system rocks The ratio ^{26}Mg/^{24}Mg is measured to have almost exactly the same value in almost all rocks found within the solar system, be they terrestrial, lunar, martian or meteoritic. Geologic chemistry has mixed the Mg isotopes thoroughly within the hot molten states that preceded the solidification of the rock. And those planetary bodies each formed from almost identical mixtures of gas and dust. Exceptions to this rule are found in the *calcium–aluminum-rich inclusions* (see **Glossary**), small rocks, typically a few mm in size, that formed very early in the process of solar-system formation. More importantly, they formed from smaller material reservoirs, either in the gaseous nebula itself or on small early planetesimals. These smaller reservoirs contained some isotopic anomalies among stable isotopes, but more to the point here, they contained radioactive ^{26}Al, often in the ratio ^{26}Al/^{27}Al $= 5 \times 10^{-5}$. That is only 50 radioactive ^{26}Al atoms per million stable Al atoms. In well-mixed matter, that would imply that only the fraction 5×10^{-5} of stable ^{26}Mg solar abundance has been the result of ^{26}Al decay in the solar-system (because ^{26}Mg and ^{27}Al have almost equal abundances). But the CAIs contain within their mineral mix some very Al-rich minerals, especially anorthite having several hundred times more Al atoms than Mg atoms. If Al/^{26}Mg $= 1000$ in the initial mineral, for example, and if the fraction ^{26}Al/^{27}Al $= 5 \times 10^{-5}$ is radioactive, a 5% excess of ^{26}Mg results after the ^{26}Al has decayed. The telltale sign of intial ^{26}Al is a *three-isotope plot* (see **Glossary**) revealing a much larger δ for ^{26}Mg than for ^{25}Mg. Almost all of the ^{26}Mg is radiogenic in the most Al-rich oxide rocks. This is commonly believed to represent ^{26}Al atoms mixed into the initial solar system from a nearby supernova, one whose pressure wave may have been responsible for the collapse of the solar cloud. But perhaps it was created in the solar disk by nuclear reactions with high-energy particles accelerated by early solar flares. (See **^{26}Al** for further implications and considerations relating to this live ^{26}Al in the early solar system.)

Excess ^{26}Mg from ^{26}Al: in presolar grains Magnesium does not, as an element, condense favorably into the known types of presolar grains, especially SiC and graphite. Because of their crystal structures, Mg is almost excluded. Because of the existence of excess ^{26}Mg in presolar grains from the decay of ^{26}Al, and because Mg does not condense easily in presolar grains, the primary role of Mg in presolar-grain studies has been the determination of the amount of ^{26}Al within the grains at the time of their condensation. In most cases the ^{26}Mg in these grains is almost entirely radiogenic, and indicates the abundance of ^{26}Al when the grain formed. The measured ^{26}Al/^{27}Al ratios are very much larger than the ratio ^{26}Al/^{27}Al $= 5 \times 10^{-5}$ that is found in solar-system rocks. Evidently the presolar grains condensed quickly as material was leaving the stars, so that none of the ^{26}Al had decayed prior to the condensation of the grains. As a result, the inferred ^{26}Al/^{27}Al ratio can be taken to be the ratio in the stellar gas from which the grain condensed. In the mainstream SiC grains from carbon-star red giants, the initial ratio was about ^{26}Al/^{27}Al $= 10^{-3}$, in good agreement with expectations for those stellar envelopes. But in the grains from supernovae, Type-X SiC and low-density graphite, larger values near ^{26}Al/^{27}Al $= 0.01$–0.1 are most often recorded. Such large ^{26}Al/^{27}Al ratios are just one of many reasons to believe that those grains are condensed portions of the interiors of expanding supernova gases, a process predicted three decades ago.

Solar high-energy particles (SEPs)

Solar high-energy particles are accelerated by the Sun. The so-called "impulsive events" are apparently accelerated by solar flares. NASA's *Advanced Composition Explorer* (ACE) has found large variations in the ^{26}Mg/^{24}Mg ratios within such events, ranging from 0.1–1. These are thought to reflect the different masses in the ratio Q/A of charge to mass for Mg isotopes, since each should have the same charge state; but the variations are much larger than would be expected from similar variations of Fe/Mg element ratio in the SEPs. Fe/Mg should vary by a factor 2 in Q/A, whereas the ^{26}Mg/^{24}Mg ratio varies by only 8% in Q/A. Making matters more puzzling, the ^{26}Mg/^{24}Mg ratio is larger in those impulsive flares yielding the highest ^{3}He/^{4}He ratios, which are augmented by up to 10 000 times! But the He results favor the lighter isotope, whereas the Mg results favor the heavier one. A similar situation is observed for neon in the SEPs. This exciting data is not well understood.

13 | Aluminum (Al)

All aluminum nuclei have charge +13 electronic units, so that 13 electrons orbit the nucleus of the neutral atom. Its electronic configuration can be abbreviated as an inner core of inert neon (a noble gas) plus three more electrons: $(Ne)3s^2 3p^1$, which locates Al in Group IIIA of the chemical periodic table, between boron (B) and gallium (Ga). Its principal oxidation state is +3, so its oxide is Al_2O_3, a very important compound in cosmochemistry. Its chloride is $AlCl_3$.

Like many elements, aluminum's name derives from the Latin because the Roman Empire used two of its naturally occurring minerals, *alum* (a potassium–aluminum sulfate) and *alumina* (aluminum oxide). It was isolated in 1825 by H. C. Oersted after decades of suspicion that it existed.

As an element, Al is the third most abundant in the Earth's crust, and the most abundant of all metals there, whereas it is but 12th in the universe. This is because its oxide, Al_2O_3, is so easily made and so tightly bound into rocks that Al remained in the crust when less reactive metals (including most Fe) sank to the molten core of Earth. Al_2O_3 is the fourth most abundant compound in the Earth's crust, following SiO_2, MgO, and FeO.

Pure Al metal is lightweight silvery white. It is highly malleable and ductile. Although too soft for construction, it becomes hard and strong when alloyed with small amounts of Si and Fe, making it useful in airplanes and in engines. It oxidizes easily but only superficially, because the thin film of Al_2O_3 that coats its surface protects it from further corrosion. Commercial Al is produced by electrolysis of Al_2O_3 which is found in the common *bauxite* mineral.

Aluminum compounds are well known and important to many commercial products.

Because aluminum has but a single stable isotope it sheds no light on isotopic anomalies within presolar material except through its long-lived radioactive isotope ^{26}Al, which has become the most common gamma-ray-line emitter in the interstellar gas, confirming the rates of continuing nucleosynthesis during the past million years. ^{27}Al nucleosynthesis derives almost entirely from a single main-line source, the fusion of carbon in massive stars that will become Type II supernovae.

Natural isotopes of aluminum and their solar abundances

A	Solar percent	Solar abundance per 10^6 Si atoms
26	extinct radioactivity in solar system and in presolar grains	
27	100	8.49×10^5

^{26}Al | $Z = 13$, $N = 13$ $t_{1/2} = 7.28 \times 10^5$ years
Spin, parity: J, $\pi = 5+$;   **Mass excess**: $\Delta = -12.210$ MeV

The two Al isotopes ^{26}Al and ^{27}Al resemble the two Na isotopes ^{22}Na and ^{23}Na in that the $Z = N$ isotope of both is a long-lived positron-emitting radioactive nucleus that appears in nature, and both can be made by adding alpha particles to ^{14}N. It is the large spin, $S = 5$, of the ^{26}Al nuclear ground state that makes it decay so slowly. But a 0+ isomeric state of low excitation decays in just 6.3 seconds, becoming an issue for ^{26}Al nucleosynthesis.

Extinct radioactivity

This isotope of aluminum has great importance for the science of *extinct radioactivity* (see **Glossary**). It appears to have had significant abundance in the early solar system, about 5×10^{-5} that of stable aluminum; but it is also of the type of extinct radioactivity that shows itself within presolar grains where it was much more abundant. The ^{26}Al created in explosive events, novae or supernovae, remains alive for the time needed to grow dust grains within the expanding and cooling gas of the explosions, but its halflife is too short for us to understand how its large abundance in the early solar system could be typical of interstellar molecular clouds, where stars and planetary systems form. If all molecular clouds contained that much ^{26}Al, the gamma-ray maps of the Milky Way would be brighter.

Presolar grains   Extinct radioactive ^{26}Al is recognized as having once been present within many different types of presolar grains found within the carbonaceous meteorites by the excess of daughter ^{26}Mg that those grains contain. The magnesium within some Al-rich grains is almost pure ^{26}Mg, arguing strongly that the condensation of ^{26}Al as one of two isotopes of chemically reactive aluminum is its reason for being so abundant in presolar grains. This expectation was described in 1975 by the writer, more than a decade prior to the actual discovery of these grains. Because aluminum is so chemically reactive, it forms very tightly bonded solids that can exist at high temperatures. This helps it to appear condensed in several types of presolar grains – in the SiC grains from AGB stars; in low-density graphite grains and Type X

SiC grains from supernovae; and in Si_3N_4 grains from supernovae. The 1975 prediction was specifically that ^{26}Al must appear condensed in corundum grains, Al_2O_3, from supernovae; but these have been very hard to find, probably because they are smaller than the large presolar grains that have been amenable to study. By comparing the daughter ^{26}Mg abundance in these grains with the abundance of stable ^{27}Al in them it is found that the ratio for ^{26}Al/^{27}Al in these grains at the time of their condensation ranges from values as large as 0.6 down to near 0.001, the lowest level detectable using SIMS (see **Glossary**). This range of initial isotopic ratios at the times that the grains condensed (usually within a few months of ejection or explosion) has set exciting challenges concerning the location and nature of nucleosynthesis sites.

Solar-system rocks Extinct radioactive ^{26}Al is also recognized as having once been present within rocks during planetary formation chemistry in the early solar system. These rocks are larger than presolar grains but still small to ordinary geologic experience, having sizes between 0.01 and 10 mm. Prior presence of ^{26}Al is again attested to by the excess of daughter ^{26}Mg that these rocks contain. But in these cases the rocks contain all three isotopes of magnesium in considerable abundance; therefore the detection of excess ^{26}Mg resulting from ^{26}Al decay rests upon studying rocks that are very rich in the element Al. If the initial chemical abundance ratio Al/Mg is large in the rock as it forms by hot chemistry, the decay of live ^{26}Al to ^{26}Mg can be detected because it produces a significant excess of ^{26}Mg. To understand how this can be detected, consider that early in solar-system history ^{26}Al persisted only in the modest isotopic ratio ^{26}Al/^{27}Al $= (0.1 - 5) \times 10^{-5}$, which is not capable of greatly enriching the ^{26}Mg unless Al is much more abundant than ordinary Mg. The types of rock in which this is easily done are aluminum oxides, corundum or hibonite, and some forsterite olivine grains that have low Mg content.

When excess ^{26}Mg was first detected in Al-rich rocks, the argument ensued whether that excess was the result of live ^{26}Al that had decayed within that very rock, or whether it was a fossil of live ^{26}Al that had previously decayed within presolar Al-rich grains that in turn were later fused together to make the rocks in the solar system. The former process, decay in the rock, is called *in-situ* decay, whereas the latter, prior decay within presolar grains, is called *fossil* ^{26}Mg. This argument was sometimes acrimonious, for a decade or more. But careful isotopic measurements within a given rock, comparing portions having locally differing chemical ratios Al/Mg, have established that ^{26}Al was alive at the level 5×10^{-5} of stable ^{27}Al at the time and place where many rocks formed. Fossil ^{26}Mg must also exist in some rocks of differing types and origins; but it has proven difficult to establish the effect owing to the large contaminating abundance of normal magnesium within them. The fossil ^{26}Mg is very clear in the presolar grains, however; but only in them.

Even the small isotopic ratio $^{26}Al/^{27}Al = 5 \times 10^{-5}$ is too great to be expected on average within molecular clouds. These clouds take 50 Myr to form from dilute interstellar gas, so that ^{26}Al should be virtually absent by that time, owing to its much shorter halflife. This conundrum has not been solved; but the leading bet is that a massive star existed quite close to the cloud core that would soon form our solar system. This massive star ejected about 10^{-4} solar masses of ^{26}Al, either in a steady wind over a million-year period or in an explosive ejection when that star became a supernova (see **Nucleosynthesis origin**, below). It is speculated that this ejected mass of ^{26}Al mixed into the cloud that was just beginning to form the solar system. This is not so great a coincidence as it might seem to be if the massive-star ejecta actually *caused* the initial collapse of the presolar cloud. The massive star is pictured as *a trigger* for the formation of the Sun. In this way one does not require the ratio $^{26}Al/^{27}Al = 5 \times 10^{-5}$ to be typical of the entire molecular cloud, but rather just of that portion that was triggered into collapse by the massive star. This is an attractive resolution in many ways, but not without its own physical difficulties and implausibilities.

The only two alternatives to the trigger picture have not won majority scientific support. The first is the suggestion that the early solar-system chemistry may be able to produce the ^{26}Mg excess within solar rocks from the fossil ^{26}Mg excess that exists in presolar grains from which the rocks were formed (fossil theories). Although the initial conditions for such a picture do exist, chemists conclude that the chemical processes that formed the CAIs would have eliminated the fossil-^{26}Mg correlation with the element Al. The second alternative to triggered solar formation is that processes within the molecular cloud or within the early solar system accelerate energetic particles (low-energy cosmic rays), and that these create the ^{26}Al by nuclear reactions during collisions with other heavy nuclei (energetic-particle theories). One leading version of that theory utilizes ^{3}He-rich solar-flare particles, which create ^{26}Al from collisions with ^{24}Mg in the forming rocks. Observations of young stars in Orion with the *Chandra X-ray Observatory* have shown that those stars experience more frequent and more violent flares than does the present Sun. Most astonishing enrichments of ^{3}He relative to ^{4}He are found in atoms accelerated by solar flares today. The charge-to-mass ratio $Z/A = 2/3$ for the bare ^{3}He nucleus is greater than any other nucleus save ^{1}H. This fact favors its acceleration by the hydromagnetic phenomenon of solar flares, or ^{3}He-rich flares. Observations of solar-flare particles at 385 keV/nucleon made from NASA's *Advanced Composition Explorer (ACE)* in January 2000 revealed an abundance ratio $^{3}He/^{4}He = 33$, a million times greater than the mean value of that ratio in the Sun! Abundant flares of this type may have produced ^{26}Al in the early solar system through collisions of accelerated ^{3}He ions with the abundant ^{24}Mg nuclei already contained in the solids in the solar disk. Unquestionably the truth can be discerned only by simultaneously understanding all of the ten *extinct radioactivities* within a self-consistent picture that includes a physical understanding of the origin of those rocks, chondrules, and Ca–Al-rich inclusions.

Live ^{26}Al globular-cluster red-giant stars   The surface compositions of the red-giant stars in the old globular star clusters of our Galaxy show a puzzling anticorrelation between aluminum abundance and oxygen abundance. These stars presumably formed at the same time (about 10 billion years ago) and from the same gas (except for some variations of initial abundances that were caused by differing contributions of nucleosynthesis from earlier stars in that globular cluster). But in those stars for which Al (see also Na) exist in greater-than-average abundance for the stars of the cluster, oxygen is less abundant than average. It was natural to think that this Al abundance was of its stable isotope, ^{27}Al; but modern arguments show that it is more likely to be of radioactive ^{26}Al, which is made continuously in its hydrogen-burning-shell from the initial ^{25}Mg (see **Nucleosynthesis origin**, below). A very subtle part of this argument is that an earlier generation of AGB stars in the globular cluster, stars that are now no longer there, had enriched the cluster gas in ^{25}Mg, because it is from that ^{25}Mg that the ^{26}Al is continuously produced by mixing of surface material into the hydrogen-burning shell. This mixing is not totally understood. But the amount of ^{26}Al that is produced in the last million years by this process, which is all that can exist in the star today, is enough to render aluminum overabundant today. The same burning processes consume oxygen. The net result is noteworthy; namely, that stars exist in which their major Al abundance is apparently this radioactive isotope. A consequence is that those atmospheres must also be enriched in ^{25}Mg and ^{26}Mg, but that can not yet be detected. For this mechanism to work, the first-generation AGB stars must be able to synthesize ^{25}Mg despite having negligible C and O abundance in their initial compositions. (See **^{25}Mg, Nucleosynthesis origin** for the reason for ^{25}Mg-richness of the AGB stars.)

Gamma-ray astronomy

Live ^{26}Al has been detected in the interstellar gas of the Milky Way by its gamma-ray line at 1.809 MeV emitted following its beta decay to ^{26}Mg. This was first discovered in 1982 by a high-resolution gamma-ray spectrometer on NASA's *HEAO-C* spacecraft. Its total flux from the Milky Way was later measured by NASA's *Solar Maximum Mission*. The number of arriving gamma rays is about 0.0004 per cm^2 per second, or about 4 per second striking a square detector of dimension 1 meter (about 3 ft) on a side. This search has culminated with NASA's *Compton Gamma Ray Observatory*, the second of NASA's planned sequence of Great Space Observatories. Its COMPTEL (COMPton TELescope) has shown the radioactivity to be concentrated to the plane of the Milky Way. There exist intense regions near the galactic center and several concentrated areas of higher flux that correlate best with the known positions of young star-forming regions of the Milky Way. These correlations suggest very strongly that interstellar ^{26}Al is produced primarily by the evolution of massive stars, which are found only in regions of current star formation. The current detections are from gas containing the ejecta of many supernovae over a few million year period prior to today, so that the concentration of ^{26}Al can build up. Each Type II supernova ejects about 0.0001 solar masses of ^{26}Al, which sounds more impressive as 20–30 times the mass of the Earth. About 2 solar masses of ^{26}Al exists in the entire interstellar medium (containing about 10 billion solar masses of gas). Hot gas between Earth and Orion seems to also contain detectable ^{26}Al. The ^{26}Al gamma ray from a single exploding star (or Wolf–Rayet star) has not yet been detected, however, because ^{26}Al decays so slowly that it emits its gamma rays grudgingly.

The ^{26}Al decay in the interstellar medium also emits positrons (antielectrons), which may be a partial source of the positron–electron annihilation into gamma rays that are also seen emanating from the interstellar gas.

Nucleosynthesis origin

^{26}Al nuclei in stars are primarily the result of hydrogen burning or explosive helium burning in either novae or in supernovae. During hydrogen burning in stars, ^{26}Al is made when stable ^{25}Mg captures a free proton. This occurs first while the star is burning hydrogen in hydrostatic equilbrium prior to the later explosion, but may also occur during the elevated temperature of the explosion in some circumstances. The ^{26}Al may be continuously ejected in a wind of outflowing gas from very massive stars, or it may be primarily ejected by the final explosion. Even more ^{26}Al may be synthesized deeper within the massive star, during burning processes called carbon burning and neon burning. Which source dominates the observed gamma-ray line from star-forming regions is still not certain. A very interesting nuclear problem for ^{26}Al synthesis in the inner burning shells is the question whether the long-lived ground state of ^{26}Al can be thermally converted to the rapidly decaying isomeric state of ^{26}Al, thereby lowering greatly its effective lifetime and the amount synthesized. Common core-collapse supernovae about 20–30 times more massive than the Sun eject about 0.0001 solar masses of ^{26}Al. These provide most of the steady interstellar anticipated ^{26}Al activity. But rarer, more massive supernovae can individually eject more in the massive winds that bring those stars down to size before they explode. Calculations imply that supernovae 100–120 times more massive than the Sun ejected about 0.001 solar masses of ^{26}Al during the few million years leading up to their explosion, about ten times more than the common supernovae. However, the 100-solar-mass star should occur only 2% as frequently as the 25-solar-mass variety, and therefore cannot dominate the yield from the full spectrum of stars. Nevertheless, two applications of significance for these rare very-massive supernovae are of high interest. Firstly, in massive associations of large numbers of new stars of types O and B (such as occurred in Orion about 10 million years ago), these brutes may cause a bright gamma-ray-emitting region for a couple of million years as they eject their ^{26}Al-laden winds. Such may play a role in the galactic "hotspots" of ^{26}Al gamma rays that have been mapped by COMPTEL on NASA's *Compton Gamma Ray Observatory*. Secondly, a single massive presupernova, called in this pre-explosive phase a "Wolf–Rayet star," may be able to pollute the surrounding molecular cloud material with an abnormally high concentration of ^{26}Al, perhaps enough to account for ^{26}Al having an abundance near 0.0001 that of ^{27}Al, as it is found to have had in the small molecular cloud that collapsed to form our own Sun (see *Solar-system rocks*, above). If so, that neighbor probably triggered the solar cloud to collapse owing to the pressure of those winds on the presolar cloud.

If helium novae exist (in which the explosive envelope on the white-dwarf star is made of He rather than H), three successive captures of free helium nuclei by ^{14}N atoms partially transmute them into ^{26}Al. This process also happens in helium caps atop Type Ia supernovae if ^{14}N still remains within He at the time of explosion. Although this would not seem likely in the He-burning shells of massive stars, wherein the ^{14}N has been destroyed prior to the explosion, it does seem quite likely in He-cap detonation models for Type I supernovae. This site has been shown to produce large amounts of ^{26}Al, approximately as much as stable ^{27}Al, giving He caps the largest ratio for ^{26}Al/^{27}Al of any known nucleosynthesis site. Such high ratios for ^{26}Al have been found in the meteoritic SiC grains of Type X, so that they may come from such He-cap Type Ia supernovae.

Some ^{26}Al may also be produced by the classical nova explosions. In these the entire star does not explode; rather, about 0.0001 solar masses of "skin" on a white dwarf experiences a thermonuclear runaway, some fraction of which is ejected in the nova outburst. This ejecta should contain live ^{26}Al, especially from a subset of Mg-rich novae. Magnesium is the parent from which ^{26}Al is created during the explosion. Dust grains that condense Al in the ejecta from the novae will carry excess ^{26}Mg from this initial content of ^{26}Al, as Fred Hoyle and I pointed out in 1975–76.

It must be appreciated that ^{26}Al production is not the major source of stable daughter ^{26}Mg in nature. Most ^{26}Mg is made in stable form in stars (see **^{26}Mg**).

Other applications

The existence of ^{26}Al in cosmic rays allows some other applications of natural interest.

Dating of sediments ^{26}Al exists in the cosmic rays owing to collisions between them and interstellar atoms. The same thing is true for radioactive ^{10}Be. The ^{26}Al/^{10}Be ratio in the cosmic rays enables one to determine that they have been traveling about 10 Myr in interstellar space before arriving at Earth.

High-energy cosmic rays reach the ground and produce measurable levels of these radioactivities on Earth. The ^{26}Al/^{10}Be ratio in buried quartz allows careful measurement of the time of sedimentary burial. The production ratio $p(26)/p(10) = 6$ from cosmic-ray interactions with abundant stable elements, so differential decay allows dating of sediments, as in fluvial deposits in Mammoth Cave, KY. This works with 100 000-yr accuracy for deposit ages up to 5 Myr.

Cosmogenic radioactivity ^{26}Al is created as a collision fragment in meteorites when cosmic rays strike the meteorite during its journey to the Earth. This radioactivity is counted in the lab, alive today in the meteorite, and gives information on its history in space.

^{27}Al $\quad$ Z $= 13$, N $= 14$
Spin, parity: J, $\pi = 5/2 +$;
Mass excess: $\Delta = -17.197$ MeV; $\qquad S_n = 13.06$ MeV

^{27}Al and ^{39}K have the largest neutron binding energies, S_n, of any stable odd-A nuclei (with ^{39}K winning by a mere 0.03 MeV).

Abundance

^{27}Al is the only stable isotope of aluminum; therefore all elemental observations of its abundance are of the abundance of ^{27}Al. This isotope has

solar abundance of ^{27}Al $= 8.49 \times 10^5$ per million silicon atoms.

^{27}Al is the 18th most abundant nucleus within the universe and 12th most abundant element.

Nucleosynthesis origin

In the mix of interstellar atoms from which the solar system formed, ^{27}Al exists primarily owing to the carbon-burning process of stellar nucleosynthesis. Put simply, during carbon burning two ^{12}C nuclei collide, causing several reactions that synthesize the isotopes of Mg (see **^{26}Mg**, *Nucleosynthesis origin*). Once ^{26}Mg is formed, its reactions with free protons and neutrons, which are sustained by the main reactions of carbon burning, easily and efficiently synthesize ^{27}Al as well. In this sense ^{27}Al synthesis is closely tied to that of the two neutron-rich isotopes of Mg. Essentially all of this happens in Type II supernovae, or, more precisely, in the burning that leads up to the supernova, which on average eject about 20–30 times as much ^{27}Al as do Type Ia supernovae and are about five times more frequent. Nature's numbers are awesome, each Type II ejecting a mass of ^{27}Al comparable to 5000 times the mass of the Earth!

A small percentage of the ^{27}Al is created instead in the hydrogen-burning shells of evolved stars, where the ^{26}Mg $+ \, ^1$H $\rightarrow \, ^{27}$Al reaction can occur. At the time of the supernova explosion, this shell burning can have almost doubled the ^{27}Al abundance locally in the hydrogen convective shell that has been exposed at the stellar surface.

The above reactions might make it seem that, on the one hand, ^{27}Al is a *primary nucleus* in nucleosynthesis parlance. That designation is given to those nuclei that are abundantly produced within stars whose initial composition is hydrogen plus helium, the two elements left from the Big Bang. Indeed, initial hydrogen and helium are converted to ^{12}C by the natural thermonuclear evolution of matter. So ^{12}C is a primary nucleus, and so therefore might be ^{27}Al, which is made from the fusion reactions ^{12}C + ^{12}C. On the other hand, calculations of ^{27}Al production show secondary behavior in that the abundance synthesized depends on the amount of ^{22}Ne burned during the carbon burning; and the amount of ^{22}Ne depends on the initial C and O content of the star (see *Secondary nucleus*, in **Glossary**).

Astronomical observations

Aluminum has been studied in stellar optical spectra. The lines at 3944 and 3961 Å have dominated Al analyses. As nucleosynthesis progressed the galactic Al/H abundance ratio in newly born observed stars increased from 10^{-4} of solar in some early stars to a bit in excess of solar Al/H today. If compared instead to Mg, the ratio Al/Mg increases by a factor of ten from the values in metal-poor stars to those in recent stars. Augmenting the coproduction of Al and Mg in massive Type II supernovae, the gradual increase in Al/Mg ratio as stellar metallicity increased is understood by the initial metallicity aiding Al production during carbon-burning (see above).

Anomalous meteoritic isotopic abundance

Because ^{27}Al is the only stable isotope, no isotopic anomalies in Al can be found except for the aforementioned abundance of radioactive ^{26}Al.

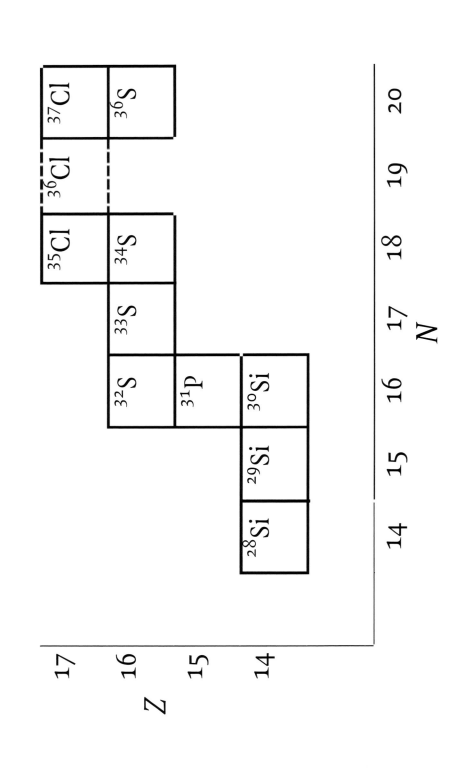

14 | Silicon (Si)

All silicon nuclei have charge $+14$ electronic units, so that 14 electrons orbit the nucleus of the neutral atom. Its electronic configuration can be abbreviated as an inner core of inert neon (a noble gas) plus four more electrons: $(Ne)3s^2 3p^2$, which locates Si beneath carbon in Group IV of the periodic table. Silicon therefore normally has valence $+4$ and combines readily with pairs of oxygen atoms in making its oxide.

Silicon is a nonmetallic element having great importance to man. It constitutes about 28% by weight of the Earth's crust, being the second most abundant element on Earth. It occurs in the form of *silica* (sand and glass) and *silicates*, a major mineral form of rocks. Because the oxygen atoms outnumber the silicon atoms in these minerals, oxygen is the most abundant element on Earth, and silica (sand), SiO_2, its most abundant compound. The Earth's crust is, to a first approximation, comprised of minerals made of Si and O atoms, with just enough Na, Mg, Al, K and Fe atoms to assemble some of them into silicates. Its high abundance in meteorites and Earth have caused silicon to be adopted as an abundance standard. The abundances of other elements are usually given *per million Si atoms*. In the wider universe, Si is very abundant but not dominantly so, taking seventh place in the cosmic list. Carbon is 3.5 times more abundant in the universe and in the Sun, but Si is more abundant than C by a large factor on Earth. In the meteorites, Si exists primarily as the silicates, $MgSiO_3$, $FeSiO_3$, Mg_2SiO_4, Fe_2SiO_4, and their solid mixtures, which reveal Mg and Fe to be only slightly less abundant than Si. The Earth's crust had given a distorted view of the relative abundances of these elements in the universe.

Silicon was first recognized as a distinct element in 1824 by Swedish chemist Jakob Berzelius. In this form it is a black to dark-gray solid, brittle, and with a metallic luster. Commercially, Si is liberated from silica through reduction by carbon: $SiO_2 + 2C \rightarrow 2CO + Si$. Although pure Si is a poor conductor of electricity, the electronic configuration of its mineral form has spawned the computer revolution (silicon computer chips). *Silicon carbide* (SiC) is a tough abrasive mineral that is used in grinding tools.

Silicon is composed of three stable isotopes, masses 28, 29 and 30, and is one of the cornerstone elements of nucleosynthesis theory. Thermal decomposition of its nuclei after first becoming dominant in the cores of stars and its subsequent reassembly into iron-group elements is called "silicon burning" in stellar evolution and nucleosynthesis. That process is important for understanding how supernova explosions assemble the relative abundances of the elements having masses between

$A = 28$ (Si) and $A = 56$ (Fe). It also provided the cornerstone to a new e-process theory in which iron is synthesized as radioactive nickel, rather than as iron itself as Hoyle initially argued.

Silicon is one of the most important elements in astronomical studies based upon the anomalous isotopic compositions of presolar grains. These include not only the presolar grains containing silicon (as SiC), but also presolar graphite particles containing trace Si compositions. These isotopic variations have helped define physically distinct families of presolar grains and helped to identify their stars of origin.

Natural isotopes of silicon and their solar abundances

A	Solar percent	Solar abundance per 10^6 Si atoms
28	92.2	9.22×10^5
29	4.67	4.67×10^4
30	3.10	3.10×10^4

^{28}Si | $Z = 14$, $N = 14$
Spin, parity: $J, \pi = 0+$;
Mass excess: $\Delta = -21.492$ MeV; $\qquad S_n = 17.17$ MeV

^{28}Si is a very tightly bound nucleus. Its neutron separation energy exceeds that of all heavier stable nuclei, and its proton and alpha particle separation energies are also impressively large. That is why ^{28}Si is very resistive to thermal disintegration, and why silicon burning in stars must overcome the problem of breaking ^{28}Si down before the nuclei can be reassembled into the iron group.

Abundance

From the isotopic decomposition of normal silicon one finds that the mass-28 isotope, ^{28}Si, is the most abundant of the three stable Si isotopes; 92.23% of all Si. Using one million silicon atoms, the common cosmochemical standard in solar-system matter,

$$\text{solar abundance of } {}^{28}\text{Si} = 0.9223 \times 10^6 \text{ per million silicon atoms.}$$

This makes ^{28}Si the eighth most abundant nucleus in the universe, between ^{2}H and ^{24}Mg.

Nucleosynthesis origin

^{28}Si exists primarily owing to the oxygen-burning process of stellar nucleosynthesis. Put simply, during oxygen burning two ^{16}O nuclei collide, causing the reaction ^{16}O $+ ^{16}$O $\rightarrow {}^{28}$Si $+ ^4$He. The alpha particles are initially consumed by ^{24}Mg, which was already abundant in the prior ashes of carbon burning: $^{24}Mg + {}^4He \rightarrow {}^{28}Si +$ gamma ray. By these two reactions one effectively gets the net reaction $2\,{}^{16}O +$ $^{24}Mg \rightarrow 2\,{}^{28}Si$. These make ^{28}Si the most abundant product of oxygen burning by a large margin. Its large neutron separation energy, $S_n = 17.17$ MeV, ensures that ^{28}Si will survive high temperatures well, somewhat in contrast to the two heavier isotopes of silicon, which are being destroyed at the same time that ^{28}Si is being created.

Primary nucleus   The above reactions are helpful in understanding that ^{28}Si is called a "primary nucleus" in nucleosynthesis parlance. That designation is given to those nuclei that are abundantly produced within stars whose initial composition is hydrogen plus helium, the two elements left from the Big Bang. Indeed, initial hydrogen and helium are converted to ^{16}O by the natural thermonuclear evolution of matter. So ^{16}O is a primary nucleus, and so therefore is ^{28}Si, which is made from the ^{16}O. ^{28}Si is distinguished from the two heavier isotopes of silicon in this way; ^{29}Si and ^{30}Si are *secondary nuclei* (see **Glossary**) because their nucleosynthesis requires that the star manufacturing them have not only hydrogen and helium in its initial composition, but also carbon and oxygen. Following changes during H and He burning, the initial carbon and oxygen provide for the excess neutrons that the heavier isotopes contain. Nucleosynthesis of ^{28}Si in this way occurs abundantly both in core-collapse Type II supernovae and in thermonuclear Type Ia supernovae. Each can eject one-tenth of a solar mass of pure ^{28}Si, which is 30 000 Earth masses!

Astronomical measurements

The Si abundance has been studied in stellar optical spectra, usually in line transitions of the neutral Si atom. Because ^{28}Si dominates the spectra of Si in stars, and especially because ^{28}Si is a primary "alpha nucleus," the galactic evolution of ^{28}Si defines the observed galactic evolution of Si elemental abundance. As nucleosynthesis progressed the galactic Si/H abundance ratio in newly born observed stars increased from 10^{-3} of solar in some stars to a bit in excess of solar Si/H. The behavior is very similar to that of oxygen, and indeed to all alpha nuclei that can be created from the fusion of He, C and O. These include the elements Ne, Mg, Si, S, Ar, Ca, and Ti. The ratio Si/Fe declines, however, from about three-times solar in metal-poor stars to solar in young stars. This is understood as the rapid coproduction of the alpha nuclei, along with O, in early massive Type II supernovae. These produce less iron; but the later onset of Type Ia supernovae explosions increases Fe more dramatically than it does the Si abundance. For that reason, interstellar Si/Fe undergoes a steady gentle decline.

X-ray emissions from young, hot, shocked supernova remnants, as observed with NASA's *Chandra X-Ray Observatory*, confirm that core-collapse Type II supernovae are prolific nucleosynthesis sources of Si; but X-ray emissions do not yet measure isotopic composition.

Anomalous meteoritic isotopic abundance

The isotopes of Si do not always occur in all natural samples from meteorites in their usual proportions. In particular, presolar grains of several different types reveal sometimes a deficiency and sometimes an excess of ^{28}Si relative to the two heavier isotopes, whose abundances are represented by their ratio to ^{28}Si (traditional because ^{28}Si is the most abundant isotope). The anomaly is defined by the fractional difference of the ^{30}Si/^{28}Si ratio (or the ^{29}Si/^{28}Si ratio) from its solar value; that is, by the ratio

$$(^{30}\text{Si}/^{28}\text{Si})_{\text{sample}}/(^{30}\text{Si}/^{28}\text{Si})_{\text{solar}} = 1 + \delta,$$

where δ is defined as the fractional deviation from the solar ratio. It may be positive or negative. Rather than giving the value of δ as a decimal fraction, however, it is more common to express it in percent, in parts per thousand, or in parts per ten thousand. All are found in the science literature. There is no way to determine whether those measured silicon isotope ratios having a value less than unity relative to solar (negative δ) are to be interpreted as an excess in the number of ^{28}Si atoms (the denominator) or a deficiency in the number of ^{30}Si atoms (the numerator). For the most part, such a distinction is not meaningful, representing more a bias in the mental picture than a physical situation. Astrophyiscs must provide the sense of the deviation from solar values.

Thermally fractionated Si isotopes   When the abundances of all three Si isotopes are measured in any sample, it is possible to locate that composition as a point on a graph whose y axis is the ratio ^{29}Si/^{28}Si and whose x axis is the ratio ^{30}Si/^{28}Si. Such a graph has been named the *three-isotope plot* (see **Glossary**), because three isotopes make only two independent ratios, the two coordinates of the composition. Each bulk sample is represented by its location point on that graph. In all bulk samples of Earth and Moon, the different samples are measured to lie very close to one another in this plot. The Si isotopic ratios vary significantly from their average solar values in the calcium–aluminum-rich inclusions in meteorites, however. The deviations are very accurately measured and give important thermal information about the formation of the CAIs. Deviations in ^{29}Si/^{28}Si can be a few percent in the FUN CAIs (see **Glossary**, *calcium–aluminum-rich inclusions*), which is quite large for thermal fractionation (fractionation owing to chemical effects). In these same samples the ^{30}Si/^{28}Si ratio deviates from average in the same direction, but almost exactly twice as much. As example, if $\delta(^{29}\text{Si}/^{28}\text{Si}) = 1\%$, then $\delta(^{30}\text{Si}/^{28}\text{Si}) = 2\%$. This correlated behavior is a well-understood consequence of thermal chemistry. These chemical effects are those that depend to some degree on the mass of the isotope that is reacting (or in this case, evaporating). The mass-30 isotope differs in mass from the mass-28 isotope by twice the amount that the mass-29 isotope differs from it. Temperature and average-velocity effects are thus twice as large for the ^{30}Si/^{28}Si as for ^{29}Si/^{28}Si, so that such chemical factors can "fractionate" one isotope from another. Such samples lie on a correlation line having slope 1/2 in the three-isotope plot. Such "chemical fractionation" can be used as a tool to study the thermal history of samples. But it is a mass-dependent chemical effect (chemical fractionation), not an abundance difference caused by some initial difference in abundance. Isotopic data of this kind seem to imply that the CAIs were partially molten rocks in the early solar disk, and that the enrichments occurred by evaporation of atoms from the small molten CAIs. This tells us something very important about the early solar system.

Mainstream SiC grains   The first presolar grains to be characterized in the laboratory were SiC (silicon carbide) grains. Isotopic abnormalities exist within all SiC grains in meteorites, showing that this mineral was not one that was commonly made within our solar system from normal silicon atom populations. All SiC grains seem to be presolar, older than the Earth, inherited from the interstellar medium. Nonetheless, the SiC presolar grains have been divided into six distinct groups, based not only on the silicon isotopic ratios within the grains but also on the isotopic ratios of the other elements (either the carbon or nitrogen isotopes condensed within the SiC). These six groups are called "mainstream, A, B, X, Y, and Z." The names are historical and have little mnemonic value, save the "mainstream," which is a correlated set of isotope ratios found in the vast majority (90–95%) of the SiC grains. The grains range in size from about 0.1 to about 10 μm. These are tiny by human experience, but larger than average size of interstellar dust particles.

^{28}Si is the most abundant isotope also in the presolar grains. But it is not quite as dominant, a statement that can be said with high precision because the isotope ratios are measured with an accuracy of one part per thousand (to accuracy 0.1%, or $\delta = 1$). The characteristic signature of the mainstream SiC grains is ^{29}Si/^{28}Si and ^{30}Si/^{28}Si isotopic ratios greater than solar. The excess ^{29}Si/^{28}Si ratio is a continuum of values up to 20% ($\delta = 200$). The excess ^{30}Si/^{28}Si ratio is a continuum of values up to 15% ($\delta = 150$). Thus ^{28}Si is a smaller fraction of all Si atoms in the mainstream presolar grains than it is in solar-system matter (sand at the beach). But one can not discuss ^{28}Si without discussing the two heavier isotopes. Plotting the isotopic ratio of each SiC grain in the three-isotope plot $\delta(^{29}$Si/^{28}Si) versus $\delta(^{30}$Si/^{28}Si) shows their values to align along a line having slope 4/3. That is, an increase of $\delta = 150$ for ^{30}Si/^{28}Si is accompanied by an increase $\delta = 200$ for ^{29}Si/^{28}Si. This linear correlation strongly suggests that the mainstream is a family, probably having a common origin. It is believed that the mainstream SiC grains condensed in dense winds leaving the surfaces of AGB stars called "carbon stars" (see **^{29}Si** or **^{30}Si** for further interpretation).

Other presolar grain groups   (See **^{29}Si** or **^{30}Si** for this information). Since ^{28}Si is used as the normalizing isotope in forming the isotopic ratios, it is more convenient to list these presolar groups under the two neutron-rich isotopes of Si. The Type X grains appear to be rich in ^{28}Si in comparison with the two heavier isotopes; but this can just as well be described as a deficiency in the two heavier isotopes. These Type X grains are especially exciting in that they appear to have condensed within the expanding interiors of supernovae, where the newly synthesized ^{28}Si can be more dominant than in the solar abundances (see **Nucleosynthesis origin**, above).

^{29}**Si** | $Z = 14, N = 15$
Spin, parity: $J, \pi = 1/2+$;
Mass excess: $\Delta = -21.895$ MeV; $\qquad S_n = 8.47$ MeV

Abundance

From the isotopic decomposition of normal silicon one finds that the mass-29 isotope, ^{29}Si, is the second most abundant of the Si isotopes; 4.67% of all Si. This makes it 19.7 times less abundant than the most abundant of the silicon isotopes, namely ^{28}Si, which constitutes 92.23% of natural Si. Considering that one million silicon atoms is the common cosmochemical standard in solar-system matter,

solar abundance of ^{28}Si $= 4.67 \times 10^4$ per million silicon atoms.

This is the 22nd most abundant isotope in the universe, comparable to ^{18}O and ^{54}Fe. Among neutron-rich nuclei having an odd number of neutrons, only ^{13}C and ^{25}Mg are more abundant.

Nucleosynthesis origin

In the mix of interstellar atoms from which the solar system formed, ^{29}Si exists primarily owing to carbon-burning and a subsequent neon-burning process of stellar nucleosynthesis. These processes liberate free neutrons into the stellar gas, and those free neutrons are captured to convert initial ^{28}Si into ^{29}Si (in part). On top of that, the neon-burning process uses hot alpha particles set free in the gas by the neon burning to convert the two heavier isotopes of Mg into the two heavier isotopes of Si. That is, in neon-burning the abundant ^{25,26}Mg are transformed into ^{29,30}Si. The ^{25,26}Mg had previously been overabundant owing to their nucleosynthesis during the preceding carbon burning. Owing to this progression of silicon isotopic abundances as the massive star core evolves, idealized computer models of that core contain regions of quite different silicon isotopic ratios. These may be instrumental in understanding the observed isotopes in presolar grains from supernovae. In a 25 solar-mass supernova the silicon is quite ^{28}Si-rich within the inner three solar masses of ejecta, which also contain telltale clues such as radioactive ^{44}Ti in supernova grains. But in the overlying three solar masses the silicon is rich in ^{29}Si and ^{30}Si, roughly threefold more so than the solar isotopic ratios. This is the part in which carbon and neon have burned, as outlined above. The next solar mass above those layers contains even more ^{29,30}Si-rich silicon; but the total amount of each ejected from that thin shell is relatively small. When the supernova grains are considered, each of these supernova shells plays a potentially important role. It is also noteworthy that the inner 7 solar masses of this supernova model are oxygen-rich in comparison with carbon, a circumstance that required new theoretical treatment of carbon condensation that circumvents the combustion (in the traditional chemical sense) of the carbon by hot oxygen. Nucleosynthesis of ^{29}Si in this way occurs abundantly only in core-collapse Type II supernovae. Each can eject 1/100th of a solar mass of pure ^{29}Si, which is 3000 Earth masses! Type Ia explosions are not efficient producers.

Although core-collapse supernovae provide in this manner the nucleosynthesis origin of the bulk of the ^{29}Si, other stars can create ^{29}Si-rich compositions locally. In particular, this occurs as a byproduct of the s-process neutron-capture nucleosynthesis. In the He-burning shells of AGB stars (see **Glossary**) where this occurs, the initial ^{28}Si is converted partially to ^{29}Si. The same thing occurs in the He-burning shells of massive stars. The result is that locally ^{29}Si-rich matter is created and ejected. That matter is even more enriched in ^{30}Si, however. Should presolar dust form from this enriched composition before it is mixed with interstellar gas, a distinct ^{29}Si-richness may be observed. Nonetheless, these are not the major bulk causes for ^{29}Si in nature, which are the C-burning and the Ne-burning shells of massive stars.

^{29}Si is a *secondary nucleosynthesis isotope* . This means that ^{29}Si can not be man-ufactured in stars from initial hydrogen and helium in the stellar gas, but in addition requires initial carbon and oxygen in the star. The separation energy, $S_n = 8.47$ MeV, is only half that of the abundant primary isotope, ^{28}Si, meaning that ^{29}Si will not survive late high-temperature events as well as does ^{28}Si. Because of their secondary nature in supernova production, the yield of ^{29}Si and ^{30}Si increases with time as the interstellar gas from which stars form becomes increasingly enriched in carbon and oxygen, the seeds for ^{29}Si and ^{30}Si synthesis. The excess neutrons needed for ^{29}Si nucleosynthesis are established during *helium burning* (see **Glossary**) in stars when ^{14}N (no excess neu-trons) is converted to ^{18}O (two excess neutrons) by an alpha capture and a beta decay.

Astronomical measurements

The Si abundance has been studied in stellar optical spectra, usually in line transi-tions of the neutral Si atom. Unfortunately, these silicon lines are dominated by ^{28}Si, so that little is known about the galactic chemical evolution of ^{29}Si and ^{30}Si except through theoretical calculations. It is important that techniques (probably molecular) be developed to measure ^{29}Si itself in stars.

Anomalous meteoritic isotopic abundance

Presolar grains of several different types reveal sometimes a deficiency and sometimes an excess of this isotope relative to ^{28}Si. It is common to represent the anomaly by the fractional difference of the ^{29}Si/^{28}Si ratio from its solar value:

$$(^{29}Si/^{28}Si)_{sample}/(^{29}Si/^{28}Si)_{solar} = 1 + \delta,$$

where δ is the fractional deviation from the solar ratio. It may be positive or negative. Rather than giving the value of δ as a decimal fraction, however, it is more common to express it in percent, in parts per thousand, or in parts per ten thousand. All are found in the literature. Only astrophysical interpretation of the site of origin can determine whether a measured ratio having a value less than unity (negative δ) is to be interpreted as an excess in the number of ^{28}Si atoms or a deficiency in the number of ^{29}Si atoms.

Mainstream SiC grains   The first presolar grains characterized in the laboratory were SiC (silicon carbide) grains. They comprise about 90–95 % of all SiC grains in meteorites, showing that this mineral was not one that was commonly made within our solar system from normal silicon atom populations. All seem to be presolar, older than the Earth, inherited from the interstellar medium. Nonetheless, the SiC presolar grains have been divided into six distinct groups, based not only on the silicon isotopic ratios within the grains but also on the isotopic ratios of the other elements (the carbon or of trace nitrogen condensed within the SiC). These groups are called "mainstream, A, B, X, Y, and Z." The names are historical and have little mnemonic value, save the "mainstream," which is a correlated set of isotope ratios found in the vast majority (more than 90%) of the SiC grains. The grains range in size from about 0.1 to about 10 μm. These are tiny by human experience, but large in terms of the average size of interstellar dust particles. The characteristic signature of the mainstream grains is ^{29}Si/^{28}Si ratios greater than solar (also for ^{30}Si/^{28}Si). The excess ^{29}Si/^{28}Si ratio is a continuum of values up to 20% (δ^{29}Si $= 200$). The excess ^{30}Si/^{28}Si ratio is a continuum of values up to 15% (δ^{30}Si $= 150$). Furthermore, the set of all points roughly align in the three-isotope plot $\delta(^{29}$Si/^{28}Si) versus $\delta(^{30}$Si/^{28}Si) along a line having slope 4/3. That is, an increase of δ^{30}Si $= 150$ for ^{30}Si/^{28}Si is accompanied by an increase $\delta(^{29}$Si) $= 200$ for ^{29}Si/^{28}Si. This linear correlation strongly suggests that the mainstream is a family, probably having a common origin. It is believed that the mainstream SiC grains condensed in dense winds leaving the surfaces of AGB stars called "carbon stars" because their surface carbon is more abundant than oxygen. This origin is attested to by trapped xenon having s-process (see **Glossary**) isotopic ratios and by neon being strongly enriched in ^{22}Ne, both of which are expected in the surfaces of carbon stars. Trapped xenon having s-process isotope ratios was a 1975 prediction as an identifier of these presolar grains, almost a dozen years before they were actually isolated. Finally, these mainstream SiC grains contain carbon isotopes in the ratio ^{12}C/^{13}C $= 40$–80, rather than the solar isotopic ratio of 89. Because these are the range of ^{12}C/^{13}C ratios observed by astronomers to exist in carbon stars, the origin of the mainstream grains within them seems secure.

But one huge problem exists for this interpretation. According to the theory for the chemical enrichment of the Galaxy with time (see *Chemical evolution of the Galaxy*, in **Glossary**), the ^{29}Si/^{28}Si and ^{30}Si/^{28}Si ratios should be increasing with time owing to their secondary-isotope nature. This means that in stars born prior to the Sun, the initial values for those ratios should be less than solar (negative δ). And yet the ratios are, in an overwhelming majority of SiC presolar grains, greater than the solar ratios in the presolar particles. Since they had to arise from stars that were not only born prior to the Sun but that lived their entire lives (about 2000 Myr) prior to the Sun's birth, a conflict with astrophysical theory exists. How can objects older than the Sun look younger than the Sun? Three answers have been given – one that has been discredited as incorrect, and two that have not yet found substantial endorsement. The incorrect idea is that the slope 4/3 correlation lines for $\delta(^{29}\text{Si})$ versus $\delta(^{30}\text{Si})$ represents a mixture between two endmembers (see *Three-isotope plot*, in **Glossary**). The low-δ endmember would be the initial compositon of the AGB star, and the high-δ endmember would be the heavier silicon isotopically that is generated in the star by the capture of free neutrons (whose presence is the signature of the s process that is observed to happen in these stars). The failure of the endmember concept is that capture of free neutrons by Si creates a shift of $\delta(^{29}\text{Si})$ that is indeed positive, but is only 0.3 times the shift of $\delta(^{30}\text{Si})$, whereas in the mainstream particles $\delta(^{29}\text{Si})$ is $4/3 \times \delta(^{30}\text{Si})$, with both values being much larger than the shifts induced by the s process. The s process creates considerably more ^{30}Si than ^{29}Si. So, generating the observed line as a mixing line within one AGB star is discredited. The remaining suggestions to resolve the crisis attribute the correlation line to a correlation between the intial isotopic compositions of a large number of presolar AGB stars from which the grains arose. This correlation between silicon isotopes might have come about in one or more of three ways: (1) the AGB stars may have been born in more central portions of the Milky Way disk and migrated out to the solar birthplace helped by a gravity assist by swerving past massive molecular clouds, much as gravity assist sends spacecraft to Saturn by swerving them close past Venus and Jupiter. In this way the donor stars of 2–3 solar masses are made to be metal-rich stars for their times and places of birth; (2) the AGB stars may have been born with isotopically heavier Si because they all happen to be born near recent supernovae whose ejecta have made their star-forming gas ^{29}Si-rich and ^{30}Si-rich. Why all stars of 2–3 solar masses should have undergone such birth is yet unanswered, and is a problem for that interpretation; (3) the AGB stars may have been born from linear mixtures of Galaxy gas with the more metal-poor gas of a galactic companion galaxy, similar to one of the Magellanic Clouds, which was merged into the Galaxy by capture about seven billion years ago. None of these three pictures has been well studied.

Type Y and Type Z presolar SiC grains   These grains fall to the right of the mainstream grains in the three-isotope plot $\delta(^{29}\text{Si}/^{28}\text{Si})$ versus $\delta(^{30}\text{Si}/^{28}\text{Si})$. That is, for a given value of $\delta(^{29}\text{Si}/^{28}\text{Si})$ these grains are more ^{30}Si-rich than are the mainstream grains. What distinguishes the Type Z grains from the Type Y grains is that the Z grains are ^{29}Si-poor even though ^{30}Si-rich; i.e. $\delta(^{29}\text{Si}/^{28}\text{Si})$ is negative for the

Z grains, despite being even more ^{30}Si-rich than the Y grains, which are themselves more ^{30}Si-rich than the mainstream grains. Differences exist for isotopes of other elements as well. Nonetheless, both groups appear to also have originated in AGB carbon stars. The difference from the mainstream grains is that the Y and Z grains appear to have condensed from atmospheric gas that has experienced much stronger dredgeup of neutron-irradiated Si from the He-burning shells of these stars, so that the surface is much more enriched in that silicon relative to the silicon that was in the initial atmosphere of these stars. This might even occur during a terminal phase of the AGB star's transition from red-giant to white-dwarf star, when most of the envelope has already been lost in the descent toward that stellar graveyard.

Type X presolar SiC grains   These grains are characterized by ^{29}Si/^{28}Si and ^{30}Si/^{28}Si ratios that are dramatically smaller than the solar values, placing their locations in the negative quadrant of the three-isotope plot $\delta(^{29}$Si/^{28}Si$)$ versus $\delta(^{30}$Si/^{28}Si$)$. They may be thought of as ^{28}Si-rich, by factors of between 2–5, considering that ^{28}Si-richness gives simultaneous deficiencies in the ^{29}Si/^{28}Si and ^{30}Si/^{28}Si ratios. In either way of stating it, these grains appear to have condensed from matter inside the expanding post-explosion supernova, and to have done so in the ashes of the oxygen-burning region where ^{28}Si is synthesized and where the heavier isotopes are destroyed. Trying to get these and other isotopic ratios right is a science frontier for supernova astronomers. Most of these grains are also richer in ^{12}C/^{13}C than is solar matter (up to 80 × solar!), further suggesting that supernova matter is involved, since the inner shells of supernovae are also big ^{12}C producing regions. But a difficult grain-growth history is implied chemically, because that supernova matter that has synthesized new ^{12}C has not synthesized ^{28}Si, and, conversely, that that has synthesized ^{28}Si tends to have little carbon. So, unexplained mixing between the time of the explosion and the time of the condensation chemistry may be required. Another historical problem is the dominance of oxygen over carbon in these supernova zones. Although it had long been believed that SiC could not condense in gas where oxygen is more abundant than carbon, important new chemical studies have shown that the radioactivity within the young supernova alters this old chemical rule, so that condensation without oxidative destruction within mixed gas can now be understood. This can only occur in supernovae, where the radioactivity destroys the CO molecule that is otherwise the inevitable consequence of oxidation of carbon. Many other dramatic isotope ratios exist for other trace elements within these SiC grains, so understanding their chemical origins is an exciting scientific frontier.

It seems also possible that the Type X presolar SiC grains originate instead in Type I supernova explosions (the exploding white dwarf). In that case it would occur in the matter within a helium cap on top of the white dwarf that would make isotopes in the correct ratios (for many elements). The trouble with this picture is that it is hard to see how grains can grow fast enough to reach the observed size in a Type I supernova, where ejecta are hot and rapidly thinning.

One unique presolar SiC grain   One SiC grain has been discovered in the Murchison meteorite and characterized by ^{29}Si/^{28}Si and ^{30}Si/^{28}Si ratios that are dramatically larger than the solar values, placing its location far into the positive quadrant of the three-isotope plot $\delta(^{29}$Si/^{28}Si) versus $\delta(^{30}$Si/^{28}Si). It may be thought of as simultaneously rich in the ^{29}Si/^{28}Si (by a factor 3.7 compared with solar) and ^{30}Si/^{28}Si (by a factor 4.3 compared with solar) ratios. Out of about 20 000 SiC grains that have been studied, this one is the only example of this unusual isotopic pattern. The grain appears to also be a very important supernova particle, having condensed from matter inside the expanding post-explosion supernova, and to have done so in the ashes of the (He, C, Ne)-burning regions where ^{29}Si and ^{30}Si are abundantly synthesized (see *Nucleosynthesis origin*, above). ^{12}C-richness also suggests its supernova origin. Similar condensation chemistry problems exist to those of the Type X grains. The C-burning matter has much more oxygen than carbon, and only the radioactivity enables the SiC to condense without destruction by oxidation. Supernovae will yield many secrets to the study of these supernova grains. It may be possible that this particle has instead been condensed in the wind of a Wolf–Rayet star that has lost its entire hydrogen and helium envelope and is ejecting carbon-rich matter from the He-burning core, within which ^{29}Si and ^{30}Si are overabundant in almost the observed ratio.

Presolar graphite grains   Graphite is one type of solid carbon. Most of these grains are characterized by ^{29}Si/^{28}Si and ^{30}Si/^{28}Si ratios that are dramatically smaller than the solar values, placing their locations in the negative quadrant of the three-isotope plot $\delta(^{29}$Si/^{28}Si) versus $\delta(^{30}$Si/^{28}Si). This makes them resemble the Type X grains of SiC (see above). It is certain that at least some of these grains condensed in supernovae because of the excess ^{44}Ca within them, which has resulted from the decay of ^{44}Ti with its 60-yr halflife. Only within a supernova interior does enough ^{44}Ti exist (see **^{44}Ti**). An eyecatching but unsolved aspect of some graphite grains is that some are deficient only in the ^{30}Si isotope; that is, $\delta(^{29}$Si/^{28}Si) is positive but $\delta(^{30}$Si/^{28}Si) is negative. This reflects some interesting decoupling of the production of the neutron-rich isotopes of Si; but it is not yet understood.

Presolar silicon nitride grains   Most of these Si_3N_4 grains are characterized by ^{29}Si/^{28}Si and ^{30}Si/^{28}Si ratios that are dramatically smaller than the solar values, placing their locations in the negative quadrant of the three-isotope plot $\delta(^{29}$Si/^{28}Si) versus $\delta(^{30}$Si/^{28}Si), with values in the range $\delta = -20\%$ to -35%. This makes them resemble the Type X grains of SiC (see above). Several other isotopic similarities suggest that supernovae are the origin of these particles as well.

Presolar oxides   The Si concentration in presolar *corundum* (Al_2O_3) is not sufficiently great that its isotopes can be measured in those particles. And presolar silicates (which would be rich in Si) have not yet been found.

^{30}Si | $Z = 14$, $N = 16$
Spin, parity: J, $\pi = 0+$;
Mass excess: $\Delta = -24.433$ MeV; $S_n = 10.62$ MeV

Abundance

From the isotopic decomposition of normal silicon one finds that the mass-30 isotope, ^{30}Si, is the least abundant of the three stable Si isotopes; 3.10% of all Si. This makes it 29.8 times less abundant than the most abundant silicon isotope, namely ^{28}Si, which constitutes 92.23% of natural Si. Using one million silicon atoms as the common cosmochemical standard in solar-system matter, this isotope has

solar abundance of ^{28}Si $= 3.10 \times 10^4$ per million silicon atoms.

This is the 25th most abundant nucleus in the universe, comparable to ^{18}O and ^{34}S.

Nucleosynthesis origin

In the mix of interstellar atoms from which the solar system formed, ^{30}Si exists primarily owing to carbon-burning and the subsequent neon-burning process of stellar nucleosynthesis. Simply put, these processes liberate some free neutrons within the stellar gas, and those free neutrons are captured to convert initial ^{28}Si into ^{30}Si (in part). More importantly than that, the neon-burning process uses hot alpha particles set free in the gas by the neon burning to convert the two heavier isotopes of Mg into the two heavier isotopes of Si. That is, in neon-burning the abundant ^{25,26}Mg are transformed into ^{29,30}Si. The ^{25,26}Mg had previously been made overabundant owing to their nucleosynthesis during the preceding carbon burning. Owing to this progression of silicon isotopic abundances as the massive star core evolves, idealized computer models of that core show regions of quite different silicon isotopic ratios. These may be instrumental in understanding the observed isotopes in presolar grains from supernovae. In a 25 solar-mass supernova the silicon is quite ^{28}Si-rich within the inner three solar masses of ejecta, which also contain telltale clues such as radioactive ^{44}Ti in supernova grains. But in the overlying three solar masses the silicon is rich in ^{29}Si and ^{30}Si, roughly threefold more than the solar isotopic ratios. This is the part in which carbon and neon have burned, as outlined above. The next solar mass above those layers contains even more ^{29,30}Si-rich silicon; but the total amount of each ejected from that thin shell is relatively small. When the supernova grains are considered, each of these supernova shells plays a potentially important role. It is also noteworthy that the inner 7 solar masses of this supernova model are oxygen-rich in comparison with carbon, a circumstance that required new theoretical treatment of carbon condensation that circumvents the combustion (in the traditional chemical sense) of the carbon by hot oxygen. Nucleosynthesis of ^{30}Si in this way occurs abundantly only in core-collapse Type II supernovae. Each can eject 1/100th of a solar mass of pure ^{30}Si, which is 3000 Earth masses! Type Ia explosions are not efficient producers.

Although this is the nucleosynthesis origin of the bulk of the ^{30}Si, other stars can create ^{30}Si-rich compositions locally. In particular, this occurs as a byproduct of the s-process neutron-capture nucleosynthesis. In the He-burning shells of AGB stars where this occurs, the initial ^{28}Si is converted efficiently to ^{30}Si. The small neutron-capture cross section of ^{30}Si causes the weak capture flow to dam up there. The same thing occurs in the He-burning cores of massive stars. In stellar regions where the s process is occurring, the large cross section of the important reaction ^{33}S + n → ^{30}Si + ^{4}He directs the capture flow through sulfur back into silicon. This is significant for the silicon isotopes in mainstream presolar SiC grains. The result is that locally ^{30}Si-rich matter is created and ejected. In presolar dust formed from this enriched composition before it is mixed with interstellar gas, a distinct ^{30}Si-richness may be observed. Nonetheless, these are not the major causes for bulk ^{30}Si in nature, which are the C-burning and the Ne-burning shells of massive stars.

^{30}Si is a *secondary nucleosynthesis isotope*. This means that ^{30}Si can not be manufactured as efficiently in stars from initial hydrogen and helium in the stellar gas, but in addition is augmented by initial carbon and oxygen in the star. The separation energy, $S_n = 10.62$ MeV, is near half that of the abundant primary isotope, ^{28}Si, meaning that ^{30}Si will not survive late high-temperature events as well as does ^{28}Si. The excess neutrons required for ^{30}Si secondary nucleosynthesis are established during *helium-burning* (see **Glossary**) in stars when ^{14}N (no excess neutrons) is converted to ^{18}O (two excess neutrons) by an alpha capture and a beta decay.

Astronomical measurements

The Si abundance has been studied in stellar optical spectra, usually in line transitions of the neutral Si atom. Unfortunately, these silicon lines are dominated by ^{28}Si, so that little is known about the galactic chemical evolution of ^{29}Si and ^{30}Si except through theoretical calculations. It is important that techniques be developed to measure ^{30}Si itself in stars.

Anomalous meteoritic isotopic abundance

Presolar grains of several different types reveal sometimes a deficiency and sometimes an excess of this isotope relative to ^{28}Si. It is common to represent the anomaly by the fractional difference of the ^{30}Si/^{28}Si ratio from its solar value; that is, by the ratio

$$(^{30}\text{Si}/^{28}\text{Si})_{\text{sample}}/(^{30}\text{Si}/^{28}\text{Si})_{\text{solar}} = 1 + \delta,$$

where δ is the fractional deviation from the solar ratio. It may be positive or negative. Rather than giving the value of δ as a decimal fraction, however, it is more common to express it in percent, in parts per thousand, or in parts per ten thousand. All are found in the literature. Only astrophysical interpretation of the site of origin can determine whether a measured ratio having a value less than unity (negative δ) is to be interpreted as an excess in the number of ^{28}Si atoms or deficiency in the number of ^{30}Si atoms. Either gives a ratio less than unity.

Mainstream SiC grains   The first presolar grains characterized in the laboratory were SiC (silicon carbide) grains. They comprise about 90% of all SiC grains in meteorites, showing that this mineral was not one that was commonly made within our solar system from normal silicon atom populations. All seem to be presolar, older than the Earth, inherited from the interstellar medium. Nonetheless, the SiC presolar grains have been divided into six distinct groups, based not only on the silicon isotopic ratios within the grains but also on the isotopic ratios of the other elements (the carbon or of trace nitrogen condensed within the SiC). These groups are called "mainstream, A, B, X, Y, and Z." The names are historical and have little mnemonic value, save the "mainstream," which is a correlated set of isotopic ratios found in the vast majority (90%) of the SiC grains. The grains range in size from about 0.1 to about 10 μm. These are tiny by human experience, but large in terms of the average size of interstellar dust particles. The characteristic signature of the mainstream grains is ^{30}Si/^{28}Si and ^{29}Si/^{28}Si ratios greater than solar. The excess ^{30}Si/^{28}Si ratio is a continuum of values up to $\delta = 150$ (15% excess). Furthermore, the set of all points roughly align in the three-isotope plot $\delta(^{29}$Si/^{28}Si$)$ versus $\delta(^{30}$Si/^{28}Si$)$ along a line having slope 4/3. Because ^{30}Si/^{28}Si and ^{29}Si/^{28}Si ratios must be discussed together, see **^{29}Si** for more details.

Type Y and Z presolar SiC grains   These grains fall to the right of the mainstream grains in the three-isotope plot $\delta(^{29}$Si/^{28}Si$)$ vs $\delta(^{30}$Si/^{28}Si$)$. Because ^{30}Si/^{28}Si and ^{29}Si/^{28}Si ratios must be discussed together, see **^{29}Si** for more details.

Type X presolar SiC grains   These grains are characterized by ^{29}Si/^{28}Si and ^{30}Si/^{28}Si ratios that are dramatically smaller than the solar values, placing their locations in the negative quadrant of the three-isotope plot $\delta(^{29}$Si/^{28}Si$)$ versus $\delta(^{30}$Si/^{28}Si$)$. They may be thought of as ^{28}Si-rich, by factors of between 2–5, considering that ^{28}Si-richness gives simultaneous deficiencies in the ^{29}Si/^{28}Si and ^{30}Si/^{28}Si ratios. Because ^{30}Si/^{28}Si and ^{29}Si/^{28}Si ratios must be discussed together, see **^{29}Si** for more details.

One unique presolar SiC grain   One SiC grain has been discovered in the Murchison meteorite and characterized by ^{29}Si/^{28}Si and ^{30}Si/^{28}Si ratios that are dramatically larger than the solar values, placing its location far into the positive quadrant of the three-isotope plot $\delta(^{29}$Si/^{28}Si$)$ versus $\delta(^{30}$Si/^{28}Si$)$. It may be thought of as simultaneously rich in the ^{29}Si/^{28}Si (by a factor 3.7 compared with solar) and ^{30}Si/^{28}Si (by a factor 4.3 compared with solar) ratios. Because ^{30}Si/^{28}Si and ^{29}Si/^{28}Si ratios must be discussed together, see ^{29}**Si** for more details.

Presolar graphite grains   Graphite is a solid form of carbon. But that carbon contains trace amounts of Si that are not small. Most of these grains are characterized by ^{29}Si/^{28}Si and ^{30}Si/^{28}Si ratios that are dramatically smaller than the solar values, placing their locations in the negative quadrant of the three-isotope plot $\delta(^{29}$Si/^{28}Si$)$ versus $\delta(^{30}$Si/^{28}Si$)$. Because ^{30}Si/^{28}Si and ^{29}Si/^{28}Si ratios must be discussed together, see ^{29}**Si** for more details.

Presolar silicon nitride grains   Most of these Si_3N_4 grains are characterized by ^{29}Si/^{28}Si and ^{30}Si/^{28}Si ratios that are dramatically smaller than the solar values, placing their locations in the negative quadrant of the three-isotope plot $\delta(^{29}$Si/^{28}Si$)$ versus $\delta(^{30}$Si/^{28}Si$)$, with values in the range $\delta = -20\%$ to -35%. This makes them resemble the Type X grains of SiC (see above). Several other isotopic similarities suggest that supernovae are the origin of these particles as well.

Presolar oxides   The Si concentration in presolar corundum (Al_2O_3) is not sufficiently great that its isotopes can be measured in those particles. And presolar silicates (which would be rich in Si) have not been found.

15 | Phosphorus (P)

Phosphorus nuclei have charge $+15$ electronic units, so that 15 electrons orbit the nucleus of the neutral atom. Its electronic configuration can be abbreviated as an inner core of inert neon (a noble gas) plus five more electrons: $(Ne)3s^2 3p^3$, which locates P in Group VA of the chemical periodic table, between nitrogen (N) and arsenic (As). Chemically, phosphorus is a very versatile element. It gives up three or five electrons respectively for its oxidation states of $+3$ and $+5$, with oxides P_4O_6 and P_4O_{10} respectively, and it takes three electrons in its -3 oxidation state. P is the 17th most abundant of the chemical elements, between chromium and manganese

Phosphorus's name derives from the Greek *phosphoros*, "bringer of light." The ancients knew of materials that glow in the dark after the exciting source has been turned off, called "phosphors." Many of these contained phosphates, the oxidation (-3) group (PO_4). In phosphorescent materials a dislodged electron requires time to return to its low-energy configuration accompanied by the emission of light. Phosphorus the element was isolated quite early for chemical elements, 1669. It is recovered from abundant *calcium phosphate* $Ca_3(PO_4)_2$ rock. Pure P has ten distinct allotropes, an unusally large number of physical solids that are distinguished by differing crystal-structure arrangements and by having different colors (white P, red P, and black P). White phosphorus is the most chemically useful allotrope, and the most chemically reactive. White phosphorus is waxy in texture. Phosphorus is used abundantly in incendiary devices, matches and fireworks. *Phosphoric acid* (H_3PO_4), is used in prepared soft drinks. Mid-20th century Americans might have gone to the soda fountain of their local drugstore for "a phosphate." Another major industrial use of P is in fertilizers.

Phosphorus compounds are well known and important to many commercial products, making it a valuable and useful element.

Because phosphorus has but a single stable isotope it sheds no light on isotopic anomalies within presolar material. P nucleosynthesis derives almost entirely from a single main-line source, neon burning, which occurs at the end of carbon burning in massive stars that will become Type II supernovae.

^{31}P | Z = 15, N = 16
Spin, parity: J, π = 1/2+;
Mass excess: $\Delta = -24.441$ MeV; $\qquad S_n = 12.31$ MeV

Abundance

^{31}P is the only stable isotope of phosphorus; therefore all elemental observations of its abundance are of the abundance of ^{31}P. This isotope has

solar abundance of ^{31}P = 1.04×10^4 per million silicon atoms.

^{31}P is the 32nd most abundant nucleus within the universe and the 17th most abundant element.

Nucleosynthesis origin

In the mix of interstellar atoms from which the solar system formed, ^{31}P exists primarily owing to the *carbon-burning* (see **Glossary**) process of stellar nucleosynthesis and to the neon burning into which carbon burning merges as carbon is exhausted. The concentration of ^{31}P jumps by a factor 50 above its concentration in the initial star within the carbon-burning zones. Once the neutron-rich isotopes of silicon have been synthesized in a star in the same burning process, capture of another free nucleon can produce ^{31}P. Core-collapse Type II supernovae are the sites of 95% of P nucleosynthesis.

Astronomical measurements

Phosphorus has not been well studied in stellar optical spectra. One anticipates a steady increase in the P/Mg ratio as stellar metallicity increases during *chemical evolution of the Galaxy* (see **Glossary**).

16 | Sulfur (S)

All sulfur nuclei have charge +16 electronic units, so that 16 electrons orbit the nucleus of the neutral atom. Its electronic configuration can be abbreviated as an inner core of inert neon (a noble gas containing ten inner electrons) surrounded by six more electrons: $(Ne)3s^2 3p^4$, which locates S beneath oxygen in Group VIA of the periodic table. Sulfur therefore normally has valence -2, like O, as in *hydrogen sulfide* (H_2S) with the smell of rotten eggs, or $+4$ and combines readily with pairs of oxygen atoms, as in *sulfur dioxide* (SO_2). It combines also with oxygen as valence $+6$ to make the very important (-2) valence sulphate radical SO_4^{2-}, which appears in *sulphuric acid* (H_2SO_4) and in mineral sulphates such as $MgSO_4$. Sulfur is thus a quite complicated element chemically owing to its diverse bonding options. It is nonmetallic and not highly reactive, which accounts for its easy existence in elemental form. But it is highly prized as a commerical mineral because of its usefulness for many commercial goals. Forty billion kilograms of it are used per year in the manufacture of sulphuric acid, the most heavily produced of all chemicals.

Sulfur is the tenth most abundant element in the universe; thus it is an important datum for nucleosynthesis theories. It is not equally significant in terrestrial abundances because, owing to its rather high volatility, much of it was lost into space as the Earth formed. Sulfur is easily found on Earth, however, even in its pure elemental form, in which it is a pale-yellow, brittle solid. It melts near 113 °C, just above the boiling point of water. Sulfur was not recognized as a distinct element until 1777, by Lavoisier. It suddenly became much more important after the discovery of its use in gunpowder.

Again because of its high volatility, sulfur is relatively undepleted from interstellar gas by condensation in interstellar grains, so that its isotopic composition has been partly measured by radioastronomers, who detect emission lines from the molecule CS in the interstellar medium. By comparing the rotational transitions of $^{13}C/^{32}S$ with those of $^{12}C/^{34}S$ the interstellar isotopic ratio has been found for sulfur (see ^{34}S, *Anomalous isotopic abundance*).

Sulfur is composed of four stable isotopes, masses 32, 33, 34, and 36, and is a highly diagnostic element for nucleosynthesis theory. Almost simultaneous formation and thermal decomposition of S nuclei occur in the cores of massive stars. Formation is mostly during oxygen burning, but eventual reassembly into iron-group elements occurs during the subsequent "silicon burning" in stellar evolution and nucleosynthesis.

Although sulfur isotopic anomalies in presolar grains should be common, this has not been measurable owing to two problems. Firstly, S is moderately volatile, so it is not easily incorporated into hot grains during thermal condensation within supernovae where these isotopes are made. Observed presolar grains therefore do not contain enough S for its isotopes to be measured easily. Secondly, S chemistry does not associate easily with common refractory materials. So the sulfur found in its few refractory forms in meteorites, for example, large 10–100-μm CaS grains in primitive nonoxidized (enstatite) meteorites, is a mix of sulfur from many diverse presolar materials, and the melting pot that grew such grains therefore contains normal S isotopes. No definite isotopic anomalies in S in either presolar grains or in meteorite solids has been found, except those attributed to mass-dependent chemical-fractionation effects rather than to nuclear formation effects.

Natural isotopes of sulfur and their solar abundances

A	Solar percent	Solar abundance per 10^6 Si atoms
32	95.0	48 900
33	0.75	3860
34	4.21	21 700
36	0.02	103

^{32}S | Z = 16, N = 16
Spin, parity: J, π = 0+;
Mass excess: $\Delta = -26.016$ MeV; $S_n = 15.09$ MeV

Abundance

From the isotopic decomposition of normal sulfur one finds that the mass-32 isotope, ^{32}S, is the most abundant of the stable S isotopes; 95.02% of all S. On the scale where one million silicon atoms is taken as the standard for solar-system matter, this isotope has

solar abundance of ^{32}S = 4.89×10^5 per million silicon atoms;

that is almost half of the abundance of silicon. ^{32}S is the eleventh most abundant isotope in the universe.

Nucleosynthesis origin

^{32}S exists primarily owing to the oxygen-burning process (both hydrostatic and explosive) of stellar nucleosynthesis. Put simply, during oxygen burning two ^{16}O nuclei collide, causing the reaction ^{16}O + ^{16}O $\rightarrow$ ^{28}Si + ^{4}He. Some alpha particles produced in that way are consumed by ^{28}Si, which is also abundant: ^{28}Si + ^{4}He → ^{32}S + gamma ray. By these two reactions one effectively gets the net reaction $2\,^{16}$O → ^{32}S with reasonable efficiency. Within the supernova structure, therefore, the ^{32}S concentration leaps to 1000 times greater than its concentration within the initial star, and it does so in the central zones that are burning oxygen at the time the star explodes. Its large neutron separation energy, $S_n = 15.09$ MeV, ensures that ^{32}S will survive high temperatures well, remaining abundant even during silicon burning until the Si is depleted. This hardiness contrasts to its heavier isotopes, which are destroyed when silicon burning commences. Almost all of this happens in Type II supernovae, which on average eject about ten times as much ^{32}S as do Type Ia supernovae and are about five times more frequent. Nature's numbers are awesome here, each Type II ejecting a mass of ^{32}S comparable to 15 000 times the mass of the Earth!

Primary nucleus   The above reactions are helpful in understanding that ^{32}S is called a "primary nucleus" in nucleosynthesis parlance. That designation is given to those nuclei that are abundantly produced within stars whose initial composition is hydrogen plus helium, the two elements left from the Big Bang. Indeed, initial hydrogen and helium are converted to ^{16}O by the natural thermonuclear evolution of matter. So ^{16}O is a primary nucleus, and so therefore is ^{32}S, which is made from the ^{16}O. The ^{32}S is distinguished from its heavier isotopes in that they are "partly secondary isotopes" because their nucleosynthesis is aided if the star manufacturing them has not only hydrogen and helium in its initial composition, but also carbon and oxygen, which provide for the excess neutrons that the heavier isotopes contain.

Astronomical measurements

The sulfur abundance has been studied in stars. The best indicator has been a doublet line of the neutral S atom in the near infrared at 8694 Å. Because ^{32}S dominates the spectra of S in stars, and especially because ^{32}S is a primary "alpha nucleus," the galactic evolution of ^{32}S is the same as the observed galactic evolution of sulfur elemental abundance. As nucleosynthesis progressed the galactic S/H abundance ratio in newly born observed stars increased from 10^{-3} of solar in some stars to a bit in excess of solar S/H. The behavior is very similar to that of oxygen, and indeed to all alpha nuclei that can be created from fusion of He, C and O. These include the elements Ne, Mg, Si, S, Ar, Ca, and Ti. When compared to the abundance of iron, however, the ratio S/Fe declines from about five times solar in metal-poor stars to less than solar in young stars. This decline is similar to that of O, but even more dramatic. This is understood as the rapid coproduction of the alpha nuclei, along with O, in massive Type II supernovae. These produce much less iron, especially if they are so massive that most of the Fe production falls back onto the neutron star. This seems to have happened in the nucleosynthesis sources for the most metal-poor stars. Later onset of Type Ia supernova explosions dramatically increases Fe nucleosynthesis and also causes the interstellar S/Fe ratio to decline to its solar value.

^{33}S | $Z = 16$, $N = 17$
Spin, parity: J, $\pi = 3/2+$;
Mass excess: $\Delta = -26.586$ MeV; $\qquad S_n = 8.64$ MeV

Abundance

From the isotopic decomposition of normal sulfur one finds that the mass-33 isotope, ^{33}S, is one of the lesser abundant of the stable S isotopes; only 0.75% of all S. On the scale where one million silicon atoms is taken as the standard for solar-system matter, this isotope has

solar abundance of ^{33}S $= 3860$ per million silicon atoms.

It is the 36th most abundant isotope in the solar system, between ^{17}O and ^{39}K.

Nucleosynthesis origin

^{33}S exists primarily owing to the oxygen-burning process (both hydrostatic and explosive) of stellar nucleosynthesis. Put simply, during oxygen burning two ^{16}O nuclei collide, causing the net reaction $2\,^{16}O \rightarrow\, ^{32}S$ to occur with reasonable efficiency. ^{33}S is a "partly secondary isotope" because its nucleosynthesis by neutron captures from newly made ^{32}S is aided if the star manufacturing them has not only hydrogen and helium in its initial composition, but also carbon and oxygen. Following H and He burning, the ^{18}O and ^{22}Ne created from the initial carbon and oxygen provide for the excess neutrons that the heavier S isotopes contain; but significant excess neutrons are also produced during the oxygen burning itself by positron emissions that occur there, so that ^{33}S is also "partly primary." Within the supernova structure, therefore, the ^{33}S concentration leaps to 1000 times greater than its concentration within the initial star, and it does so in the central zones that are burning oxygen at the time the star explodes. When the isotopic composition of the entire spectrum of massive supernova explosions is computed, they show that all isotopes of sulfur are created by those supernova explosions, which are their source. Even for this minor isotope nature's numbers are awesome; each Type II ejects a mass of ^{33}S comparable to 150 times the mass of the Earth!

In stellar regions where the s process is occurring, the large cross section of the important reaction $^{33}S + n \rightarrow\, ^{30}Si + {}^{4}He$ directs the capture flow through sulfur back into silicon. This is significant for the silicon isotopes (see *Nucleosynthesis origin* in ^{30}**Si**).

Astronomical measurements

The S abundance has been studied in stellar optical spectra, usually in line transitions of the neutral S atom. Unfortunately, these sulfur lines are dominated by ^{32}S, so that little is known about the galactic chemical evolution of ^{33}S and ^{34}S except through theoretical calculations. It is important that techniques be developed to measure ^{33}S itself in stars.

Anomalous isotopic abundance

Sulfur is relatively undepleted by condensation into interstellar grains, so that its isotopic composition has been partly measured by radioastronomers, who detect emission lines from the molecule CS in the interstellar medium. By comparing rotational transition lines from ^{13}C^{34}S with those from ^{12}C^{33}S it has been found that the sulfur isotopic ratio in the interstellar gas is ^{34}S/^{33}S $= 6.3 \pm 1$, which is in reasonable agreement with the value 5.5 in the solar system. One will certainly hope to check this with measurements of S isotopes in interstellar grains that today drift into the solar system and that can be captured by NASA's STARDUST mission.

Isotopic variations of ^{33}S have important application to the chemistry of the Earth's atmosphere during the Archean period and to the geologic history of the Earth. The variations are caused by a photochemical selectivity for the ^{33}S isotope during the destruction of atmospheric SO_2 molecules. This mass-independent mechanism selects ^{33}S owing to its low abundance, in analogy to such fractionation of oxygen isotopes (see ^{16}O, *Astronomical measurements*, *Isotopes in diffuse interstellar clouds*). Consequently for geology, it can be inferred that sulfur-bearing material must have been transferred from atmosphere to mantle during the Archean.

The isotopic composition of sulfur in individual presolar grains has not yet been measured owing to the very small concentration of S in those grains.

---

^{34}S $\quad$ Z $= 16$, N $= 18$
$\qquad$ **Spin, parity:** J, $\pi = 0+$;
$\qquad$ **Mass excess:** $\Delta = -29.932$ MeV; $\qquad S_n = 11.42$ MeV

Abundance

From the isotopic decomposition of normal sulfur one finds that the mass-34 isotope, ^{34}S, is the second most abundant of the stable S isotopes; 4.21% of all S. On the scale where one million silicon atoms is taken as the standard for solar-system matter, this isotope has

solar abundance of ^{34}S $= 2.17 \times 10^4$ per million silicon atoms.

It is the 26th most abundant isotope in the solar system, between ^{30}Si and ^{57}Fe.

Nucleosynthesis origin

^{34}S exists primarily owing to the oxygen-burning process (both hydrostatic and explosive) of stellar nucleosynthesis. Put simply, during oxygen burning two ^{16}O nuclei collide, the net reaction $2\,^{16}\mathrm{O} \rightarrow\,^{32}\mathrm{S}$ occurs with reasonable efficiency. ^{34}S is a "partly secondary isotope" because its nucleosynthesis from newly made ^{32}S and ^{33}S by neutron captures is aided if the star manufacturing them has not only hydrogen and helium in its initial composition, but also carbon and oxygen. Following H and He burning, the the ^{18}O and ^{22}Ne created from the initial carbon and oxygen provide for the excess neutrons that the heavier S isotopes acquire; but some significant excess neutrons are also created during the oxygen burning itself by positron emissions that occur by radioactive nuclei there, so that ^{34}S is also partly a primary isotope. When oxygen burning occurs the excess neutrons are initially housed in neutron-rich isotopes of silicon, but the charged particles reactions liberate those neutrons for relocation into the increasing sulfur abundance. Within the supernova structure, therefore, the ^{34}S concentration leaps to 1000 times greater than its concentration within the initial star, and it does so in the central zones that are burning oxygen at the time the star explodes. When the isotopic composition of the entire spectrum of massive supernova explosions is computed, they show that all isotopes of sulfur are created by those supernova explosions, which are their source. Nature's numbers are awesome here, each Type II ejecting a mass of ^{34}S comparable to 660 times the mass of the Earth!

Astronomical measurements

The S abundance has been studied in stellar optical spectra, usually in line transitions of the neutral S atom. Unfortunately, these sulfur lines are dominated by ^{32}S, so that little is known about the galactic chemical evolution of ^{33}S and ^{34}S except through theoretical calculations. It is important that techniques be developed to measure ^{34}S itself in stars.

Anomalous isotopic abundance

Sulfur is relatively undepleted by condensation into interstellar grains, so that its isotopic composition has been partly measured by radioastronomers, who detect emission lines from the molecule CS in the interstellar medium. By comparing rotational transition lines from ^{13}C^{32}S with those from ^{12}C^{34}S and using the known radial relation of ^{13}C^{12}C in the interstellar gas (see ^{12}C, *Astronomical measurements*, *Isotopes in interstellar clouds*), it has been found that the sulfur isotopic ratio varies with distance R_{GC} (in kiloparsecs, kpc) from the galactic center according to $^{32}\mathrm{S}/^{34}\mathrm{S} = (3.3 \pm 0.5)R_{GC} + 4$. This observed relation yields the value $^{32}\mathrm{S}/^{34}\mathrm{S} = 32$ at the solar position $R_{GC} = 9$ kpc, which is 40% greater than the value in the solar system, $^{32}\mathrm{S}/^{34}\mathrm{S} = 22.6$. One will certainly hope to check this with measurements of S isotopes in interstellar grains that today drift into the solar system and that can be captured by NASA's STARDUST mission.

The isotopic composition of sulfur in individual presolar grains has not yet been measured owing to the very small concentration of S in those grains.

^{36}S | $Z = 16$, $N = 20$
Spin, parity: J, $\pi = 0+$;
Mass excess: $\Delta = -30.664$ MeV; $S_n = 9.88$ MeV

^{36}S is the lightest nucleus of all nuclei that are stable against beta decay and that contain four more neutrons than protons.

Abundance

From the isotopic decomposition of normal sulfur one finds that the mass-36 isotope, ^{36}S, is least abundant of the stable S isotopes; only 0.02% of all S. On the scale where one million silicon atoms is taken as the standard for solar-system matter, this isotope has

solar abundance of ^{36}S $= 103$ per million silicon atoms.

It is but the 64th most abundant isotope in the solar system, comparable to minor calcium isotopes or to ^{7}Li, the most abundant Big-Bang product outside of H and He.

Nucleosynthesis origin

The exact site of origin for ^{36}S has been somewhat unclear. ^{36}S appears to exist primarily owing to the intense neutron flux that is set free during explosive carbon burning in the shocked carbon shell of supernova explosions. This renders ^{36}S a "secondary isotope" because its nucleosynthesis by neutron captures from lighter S isotopes is aided if the star manufacturing them has not only hydrogen and helium in its initial composition, but also carbon and oxygen. The ^{18}O and ^{22}Ne created from the initial carbon and oxygen provide for the excess neutrons that ^{36}S contains. This must be an ejected carbon shell, because subsequent silicon burning in the star would destroy the ^{36}S. Within the supernova structure, therefore, the ^{36}S concentration leaps to 100 times greater than its concentration within the initial star, and it does so in the zones that are burning carbon at the time the star explodes. When the isotopic composition of the entire spectrum of massive supernova explosions is computed, they show that all isotopes of sulfur are created by those supernova explosions, which are their source.

Astronomical measurements

The S abundance has been studied in stellar optical spectra, usually in line transitions of the neutral S atom. Unfortunately, these sulfur lines are dominated by ^{32}S, so that little is known about the galactic chemical evolution of ^{36}S. It is so low in abundance that techniques to measure ^{36}S itself in stars may not be possible.

17 | Chlorine (Cl)

All chlorine nuclei have charge +17 electronic units resulting from their 17 protons, so that 17 electrons orbit the nucleus of the neutral atom. Chemically, it behaves as if it wants one more electron to fill the six states of the 3p shell. Its configuration can be abbreviated as an inner core of inert neon (a noble gas) plus seven more electrons: $(Ne)3s^2 3p^5$, which locates it in Group VIIA of the periodic table. The elements in this group are known collectively as *halogens*, and include fluorine, bromine and iodine. Chlorine is a greenish-yellow gas (of Cl_2 molecules), taking its name from the Greek word for that color, *chloros*. Its oxidation state is −1 because one more electron will close the 3p shell. Cl is a powerful oxidizer (electron grabber) in chemical reactions. All metals form ionic bonds with chlorine's oxidation state (−1) ion. Familiar salts are NaCl (table salt) with (+1) Na, $MgCl_2$ with (+2) Mg, and so on. Metals having multiple oxidation states form multiple chlorides.

Elemental Cl is a corrosive and toxic gas, and care should be taken to avoid breathing it. We recognize its smell from summer swimming pools, to which it is added owing to its lethality to bacteria, and from chlorine bleach. The principal mineral bearing Cl is NaCl, which is 80% of the dissolved mass in common salty sea water, and is also abundant in salt mines. The element Cl is liberated from NaCl by electrolysis in water. During this current-carrying process the Cl^- ions in solution give up their electrons at the anode and appear as Cl_2 gas. It is one natural resource for which no shortage is envisioned.

Chlorine is much the most abundant halogen. It is six times more abundant than fluorine and 450 times more abundant than bromine. This is understood, because both stable Cl isotopes are produced in the main line of nuclear reactions during oxygen burning in stars. Cl is the 20th most abundant element in the universe, being almost identical in abundance to potassium and 1.5 times more abundant than titanium. But measurements of Cl abundance in stars are few, because it is rare and its emission lines are unfavorable.

Natural isotopes of chlorine and their solar abundances

A	Solar percent	Solar abundance per 10^6 Si atoms
35	75.8	2860
36	extinct radioactivity in meteorites	
37	24.2	913

^{35}Cl | Z = 17, N = 18
  Spin, parity: J, π = 3/2+;
  Mass excess: Δ = −29.014 MeV; S_n = 12.63 MeV

Abundance

^{35}Cl is 75.77% of Cl atoms on Earth. Using the observed solar-system abundance, which is uncertain to perhaps 50%, Cl = 3773 per million silicon atoms, so

solar abundance of ^{35}Cl = 2860 per million silicon atoms.

^{35}Cl is the 38th most abundant isotope in nature.

Nucleosynthesis origin

During oxygen burning in stars, free neutrons, protons and alpha particles are abundant. These produce both ^{35}Cl and ^{37}Cl from the more abundant alpha nuclei that dominate the abundance pattern during oxygen burning. Once ^{34}S is established during that process, for example, a single additional proton capture yields ^{35}Cl. This and related reaction sequences well account for the solar ^{35}Cl abundance. When oxygen burning occurs the excess neutrons are initially housed in neutron-rich isotopes of silicon, but the charged particle reactions liberate those neutrons for relocation into the growing sulfur, chlorine and argon abundances. Within the supernova structure, therefore, the ^{35}Cl concentration leaps to 500 times greater than its concentration within the initial star, and it does so in the central zones that are burning oxygen at the time the star explodes. In zones where silicon burning follows oxygen burning, the neutron-rich isotopes such as ^{35}Cl disappear as the excess neutrons shift upward into the iron peak. Both types of supernovae experience explosive O burning and may contribute to the interstellar ^{35}Cl abundance; but core-collapse supernovae produce the lion's share.

Astronomical measurements

The two isotopes of Cl have been detected in interstellar clouds in Orion via the radiofrequency emission of the HCl molecule. The hyperfine splittings seen in the J = 1 to 0 transition for both molecular isotopomers, H^{35}Cl and H^{37}Cl, enable radioastronomers to identify the lines unambiguously. Because of the chemistry of the formation of the HCl molecule, one believes that the H^{35}Cl/H^{37}Cl intensity ratio is a good measure of the relative abundances ^{35}Cl/^{37}Cl of the two isotopes in these clouds. The observed abundance ratio in two Orion clouds in ^{35}Cl/^{37}Cl = 4–6, depending on the uncertainties associated with line saturation, whereas the solar ratio is but ^{35}Cl/^{37}Cl = 3.1. The Orion clouds are more ^{35}Cl-rich than the Sun by the factor 1.3–2. The explanation of this is not known, but reinforces other differences between the Orion clouds and the Sun. Interestingly, the observed ^{35}Cl/^{37}Cl ratio holds good in both Orion clouds despite factor of five differences in the ratio of Cl/H in the gas, a ratio that depends on the extent of depletion of gaseous Cl onto dust grains.

Anomalous isotopic abundance

Only modest isotopic variations of ^{35}Cl are expected in various astrophysical samples, and these have not been measurable to date. Presolar grains of the types already known contain little Cl.

^{36}Cl $\mid$ $Z = 17$, $N = 19$ $\quad$ $t_{1/2} = 300\,000$ years
$\phantom{^{36}Cl \mid}$ **Spin, parity:** $J, \pi = 2-$; $\quad$ **Mass excess:** $\Delta = -29.798$ MeV

Abundance

Radioactive ^{36}Cl does not exist on Earth except for tiny amounts created by cosmic rays. Its earlier presence in meteorites makes it an *extinct radioactivity* (see **Glossary**). Preliminary evidence indicates that about one Cl atom in a million was of this form in the gas from which the solar system formed. If that can be substantiated, it provides either additonal information about nucleosynthesis and ejection into the molecular cloud from which the Sun formed or about energetic charged particle reactions in the early solar system.

Nucleosynthesis origin

Odd–odd ^{36}Cl has little chance for significant nucleosynthesis during the explosive oxygen burning that is responsible for the stable Cl isotopes. Its binding energy competes unfavorably in *quasiequilibrium* (see **Glossary**). Its natural abundance is more likely the result of a single neutron capture on intial ^{35}Cl in a star that ejected matter into the presolar cloud. This might be a supernova or an AGB red-giant star.

^{37}Cl | $Z = 17, N = 20$
Spin, parity: J, $\pi = 3/2+$;
Mass excess: $\Delta = -31.761$ MeV; $\qquad S_n = 10.32$ MeV

> ^{37}Cl is the lightest stable nucleus supporting three excess neutrons; however, it is synthesized as radioactive ^{37}Ar having but one excess neutron. At ^{37}Cl a nucleosynthesis trend first appears in which stable neutron-rich isotopes are made as radioactive progenitors having fewer excess neutrons. The reason is this – that at temperatures higher than 3×10^9 K, where oxygen burns explosively, the burning reactions are too fast to allow the slower beta decays to occur until after the burning is finished. In the supernova the burning ends owing to the expansion of the star, which cools the interior. Then ^{37}Ar decays to ^{37}Cl.

Abundance
^{37}Cl is 24.23% of Cl atoms on Earth. Using the observed solar-system abundance, which is uncertain to perhaps 50%, Cl= 3773 per million silicon atoms, so

solar abundance of ^{37}Cl = 913 per million silicon atoms.

^{35}Cl is the 45th most abundant isotope in nature.

Nucleosynthesis origin
During oxygen burning in stars, free neutrons, protons and alpha particles are abundant. These produce both ^{35}Cl and ^{37}Cl from the more abundant alpha nuclei that dominate the abundance patterns at that time. Most of the ^{37}Cl is produced as radioactive ^{37}Ar, which has a 35-day halflife against decay to ^{37}Cl. Once ^{36}Ar is established during the burning process, for example, a single additional neutron capture yields ^{37}Ar. When oxygen burning occurs the excess neutrons are initially housed in neutron-rich isotopes of silicon, but the charged particles reactions liberate those neutrons for relocation into the growing sulfur and argon abundances. Within the supernova structure, therefore, the ^{37}Ar concentration leaps to 500 times greater than the ^{37}Cl concentration within the initial star, and it does so in the central zones that are burning oxygen at the time the star explodes. In zones where silicon burning follows oxygen burning, the neutron-rich isotopes such as ^{37}Ar disappear as the excess neutrons shift upward into the iron peak. A significant fraction of ^{37}Cl is already transmuted from ^{37}Ar during the presupernova oxygen and neon burning, however; but it too disappears during silicon burning. These and related reaction sequences well account for the solar ^{37}Cl abundance. Both types of supernovae (I and II) experience explosive O burning and may contribute to interstellar ^{37}Cl abundance.

Astronomical measurements

The two isotopes of Cl have been detected in interstellar clouds in Orion via the radiofrequency emission of the HCl molecule (see **^{35}Cl**).

Anomalous abundance

Only modest isotopic variations of ^{35}Cl are expected in various astrophysical samples because both stable isotopes are made in the same burning zones; but these have not been measurable to date. Known types of presolar grains contain little Cl.

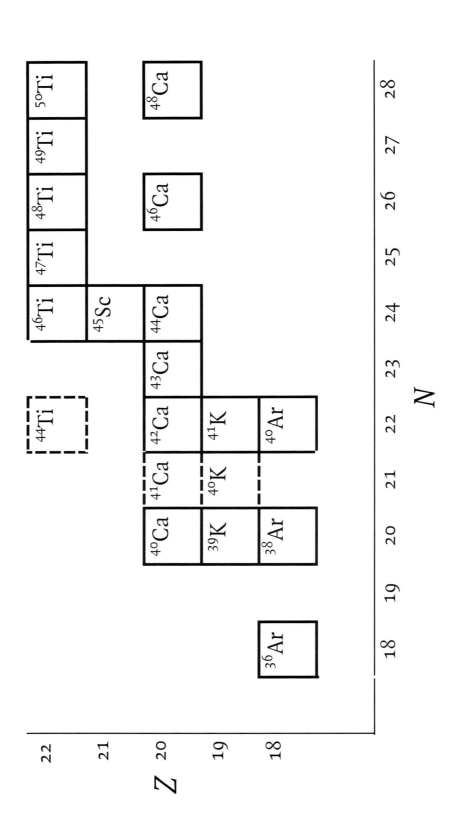

18 | Argon (Ar)

Argon is the third member of the family of inert (or *noble*) gases, coming below He and Ne is the periodic table. The gases are called inert because they react very weakly with other elements. Argon is a colorless, odorless, tasteless gas under ordinary conditions. It is a common gaseous element in the Earth's atmosphere, comprising about 1% of air molecules. It was discovered in England over a 100-year period, starting with the observation by Henry Cavendish in 1784 that atmospheric nitrogen could not be made to react completely with oxygen. About 1% of the molecules were unreactive and would not combust. In 1884 the emission spectrum in a gas-light tube showed lines of an element never before known. Its name derives from the Greek word *argos*, meaning "inactive." During the fractional distillation from a sample of liquid air, Ar gas will separate from liquid air at a temperature of $-122.3°$C, just below the critical temperature of oxygen ($-118°$C) but above that of nitrogen ($-150°$C). Noble gases, including Ar, are very useful in lamps that are excited to emit light by electron collisions, commonly called "neon lights."

All argon nuclei have charge $+18$ electronic units, so that 18 electrons orbit the nucleus of the neutral atom. Its electronic configuration is $1s^2 2s^2 2p^6 3s^2 3p^6$, which fills each of these shells. It is these closed, or full, s and p shells of electrons that make Ar and all noble gases inert. Before 1962 it was thought that these elements were absolutely inert, but at that time the first stable compound of xenon, Xe, was discovered. The newly discovered compounds of krypton, Kr, and Xe are extremely interesting to chemists.

Argon has three stable isotopes, of masses A $= 36$, 38, and 40. These nuclei are primarily produced within the cores of evolved massive stars when they explode as supernovae, but in solids, such as meteoritic rocks, they may exist owing to interactions of cosmic rays with stable atoms, such as a cosmic-ray proton colliding with calcium in the solid. Its lack of chemical affinity gives argon some unusual isotopic patterns in special objects. One of the great cosmic jokes is that in air the most abundant of Ar isotopes, ^{40}Ar, comprising 99.6% of the argon in the Earth's atmosphere, is actually its least abundant isotope in the universe. The feature that makes Ar isotopes so useful for the study of solids is argon's high volatility, causing Ar to have very low abundance within both Earth rocks and meteorites. Its low abundance allows an anomalous isotopic component of Ar to be easily detected when trapped within a solid. Harder to find is the reason why each anomalous isotopic component should be contained within that solid, but that question speaks to scientific discovery of the age and origin of the rocks. The radioactive decay of ^{40}K to ^{40}Ar has made possible the development of "potassium–argon dating" of ancient solids.

Argon is an important element for nucleosynthesis, for the discovery and interpretation of isotopic anomalies, and for the interpretation of the origin of planetary atmospheres. Argon is the eleventh most abundant element in the universe; thus it is an important datum for nucleosynthesis theories. It is much rarer in terrestrial abundances because most Ar was lost into space (owing to its inertness and high volatility) as the Earth formed.

Natural isotopes of argon and their solar abundances

A	Solar percent	Solar abundance per 10^6 Si atoms
36	84.2	8.50×10^4
38	15.8	1.60×10^4
40	small but unknown	25 (estimated)

^{36}Ar $Z = 18$, $N = 18$
Spin, parity: $J, \pi = 0+$;
Mass excess: $\Delta = -30.230$ MeV; $S_n = 15.25$ MeV

That ^{36}Ar is as abundant in the universe as the element aluminum is stunning to Earth scientists, who know ^{36}Ar as very rare and Al as a dominating chemical presence in the crust of the Earth. Furthermore, on Earth ^{36}Ar is much less abundant than the heaviest isotope of argon, ^{40}Ar, another accidental distortion of the true cosmic abundances by Earth's history. The number of nucleons in the ^{36}Ar nucleus can be counted as nine ^{4}He nuclei, causing it to be referred to as an "alpha element". Nuclei that may be assembled from an integral number n, of alpha particles are the most abundant isotopes of their respective elements between n = 3 (carbon) and n = 10 (calcium).

Abundance

Argon is the third most abundant of the noble gases. From the isotopic decomposition of solar-wind Ar one finds that the mass-36 isotope, ^{36}Ar, is the most abundant of all Ar isotopes in the Sun, and therefore also in the universe. In the Earth's atmosphere, ^{40}Ar happens to be the most abundant owing to its peculiar origin on Earth as the daughter of radioactive ^{40}K. But in the Sun and stars ^{36}Ar is overwhelmingly the most abundant of the three isotopes. ^{36}Ar comprises 84.2% of all the Ar isotopes in the Sun. Using the total abundance of elemental Ar $= 1.01 \times 10^5$ per million silicon atoms in the Sun, this isotope has

solar abundance of ^{36}Ar $= 8.50 \times 10^4$ per million silicon atoms.

This makes ^{36}Ar a quite abundant nucleus, being comparable to the abundance of ^{22}Ne or to the element aluminum. It is 17th most abundant of all nuclei. Its high abundance requires a favored position in the nuclear networks of production.

Nucleosynthesis origin

^{36}Ar exists primarily owing to the explosive *oxygen-burning* (see **Glossary**) process of stellar nucleosynthesis. Put simply, during oxygen burning two ^{16}O nuclei collide, causing the *net* reaction $2^{16}O \rightarrow {}^{32}S$ to occur with reasonable efficiency. ^{36}Ar is then created by the capture of an alpha particle by ^{32}S. Within the supernova structure, therefore, the ^{36}Ar concentration leaps to 1000 times greater than its concentration within the initial star, and it does so in the central zones that are burning oxygen at the time the star explodes. It remains abundant even during silicon burning, at least until the silicon is exhausted, because of its robust binding energy, unlike the heavier isotopes of Ar which give up their excess neutrons at the temperature of silicon burning. ^{36}Ar is a "primary isotope" because its nucleosynthesis proceeds easily in stars having only hydrogen and helium in their initial composition. Almost all of this happens in Type II supernovae, which on average eject about ten times as much ^{36}Ar as do Type Ia supernovae and are about five times more frequent. Nature's numbers are awesome here, each Type II ejecting a mass of ^{36}Ar comparable to 3000 times the mass of the Earth!

Astronomical measurements

The abundance of argon is not normally measured in stars owing to atomic physics complications. Its lines involve high excitations and are not easily detected. The most useful observations of argon abundance arise in ionized nebulae, either planetary nebulae or the HII regions surrounding young O stars where lines of ionized Ar are detectable. The latter show the composition of the interstellar medium near the O stars, and measurements agree with the accepted solar abundance for Ar. Since ^{36}Ar dominates that abundance, such measurements are effectively of the abundance of ^{36}Ar. But we have no measurements of Ar abundance in old stars, which are cool and red. Because Ar is inert, moreover, it does not make compounds that might allow isotopic compositions to be inferred from molecular lines. X-ray emission by ionized argon from young, hot, shocked supernova remnants, as observed with NASA's *Chandra X-Ray Observatory*, confirm that core-collapse Type II supernovae are prolific nucleosynthesis sources of Ar.

Anomalous isotopic abundance

Certain meteoritic samples reveal either a deficiency or an excess of this isotope. However, the isotopic ratios for argon are usually stated with ^{36}Ar as the denominator, so that variations in isotopic ratios are usually attributed to variations in the abundance of the lesser isotopes, ^{38}Ar and ^{40}Ar. Variations of these isotopic ratios are found because of (1) radioactive decay of ^{40}K, which enables the ages of minerals to be determined; (2) small isotopic fractionations that come about because of differing weights of the Ar isotopes, as in the Earth's atmosphere; and (3) differing nucleosynthesis components in the materials from which small meteoritic samples were assembled.

Solar argon in the Earth's mantle   A standing science question concerning the origin of the Earth and of the Earth's atmosphere is whether the original noble gases can still be found bubbling out of the mantle. Such primordial gases would be the small remnants of the original solar gases within the matter from which the Earth formed, and are found for He isotopes. (For Ar evidence of this see ^{38}**Ar**.)

^{38}**Ar** $\quad$ Z $= 18$, N $= 20$
$\qquad$ **Spin, parity:** J, $\pi = 0 +$;
$\qquad$ **Mass excess:** $\Delta = -34.175$ MeV; $\qquad S_n = 11.84$ MeV

Abundance

Argon is the third most abundant of the noble gases. From the isotopic decomposition of solar-wind Ar one finds that the mass-38 isotope, ^{38}Ar, is the second most abundant of Ar isotopes in the universe. In the Earth's atmosphere, ^{40}Ar happens to be the most abundant owing to its peculiar origin on Earth as the daughter of radioactive ^{40}K. ^{38}Ar comprises 15.8% of all the Ar isotopes in the Sun. Using the total abundance of elemental Ar $= 1.01 \times 10^5$ per million silicon atoms in the Sun, this isotope has

solar abundance ^{38}Ar $= 1.60 \times 10^4$ per million silicon atoms.

This makes ^{38}Ar a moderately abundant nucleus, being comparable to the abundance of ^{17}O or to the two heavy isotopes of Si. It is the 28th most abundant of all nuclei, requiring that it occur naturally in the main-line nuclear networks of element production in stars.

Nucleosynthesis origin

^{38}Ar exists primarily owing to the oxygen-burning process (both hydrostatic and explosive) of stellar nucleosynthesis. Put simply, during oxygen burning two ^{16}O nuclei collide, and the *net* reaction $2 \, ^{16}$O $\rightarrow \, ^{32}$S and subsequent production of ^{36}Ar by capture of an alpha particle by ^{32}S both occur with reasonable efficiency. ^{38}Ar is then produced by capturing excess neutrons in the thermonuclear bath. It is a "partly secondary isotope" because its nucleosynthesis by neutron-captures from newly made ^{36}Ar is enhanced if the star has not only hydrogen and helium in its initial composition but also carbon and oxygen; the ^{18}O and ^{22}Ne created from initial carbon and oxygen provide for the excess neutrons that the heavier Ar isotopes contain. However, significant excess neutrons are also produced during the oxygen burning itself owing to positron emissions by radioactive nuclei that occur there, so that ^{38}Ar is also partly primary. Within the supernova structure, therefore, the ^{38}Ar concentration leaps to 1000 times greater than its concentration within the initial star, and it does so in the central zones that are burning oxygen at the time the star explodes. Its excess neutrons are yielded up during the subsequent silicon burning, however, during which only ^{36}Ar survives until the silicon is exhausted. When the isotopic composition of the entire spectrum of massive supernova explosions is computed, they show that all isotopes of argon are created by those supernova explosions, which are their source. Almost all of this emerges from core-collapse Type II supernovae.

Anomalous isotopic abundance

Certain meteoritic samples reveal either a deficiency or an excess of ^{38}Ar with respect to ^{36}Ar. Variations are found owing to small fractionations that come about only because of differing weights of the Ar isotopes, as in the Earth's atmosphere. And they are found in meteorites owing to differing nucleosynthesis components in the materials from which small meteoritic samples were assembled.

Solar argon in the Earth's mantle   A standing science question concerning the origin of the Earth and of the Earth's atmosphere is whether the original noble gases can still be found bubbling out of the mantle. Such primordial gases would be the small remnants of the original solar gases within the matter from which the Earth formed. The ^{38}Ar/^{36}Ar isotopic ratio has the value 0.188 in the Earth's atmosphere (leading to the percentages given above under **Abundance**), but considerably smaller values in the range 0.172–0.179 are now inferred in the solar wind (an improved estimate of the Sun's ^{38}Ar/^{36}Ar ratio). Ratios smaller than atmospheric, down to ^{38}Ar/^{36}Ar = 0.182 have recently been found in basaltic glasses along the mid-ocean ridges where volcanos spew up new lava. These ratios suggest that the measured gases arising from the mantle are mixes between the original solar ^{38}Ar/^{36}Ar ratio and that in the Earth's atmosphere today. Correlations with the ^{22}Ne/^{20}Ne ratios in these samples confirm the same conclusion for neon. It may thus be concluded that the processes that formed the Earth were not able to bubble out all of the solar gases that were initially present. A consequence of this is that the larger value ^{38}Ar/^{36}Ar = 0.188 in the atmosphere today represents a selective alteration of the initial solar gases whereby the atmosphere retained more ^{38}Ar than ^{36}Ar.

Cosmic-ray exposure age   The retention of ^{38}Ar by solids enables a measure of the duration for which cosmic solids have been exposed to cosmic rays. This is useful in studies of meteorites and of lunar samples. Cosmic rays striking calcium in the rocks leave fragments of ^{36}Ar and ^{38}Ar in the ratio about 1.5:1, which is ^{38}Ar-rich in comparison with solar Ar. Since the rate of arrival of cosmic rays is known, the time required for the buildup of the excess ^{38}Ar can be computed. For the lunar soils the answers lie in the range of hundreds of millions of years, which, though great, is much less than the age of the Moon itself. This means that the lunar soils were not always exposed on the surface, but are "gardened" – turnovers caused by impacts of meteorites.

It is quite expected that in the surfaces of some stars and in the nucleosynthesis ejecta of certain stars, this isotope will have nonsolar proportions. However, its isotopic composition has not been measurable by astronomers owing to its small abundance and to the small isotopic differences in the wavelenghts of radiation emitted by this element.

^{40}**Ar** $\mid$ $Z = 18, N = 22$
$\mid$ **Spin, parity:** $J, \pi = 0+$;
$\mid$ **Mass excess:** $\Delta = -35.040$ MeV; $\qquad S_n = 9.87$ MeV

Abundance

Argon is the third most abundant of the noble gases. From the isotopic decomposition of solar-wind Ar one finds that the mass-40 isotope, ^{40}Ar, is much the rarest of Ar isotopes in the universe. In the Earth's atmosphere, however, ^{40}Ar happens to be the most abundant owing to its peculiar origin on Earth as the daughter of radioactive ^{40}K. The Earth formed with a substantial store of radioactive ^{40}K, and its decay over geologic time has caused ^{40}Ar gas, its daughter nucleus, to be outgassed from the solid Earth and to enrich the Earth's atmosphere, which would otherwise have contained only tiny quantities of argon. ^{40}Ar makes Ar the third most abundant element in the Earth's atmosphere; but that is quite deceptive of its universal abundance. In the Sun and stars ^{40}Ar is overwhelmingly the least abundant of the three isotopes. The ^{40}Ar abundance has not been measurable in the Sun and stars because it is too small; but there is evidence to suggest it to be of order 0.1% of argon.

Using the total abundance of elemental Ar $= 1.01 \times 10^5$ per million silicon atoms in the Sun, this isotope has

solar abundance of ^{40}Ar $= 25$ per million silicon atoms.

This makes ^{40}Ar a rare nucleus, 72nd in the abundance list, much less than the two lighter Ar isotopes.

Nucleosynthesis origin

The exact site of origin for ^{40}Ar has been somewhat unclear. A cosmoradiogenic portion of ^{40}Ar has appeared as the daughter of radioactive ^{40}K produced by explosive oxygen burning in supernovae. A larger portion appears to exist primarily owing to the intense neutron flux that is set free during explosive carbon burning in the shocked carbon shell during supernova explosions. Capture of those neutrons converts part of the ^{38}Ar to ^{40}Ar. Yet another portion arises in any s process, when ^{40}Ar is made from ^{39}K by neutron capture.

Anomalous isotopic abundance

Variations of the isotopic ratio are found because of radioactive decay of ^{40}K, which enables the ages of minerals to be determined. The variations also occur due to small fractionations that come about only because of the differing weight of the Ar isotopes, as in the Earth's atmosphere. And variations are found in meteorites owing to differing nucleosynthesis components in the materials from which small meteoritic samples were assembled.

Solar argon in the Earth's mantle   A standing science question concerning the origin of the Earth and of the Earth's atmosphere is whether the original noble gases can still be found bubbling out of the mantle. Such primordial gases would be the small remnants of the original solar gases within the matter from which the Earth formed. Data from ^{38}Ar and He yield information on this topic but ^{40}Ar can not be used in this way. A huge contribution to atmospheric ^{40}Ar has been made over geologic time by ^{40}K decay within the Earth. Because potassium is chemically reactive, the initial Earth retained far more K than Ar. As a result, the ^{40}Ar that is produced over time as most of the ^{40}K decays vastly outnumbers the original ^{40}Ar within the solar gases. This is the reason for ^{40}Ar being almost 1% of the atoms in the Earth's atmosphere.

Radioactive ages of rocks   The same ideas concerning production of ^{40}Ar by radioactive decay enable a technique for determining how long rocks have existed. Those that solidified recently have not had time to produce much ^{40}Ar within because not much ^{40}K has decayed in the time available. The halflife of ^{40}K is 1.28 billion years (about one-quarter of the Earth's total age). Very old rocks, by contrast, have generated much ^{40}Ar in this way. So the ratio of extra ^{40}Ar in the rock to remaining ^{40}K, provided that both can be measured accurately, reveals the age of the rock. This technique has been especially useful for the dating of lunar rocks from the *Apollo* sample returns. The Lunar Highland rocks (the oldest) are about 4.1 billion years old, older than rocks on Earth but not as old as the Earth itself.

^{40}Ar on the Moon   An early puzzle when the *Apollo* lunar samples were returned was the large amount of ^{40}Ar in the lunar soil. Since the Moon has no appreciable atmosphere, the ^{40}Ar released by radioactive decay within the Moon can not build up as on Earth. The soil on the Moon was found to contain ^{40}Ar to depths of about 0.1 μm in the dust grains, as if implanted by solar wind. And so it was. As radiogenic ^{40}Ar left the Moon it was ionized by radiation, after which it was picked up by the magnetic field of the solar wind and slammed back into the soil, explaining that lunar isotopic anomaly.

19 | Potassium (K)

All potassium nuclei have charge $+19$ electronic units, so that 19 electrons orbit the nucleus of the neutral atom. Its electronic configuration can be abbreviated as an inner core of inert argon (a noble gas) plus one more electron: $(Ar)4s^1$, which locates potassium in Group IA of the chemical periodic table, after Li and Na. The elements K and Na are so much alike that they were not recognized as distinct in the eighteenth century before both were isolated. The metals in this group are called *alkali metals*. They form strong alkaline caustic compounds with the OH^- ion. Household lye is a mix of NaOH with KOH. Leeching water through wood ashes produces alkaline potassium compounds. Potassium has valence $+1$ and combines readily with oxygen atoms as K_2O. Potassium cannot be found free because it is too reactive, so reactive that in pure form it must be stored in a nonreactive container. Potassium metal is soft and silver-white. If placed in water K catches fire to release hydrogen gas and make potassium hydroxide: $2K + 2H_2O \rightarrow H_2 + 2KOH$. Interestingly, H is at the top of Group IA, so in that sense water and K_2O are similar compounds; but this reaction shows that K can take O away from H. The pure metal was first isolated by Sir Humphrey Davy in 1807 by electrolysis of potassium hydroxide. At the same time he isolated sodium by the same method.

Like many elements, potassium's chemical symbol seems to bear no relation to its name. In such cases the symbol often derives from the Latin if the Roman Empire used one of its minerals, as in this case *kalium*, from which the word *alkali* derives. The name potassium comes from the word potash, wood ashes from which the salts of K have been recovered from antiquity. From the 7th century BC the Old Testament's Book of Jeremiah 2:22 refers to *neter*, which is potash.

As an element, K is eighth most abundant in the Earth's crust, whereas it is but 19th in the universe. This is because its oxide, K_2O, is so easily made and so tightly bound into rocks that K remained in the crust when many metals sank to the molten core of Earth. In flame-spectroscopy, K can be recognized by its spectral lines, a doublet in the far red with wavelengths 7697 and 7663 Å and a blue line at 4044 Å . Together they produce a violet hue.

Potassium compounds are in common use. A sodium-free table salt is *potassium chloride* (KCl), which can be recovered from the material dissolved in sea water. Within a cubic crystal the ions of K^+ and Cl^- are arranged alternately, one ion at each corner of the cubic cells of which the crystal is composed. Human blood is a saline solution, making K important to body chemistry. The K^+ ion plays a role in nerve tissues and in muscles. *Potassium hydroxide* (KOH) is used in soaps and in household drain cleaners; its lye is a useful drain cleaner because it reacts with animal fats to make a soap that can be dissolved. Indeed, this is the historic beginning of soap. *Potassium carbonate* ($KHCO_3$), called *pearl ash*, is used as a cleaning agent and in some glass products. *Potassium nitrate* (KNO_3) is known as *nitre* or *saltpeter* and is a common component of fertilizers and of matches.

Potassium has two stable isotopes (^{39}K and ^{41}K) and one naturally occurring radioactive isotope (^{40}K). The last, via its decay, accounts for the high abundance of ^{40}Ar in the Earth's atmosphere. It also provides a valuable means of dating ancient rocks. Potassium nucleosynthesis derives almost entirely from a single main-line source, the fusion of oxygen in massive stars that will become Type II supernovae.

Natural isotopes of potassium and their solar abundances

A	Solar percent	Solar abundance per 10^6 Si atoms
39	93.26	3516
40	0.0117 today	5.48 at solar beginning
41	6.73	254

^{39}K | $Z = 19, N = 20$
Spin, parity: $J, \pi = 3/2+$;
Mass excess: $\Delta = -33.807$ MeV; $S_n = 13.09$ MeV

^{39}K has larger neutron binding energy, S_n, than any other stable odd-A nucleus.

Abundance

^{39}K constitutes 93.26% of natural potassium. The elemental K abundance in the solar photosphere and in meteorites are in good agreement: 3770 atoms of K per million Si atoms. This isotopic abundance is then

solar abundance of $^{39}K = 3516$ per million silicon atoms.

^{39}K is the 37th most abundant nucleus within the universe. This makes ^{39}K a very abundant nucleus, comparable to chlorine or manganese.

Nucleosynthesis origin

^{39}K is primarily the result of the burning of oxygen in massive stellar explosions. It is coproduced with the other K isotopes there. The concentration of ^{39}K jumps by a factor 1000 above its concentration in the initial star within the oxygen-burning zones of the Type II *supernovae* (see **Glossary**). But it disappears if the burning progresses into silicon burning before the matter is ejected. A lesser amount is, however, included in the *alpha-rich freezeout* (see **Glossary**) material nearer the center. ^{39}K is made as itself during these burning reactions. It is not radiogenic. Some ^{39}K is also made by the *s process* (see **Glossary**) in zones farther from the supernova center, but not as much in bulk as the oxygen burning.

Astronomical measurements

Potassium has been studied in stellar optical spectra. The famous red K lines at 7663 and 7697 Å have been used. This pair of lines played a role in the history of K flame spectroscopy. As nucleosynthesis progressed the galactic K/H abundance ratio in newly born observed stars increased from 10^{-3} of solar in some early stars to a bit in excess of solar K/H today. If compared instead to Mg, the ratio K/Mg remains near the solar ratio in stars of all metallicities. This is understood as the coproduction of K and Mg in massive Type II supernovae.

It is quite expected that in the surfaces of some stars and in the nucleosynthesis ejecta of certain stars, this isotope will have nonsolar proportions. However, its isotopic composition has not been measurable by astronomers owing to its small abundance and to the small isotopic differences in the wavelengths of radiation emitted by this element.

Anomalous meteoritic isotopic abundance

Although variations of K isotopes exist, the most abundant isotope, ^{39}K, is used to normalize the abundances of the other two isotopes. Variations of the ^{41}K/^{39}K ratio are therefore interpreted as variations of ^{41}K.

^{40}K	$Z = 19$, $N = 21$ $\qquad t_{1/2} = 1.277 \times 10^9$ years
	Spin, parity: J, $\pi = 4-$; $\qquad$ **Mass excess:** $\Delta = -33.535$ MeV

Among isotopes produced primarily by stars, ^{40}K is the lightest odd-Z and odd-N nucleus having excess neutrons to exist in measurable quantities on Earth. All odd–odd nuclei bearing excess neutrons are unstable. But the large spin and parity change makes it decay so slowly that significant amounts remain on Earth, giving rise to important geologic markers.

Abundance

From the isotopic decomposition of normal K one finds that the mass-40 isotope, ^{40}K, is much the rarest of K isotopes in the universe. It is about 8000 times less abundant than ^{39}K, the most abundant of the three naturally occurring isotopes. But it was not always so rare. At the birth of the solar system, ^{40}K was 12.4 times more abundant than it is today on Earth, where its radioactive halflife has reduced its abundance by half every 1.28 billion years. In the Earth's atmosphere, ^{40}Ar happens to be the most abundant argon isotope owing to its peculiar origin on Earth as the daughter of radioactive ^{40}K. The Earth formed with a substantial store of radioactive ^{40}K, and its decay over geologic time has caused ^{40}Ar gas, its daughter nucleus, to be outgassed from the solid Earth and to enrich the Earth's atmosphere (see **^{40}Ar, Abundance**). Using the total abundance of elemental K = 3770 per million silicon atoms in the Sun, this isotope has

solar abundance of ^{40}K = 0.44 per million silicon atoms today.

This makes ^{40}K a rare nucleus, being comparable to the abundance of the heavy metals rhodium or gold. But at the birth of the solar system the abundance was

initial solar abundance of ^{40}K = 5.48 per million silicon atoms.

Nucleosynthesis origin

^{40}K is primarily the result of two processes of stellar nucleosynthesis. All three K isotopes are synthesized during the explosive burning of oxygen in massive stellar explosions. But because of the relatively weak binding energy (making its mass excess Δ about 1.3 MeV greater than that of ^{40}Ca), not very much ^{40}K can survive production in this setting. This causes it to be much rarer than it could have been if the nuclear force had bound it more strongly. Some ^{40}K is also produced by the s process. Any time that free neutrons are made available in the stars, the ^{40}K can be produced from the ^{39}K(n, gamma)^{40}K reaction. This is also not a very efficient producer of ^{40}K. So on both counts, ^{40}K is a rare but nonetheless very significant nucleus.

Anomalous isotopic abundance

The three isotopes of K certainly should not always occur in all natural samples in their usual proportions, because they are produced with widely differing ratios in differing stellar settings. Solid grains condensing in the ejecta of the stars will therefore provide isotopically variable components of the interstellar medium. In particular, certain meteoritic samples could reveal either a deficiency or an excess of ^{40}K. However, its abundance is so small as to be hard to measure in single presolar grains, where the isotopic ratios certainly must vary. And in larger samples, the isotopes tend to assume their average values for initial solar material, probably reflecting vaporization and recondensation of this fairly volatile element. Therefore, isotopic anomalies in K are not common at the present state of knowledge, although they may one day be quite important. An exception is the mass-dependent isotopic fractionation that biases K isotopes because of their different masses in thermal chemistry.

Radioactive ages of rocks   Variations of argon isotopic ratios are found within solids because of radioactive decay of ^{40}K, which also enables the ages of minerals to be determined. This works as follows. If rocks solidify with all three K isotopes present in the proportions they should have had at that time, with the abundance of ^{40}K slowly decaying with time, the abundance of daughter ^{40}Ar in the rock today depends upon the intial abundance of ^{40}K in that rock and the time that the rock has existed. Those that solidified recently have not had time to produce much ^{40}Ar within because not much ^{40}K has decayed in the time available. Measurement of the amount of ^{40}K remaining in the rock and of the amount of daughter ^{40}Ar within the rock suffice to give the time at which the rock solidified (i.e. the time at which it began to trap noble gas ^{40}Ar). This technique was particularly powerful in the lunar science studies of the returned samples from the *Apollo* missions. They showed that the oldest rocks on the Moon lie in the Lunar Highlands, and are about 4.1–4.2 billion years old. This is about 91% of the age of the Earth. It is almost certain that the Moon was broken off from the Earth very early in the Earth's history, so that no rocks can be found that survived from the first 300 million years of the Moon's existence. This is commonly believed to be due to the terribly destructive bombardment of Moon and Earth by meteorites for almost 400 million years after the Earth–Moon system first formed.

Radiogenic argon in the Earth's mantle   A huge contribution to atmospheric ^{40}Ar has been made over geologic time by ^{40}K decay within the Earth. Because potassium is chemically reactive, the initial Earth retained far more K than gaseous Ar. As a result, the ^{40}Ar that is produced over time as most of the ^{40}K decays to ^{40}Ar vastly outnumbers the original ^{40}Ar trapped from the original solar gases. This is the reason for argon being almost 1% of the atoms in the Earth's atmosphere, and for atmospheric argon being almost pure (99.6%) ^{40}Ar. It is the decay product within a large rock (the Earth) that initially contained a lot of radioactive ^{40}K, and from within which that argon has diffused out into the atmosphere.

^{41}**K** | $Z = 19$, $N = 22$
Spin, parity: J, $\pi = 3/2+$;
Mass excess: $\Delta = -35.559$ MeV; $S_n = 10.09$ MeV

^{41}K and ^{37}Cl are the lowest-Z stable isotopes to have three excess neutrons; but that is deceptive, for, like ^{37}Cl, ^{41}K is synthesized primarily as a radioactive progenitor, ^{41}Ca, having only one excess neutron.

Abundance

^{41}K constitutes 6.73% of natural potassium. The elemental K abundance in the solar photosphere and in meteorites are in good agreement: 3770 atoms of K per million Si atoms. This isotope's abundance is then

solar abundance of ^{41}K $= 254$ per million silicon atoms.

^{41}K is the 56th most abundant nucleus within the universe. Of the lighter stable isotopes that are synthesized in stars, only ^{40}Ar and ^{36}S are less abundant. On the other hand, ^{41}K is more abundant than any stable isotope heavier than zinc. In that sense then, ^{41}K may be thought of as among those few dividing the very abundant isotopes from the rare ones.

Nucleosynthesis origin

^{41}K is primarily the result of the burning of oxygen in massive stellar explosions. It is coproduced with the other K isotopes there. During oxygen burning in stars, free neutrons, protons and alpha particles are abundant. These produce both ^{39}K and ^{41}K from the more abundant alpha nuclei that dominate the abundance patterns at that time. ^{41}K is synthesized primarily in the form of radioactive ^{41}Ca, with halflife 1.03×10^5 yr, and which decays primarily after ejection by the supernova event within which oxygen burning occurs. It must be appreciated that ^{41}Ca production is the major source of stable daughter ^{41}K in nature. This makes ^{41}K one of the most radiogenic of all natural abundances. ^{41}Ca is primarily the result of neutron capture by abundant ^{40}Ca . This occurs whenever free neutrons exist in the stellar gas, since ^{40}Ca is almost always present to some degree.

The required neutron liberation for ^{41}Ca production also occurs farther out in the star while it is burning helium (see *Helium burning*, in **Glossary**) in hydrostatic equilibrium. That portion of ^{41}Ca may be continuously ejected in a wind of outflowing gas from massive stars (the Wolf–Rayet phase), or it may be primarily ejected by the final explosion (the supernova). But most ^{41}Ca is synthesized deeper within the massive star, during oxygen burning. Which of these sources dominates the observed extinct ^{41}Ca in presolar grains and in the early solar system is not clear (see **^{41}Ca**).

Astronomical measurements

See **^{39}K**.

Anomalous meteoritic isotopic abundance

The three isotopes of K certainly should not always occur in all natural samples in their usual proportions, because they are produced with widely differing ratios in differing stellar settings. Solid grains condensing in the ejecta of the stars will therefore be isotopically variable components of the interstellar medium. In particular, certain meteoritic samples could reveal either a deficiency or an excess of ^{41}K. However, its abundance is so small as to be hard to measure in single presolar grains, where the isotopic ratios certainly must vary. And even in larger samples, the isotopes tend to assume their average values for initial solar material. Therefore, isotopic anomalies in K are not common at the present state of knowledge, although they may one day be quite important. An exception is the mass-dependent isotopic fractionation that biases K isotopes because of their different masses in thermal chemistry. Especially interesting is the large ^{41}K excess found in glassy beads on the Moon. These are interpreted as melted drops created by meteorite impacts that evaporate ^{39}K faster than ^{41}K.

Presolar grains   Excess ^{41}K exists in Ca-rich presolar grains, in proportion to the amount of calcium. This identifies extinct ^{41}Ca is the source of these ^{41}K excesses. That ^{41}Ca was alive when the stellar grains condensed rather than in the early solar system. See **^{41}Ca** for more on that *extinct radioactivity* (see **Glossary**).

Ca–Al-rich inclusions from meteorites   Many of these CAIs reveal excess ^{41}K in the most Ca-rich minerals within them. This shows that ^{41}Ca was alive in the early solar system, qualifying it as an *extinct radioactivity*. The amount in the CAIs was small; but that small amount causes severe restrictions on the timing of the events that placed ^{41}Ca in the solar system. See **^{41}Ca** for more on that *extinct radioactivity*.

20 | Calcium (Ca)

All calcium nuclei have charge +20 electronic units, so that 20 electrons orbit the nucleus of the neutral atom. Its electronic configuration can be abbreviated as an inner core of inert argon (a noble gas containing eighteen inner electrons) surrounded by two more electrons: $(Ar)4s^2$, which locates calcium beneath magnesium in Group IIA of the periodic table. Calcium has valence +2 owing to its two 4s electrons, and combines readily with single oxygen (−2) atoms as CaO.

Calcium is a metal, hard, lustrous and silver gray when freshly cut, but tarnished by reaction with oxygen and nitrogen when exposed to air. It was traditionally called an "alkaline earth" element owing to its high abundance in alkaline rocks. Its name derives from the Latin *calx*, the name by which lime was known to the early Romans. Lime (CaO) is added to some soils to reduce their acidity. Ca is a very active metal utilized in bones and teeth. We brush our teeth to remove the acidity that dissolves Ca. Statues made of *limestone* ($CaCO_3$) are attacked by acid rain for the same reason. Ca metal placed in water bubbles off hydrogen gas and leaves calcium hydroxide in the water. But Ca does not exist in free elemental form because of its high reactivity. Few people have seen metallic pure Ca because the pure metal must be chemically isolated from naturally occurring minerals, such as lime or *gypsum* ($CaSO_4$). Although its minerals had many ancient uses in drywall construction and in casts for statues and for broken bones, calcium was not isolated as an element until Sir Humphrey Davy did so by electrolysis of $Ca(OH)_2$ in 1808.

Calcium is the thirteenth most abundant element in the universe, following aluminum and before sodium; thus it is a significant datum for nucleosynthesis theories. It is even more significant in terrestrial abundances, where its high reactivity incorporates it into many common minerals, making it the fifth most abundant element in Earth's rocky crust. Calcium atoms are highly depleted from interstellar gas by being condensed in interstellar grains, many of which were formed when the atoms first left the stars. So it is important in presolar stardust.

Calcium is composed of six stable isotopes, far more than any lighter element. Such fecundity owes to the great stability of twenty protons (a "magic" nuclear number). It is the only element having two doubly-magic isotopes (^{40}Ca and ^{48}Ca). Its stable masses are 40, 42, 43, 44, 46 and 48; and one long-lived radioactive isotope (^{41}Ca, $t_{1/2} = 103\,000$ yr) was demonstrably present in the early solar system and in presolar grains and may eventually be detectable by its X-ray line in supernova remnants. Its many isotopes renders calcium highly diagnostic for nucleosynthesis theory. Formation and thermal decomposition of calcium nuclei occur almost simultaneously in the cores of massive stars, so that calcium's abundance there reflects its steady-state population in that very competitive setting. Formation occurs mostly during oxygen burning and silicon burning, whereas eventual reassembly into iron-group elements occurs during the exhaustion of ^{28}Si.

Calcium isotopic anomalies in presolar grains and in *calcium–aluminum-rich inclusions* (see **Glossary**) from meteorites are common. One of the largest anomalies and the most stunning is the excess ^{44}Ca that exists in presolar grains from supernovae. Placed there by the radioactive decay of ^{44}Ti, this excess proves that those grains condensed within a few years after the supernova exploded, a sequence predicted in 1975 by the writer. Calcium gives the CAIs their name, because these rocks, made in the early solar system, are so heavily enriched in oxidized calcium. They also bear anomalous isotopic ratios in other Ca isotopes. Initially unexpected for rocks made in the early solar system, these isotopic anomalies are probably to be understood within the *cosmic-chemical-memory* (see **Glossary**) interpretation of isotopic variations inherited from preexisting interstellar material.

Natural isotopes of calcium and their solar abundances

A	Solar percent	Solar abundance per 10^6 Si atoms
40	96.94	59 200
41	extinct radioactivity in meteorites	
42	0.647	395
43	0.135	82.5
44	2.086	1275
46	0.004	2.4
48	0.187	114

^{40}Ca | $Z = 20, N = 20$
Spin, parity: $J, \pi = 0+$;
Mass excess: $\Delta = -34.846$ MeV; $\quad S_n = 15.62$ MeV

The number 20 conveys such stability to nucleons that it was called a "magic" number. This isotope is special in having magic numbers of both neutrons and protons. It has strong nucleon bonding, as evidenced by its neutron separation energy, S_n. It is the most massive of the stable "alpha nuclei," those that could be assembled from integer numbers of ^4He nuclei (ten in this case).

Abundance

From the isotopic decomposition of normal calcium one finds that the mass-40 isotope, ^{40}Ca, is the most abundant of the stable calcium isotopes; 96.94% of all Ca. On the scale where one million silicon atoms is taken as the standard for solar-system matter, this isotope has

$$\text{solar abundance of } ^{40}\text{Ca} = 5.92 \times 10^4 \text{ per million silicon atoms};$$

i.e. about 5% of the abundance of silicon. ^{40}Ca is the nineteenth most abundant isotope in the universe, lying between the elemental abundances of Al and Na.

Nucleosynthesis origin

^{40}Ca exists primarily owing to the oxygen-burning and silicon-burning processes (both hydrostatic and explosive) of stellar nucleosynthesis. The second follows the first quite quickly in the evolution of massive stars. Put simply, during oxygen burning two ^{16}O nuclei collide, causing the reaction $^{16}\text{O} + {}^{16}\text{O} \rightarrow {}^{28}\text{Si} + {}^{4}\text{He}$. Some alpha particles produced in that way are consumed by ^{28}Si, which is also abundant: $^{28}\text{Si} + {}^{4}\text{He} \rightarrow {}^{32}\text{S} + \text{gamma ray}$. Capture of two more alpha particles produces ^{40}Ca. Its steady-state abundance there is maintained by a balance between alpha-particle capture by ^{36}Ar and thermal ejection of an alpha particle from ^{40}Ca to restore ^{36}Ar. This stationary state is called *quasiequilibrium* (see **Glossary**). Within the supernova structure, therefore, the ^{40}Ca concentration leaps to 1000 times greater than its concentration within the initial star, and it does so in the central zones that are burning oxygen at the time the star explodes. Its large neutron separation energy, $S_n = 15.62$ MeV, ensures that ^{40}Ca will survive high temperatures better than its heavier isotopes, which are destroyed when silicon burning commences. The nuclei having $Z = N = A/2$ in this mass region, ^{28}Si, ^{32}S, ^{36}Ar, and ^{40}Ca, are spoken of as "alpha nuclei."

Primary nucleus   The above reactions are helpful in understanding that ^{40}Ca is called a "primary nucleus" in nucleosynthesis parlance. That designation is given to those nuclei that are abundantly produced within stars whose initial composition is hydrogen plus helium, the two elements left from the Big Bang. Indeed, initial hydrogen and helium are converted to ^{16}O by the natural thermonuclear evolution of matter. So ^{16}O is a primary nucleus, and so therefore is ^{40}Ca, which is made from the ^{16}O. ^{40}Ca is distinguished from its heavier isotopes in that they are "partly secondary isotopes" because their nucleosynthesis is aided if the star manufacturing them has not only hydrogen and helium in its initial composition, but also carbon and oxygen, which provide in part for the excess neutrons that the heavier isotopes contain.

Astronomical measurements

The Ca abundance has been studied in stars by optical spectra. Because ^{40}Ca dominates the spectra of Ca in stars, and especially because ^{40}Ca is a primary "alpha nucleus," the galactic evolution of ^{40}Ca is the same as the observed galactic evolution of calcium elemental abundance. As nucleosynthesis progressed in the galaxy the Ca/H abundance ratio in newly born observed stars increased from 10^{-3} of solar in some stars to a bit in excess of solar Ca/H. The behavior is very similar to that of oxygen, and indeed to that of all alpha nuclei that can be created from fusion of He, C and O. These include the most abundant isotopes of the elements Ne, Mg, Si, S, Ar, Ca, and Ti. The ratio Ca/Fe declines, however, from about three-times solar in metal-poor stars to solar in young stars. This is understood as the rapid coproduction of the alpha nuclei, along with O, in massive Type II supernovae. These produce less iron; but the later onset of Type Ia supernova explosions dramatically increases Fe nucleosynthesis in the Galaxy and causes the interstellar Ca/Fe ratio to decline to its solar value.

Anomalous isotopic abundance

Calcium–aluminum-rich inclusions   The CAIs reveal a lot of isotopic action in calcium. But ^{40}Ca itself is taken as a reference for other isotopes because it is the most abundant. What is measured is the ratio of the other Ca isotope abundances to that of ^{40}Ca; therefore attribution of an anomalous isotopic ratio to an anomaly in ^{40}Ca abundance is seldom made. Because of systematic linear variations with isotopic mass of the measured isotopic abundances, moreover, it is common to remove that effect by defining the ratio $^{44}Ca/^{40}Ca$ as normal, a doubtful procedure, but common. Anomalies in other Ca isotopes thus appear as deviations from the linear slope set by the $^{44}Ca/^{40}Ca$ ratio. All this is to say that a self-consistent analysis of the entire Ca suite is required.

Presolar grains ^{40}Ca itself is taken as a reference for the other Ca isotopes because it is the most abundant.

---

^{41}Ca | $Z = 20$, $N = 21$ $t_{1/2} = 1.03 \times 10^5$ years
Spin, parity: $J, \pi = 7/2-$;
Mass excess: $\Delta = -35.138$ MeV; $S_n = 8.36$ MeV

Extinct radioactivity

This isotope of calcium has special significance for the science of *extinct radioactivity* (see **Glossary**). It has the shortest halflife of any heavy isotope that gives demonstrable evidence of having existed in the early solar-system cloud. It appears to have had small but measurable abundance in the early solar system, about 1.5×10^{-8} of stable calcium; but it is also of the type of extinct radioactivity that shows itself within presolar grains where it was much more abundant. ^{41}Ca created in explosive events, novae or supernovae, remains alive for the time needed to grow grains within the expanding and cooling gas of the explosions, but its halflife is too short to understand how its abundance in the early solar system could be typical of interstellar molecular clouds, where stars and planetary systems form. Some event within the molecular cloud had to have produced it, either there or even later in the planetary disk around the emerging young Sun.

Presolar grains   Extinct radioactive ^{41}Ca is recognized as having once been present within presolar grains found within the carbonaceous meteorites by the excess of daughter ^{41}K that those grains contain. The potassium within some grains is quite enriched in ^{41}K, arguing strongly that the condensation of chemically reactive calcium is its reason for being there. This possibility was predicted in 1975 by the writer. Because calcium is so chemically reactive, it forms very tightly bonded solids that can exist at high temperatures. This helps it to condense in presolar grains – in the SiC grains from AGB stars, in low-density graphite grains and Type X SiC grains from supernovae, and in Si_3N_4 grains from supernovae. Five low-density graphite grains in particular show the expected large excess of daughter ^{41}K of ^{41}Ca. By comparing its abundance with that of the stable ^{39}K in the same grains it is found that the production ratio for ^{41}Ca/^{40}Ca in these grains ranges from values as large as 1.7% down to the limits measurable. This is a million times greater than the ^{41}Ca concentration in the solar-system rocks (see below), proving beyond reasonable doubt the supernova origin of these grains. This range of initial isotopic ratios at the times that the grains condensed (usually within a few months of ejection or explosion) has set exciting challenges to discover the nature and location of their nucleosynthesis sites. One would like to determine if this ^{41}Ca originated in the supernova envelope by neutron captures or within the supernova core in oxygen burning.

Solar-system rocks   Extinct radioactive ^{41}Ca is recognized as having once been present within rocks formed by chemical processes on protoplanetary bodies in the early solar system. These are the calcium–aluminum-rich inclusions (CAIs). CAIs are larger than presolar grains but still small to ordinary geologic experience, having sizes between 0.01 and 10 mm. Prior presence of ^{41}Ca is again attested to by the excess of daughter ^{41}K that these rocks contain. But in these cases the rocks contain both isotopes of potassium in considerable abundance; therefore the detection of excess ^{41}K resulting from ^{41}Ca decay rests upon studying rocks that are very rich in the element Ca. If the initial chemical abundance ratio Ca/K is large, as it fortunately is in several refractory high-temperture minerals, the decay of the ^{41}Ca to ^{41}K can be detected because it produces a significuant excess of ^{41}K. To see the problem consider that at that time in early solar-system history ^{41}Ca existed only in the small isotopic ratio ^{41}Ca/^{40}Ca $= 1.5 \times 10^{-8}$, which is not capable of greatly enriching the ^{41}K unless Ca is much more abundant than ordinary K.

When excess ^{41}K was first detected in Ca-rich rocks, the argument ensued whether that excess was the result of live ^{41}Ca that had decayed within that very rock, or whether it was a fossil of live ^{41}Ca that had previously decayed within presolar

Ca-rich grains that in turn were later fused together to make the rocks in the solar system. The former, decay in the rock, is called *in-situ* decay, whereas the latter, prior decay within presolar grains, is called *fossil* ^{41}K. But careful isotopic measurements within a given rock, comparing portions having locally differing chemical ratios Ca/K, have suggested that because excess ^{41}K correlates with the element calcium (^{40}Ca), the ^{41}Ca was alive at the level 1.5×10^{-8} of stable ^{40}Ca at the time and place where many rocks formed. Fossil ^{41}K must also exist in some rocks of differing types and origins; but it has proven difficult to establish the effect owing to the large contaminating abundance of normal potassium within them. The fossil ^{41}K is very clear in presolar grains, but only in them.

Even the small isotopic ratio ^{41}Ca/^{40}Ca $= 1.5 \times 10^{-8}$, is too great to be expected on average within molecular clouds. These clouds take 50 Myr to form from dilute interstellar gas, so that ^{41}Ca should be virtually absent by that time, owing to its much shorter halflife. This conundrum has not been solved; but the leading bet is that a massive star existed quite close to the cloud core that would soon form our solar system. This massive star ejected and admixed into the solar cloud enough atoms of ^{41}Ca, either in a steady wind over a million-year period or in an explosive ejection when that star became a supernova (see **Nucleosynthesis origin**, below). Because the ratio ejected from supernovae is near ^{41}Ca/^{40}Ca $= 10^{-2}$ to 10^{-3}, about 10^{-5} of the solar-system calcium must have originated in this supernova to achieve the final ratio ^{41}Ca/^{40}Ca $= 1.5 \times 10^{-8}$, with the fraction required being even greater if allowance is made for some fraction of the ^{41}Ca to decay. It is speculated that this ejected mass of ^{41}Ca mixed into the cloud that was just beginning to form the solar system. This is not so great a coincidence as it might seem to be, if the massive-star ejecta actually *caused* the initial collapse of the presolar cloud. The massive star is pictured as *a trigger* for the formation of the Sun. In this way one does not require the ratio ^{41}Ca/^{40}Ca $= 1.5 \times 10^{-8}$ to be typical of the entire molecular cloud, but rather just of that portion that was triggered into collapse by the massive star. This is an attractive resolution in many ways, but not without its own physical difficulties and implausibilities. It has been especially difficult to see how supernova ejecta are admixed at all into molecular-cloud cores.

The only two alternatives to the trigger picture have not won scientific support. The first is the suggestion that the early solar-system chemistry may be able to produce the ^{41}K excess within solar rocks from the fossil ^{41}K excess that exists in presolar grains from which the rocks were formed. The second is that processes within the molecular cloud or within the early solar system accelerate energetic particles (low-energy cosmic rays), and that these create the ^{41}Ca by nuclear reactions during collisions with other heavy nuclei. Unquestionably the true answer can be discerned only by simultaneously understanding all of the ten extinct radioactivities within a self-consistent picture.

Cosmogenic radioactivity ^{41}Ca is created as a collision fragment in meteorites when cosmic rays strike the meteorite during its journey to the Earth. This radioactivity is counted in the lab, alive today in the meteorite, and gives information on its history in space.

Nucleosynthesis origin

^{41}Ca is primarily the result of neutron capture by abundant ^{40}Ca. This occurs whenever free neutrons exist in the stellar gas, since ^{40}Ca is almost always present to some degree. The required neutron liberation occurs first while the star is burning helium (see *Helium burning*, in **Glossary**) in hydrostatic equilibrium, but may also occur during the elevated temperature of the explosion in some circumstances. The ^{41}Ca may be continuously ejected in a wind of outflowing gas from massive stars (the Wolf–Rayet phase), or it may be primarily ejected by the final explosion (the supernova). Even more ^{41}Ca is synthesized deeper within the massive star, during oxygen burning. This comes about because most of the bulk nucleosynthesis of potassium isotopes occurs during oxygen burning, but the ^{41}K isotope is synthesized as ^{41}Ca. Which source dominates the observed extinct ^{41}Ca is not clear; but the latter portion of the ^{41}Ca does not get out until the final explosion ejects it, whereas the former may be ejected in winds from massive stars prior to their explosions. For the presolar grains what matters is identifying the material from which the ^{41}Ca-bearing grains condense. For the live ^{41}Ca in the early solar system what matters is whether the entire supernova trigger of solar birth (if that picture be correct) was admixed into the solar cloud, or if only the outer layers of that triggering star were admixed.

It must be appreciated that ^{41}Ca production is the major source of stable daughter ^{41}K in nature. This makes ^{41}K one of the most radiogenic of all natural abundances.

X-ray-line astronomy

The decay of ^{41}Ca is accompanied 11% of the time by an X-ray line having energy 3.31 keV. So although it is not accompanied by a gamma ray, the decay may nonetheless one day be observed in a nearby supernova remnant.

^{42}Ca | $Z = 20$, $N = 22$
Spin, parity: J, $\pi = 0+$;
Mass excess: $\Delta = -38.547$ MeV; $\qquad S_n = 11.48$ MeV

Abundance

From the isotopic decomposition of normal calcium one finds that the mass-42 isotope, ^{42}Ca, is third most abundant of the stable Ca isotopes; 0.647% of all calcium. On the scale where one million silicon atoms is taken as the standard for solar-system matter, this isotope has

solar abundance of ^{42}Ca = 395 per million silicon atoms.

Nucleosynthesis origin

^{42}Ca exists primarily owing to the oxygen-burning process (both hydrostatic and explosive) of stellar nucleosynthesis. Put simply, during oxygen burning (fusion) two ^{16}O nuclei collide, the net reaction $2\,^{16}$O$ + 2\,^4$He $\rightarrow$ ^{40}Ca occurs with reasonable efficiency. Within the supernova structure, therefore, the ^{42}Ca concentration leaps to 200 times greater than its concentration within the initial star, and it does so in the central zones that are burning oxygen at the time the star explodes. ^{42}Ca is a partly *secondary nucleus* (see **Glossary**) because its nucleosynthesis by neutron captures from newly made ^{40}Ca is aided if the star manufacturing them has not only hydrogen and helium in its initial composition, but also carbon and oxygen. Following H and He burning, the ^{18}O and ^{22}Ne created from the initial carbon and oxygen provide for the excess neutrons that the heavier Ca isotopes contain; but significant excess neutrons are also produced during the oxygen burning itself by positron emissions that occur there by radioactive nuclei, so that ^{42}Ca is also a partly primary nucleus. The ^{42}Ca does not survive when silicon burning begins, unlike ^{40}Ca.

Although the oxygen-burning process in the core accounts for the bulk of ^{42}Ca production in stars, a larger isotopic ratio ^{42}Ca/^{40}Ca is created in the star's mantle by neutrons liberated during helium burning (the s process, see **Glossary**). The neutrons deplete ^{40}Ca by turning it in part into ^{42}Ca. Inverse ratios as low as ^{40}Ca/^{42}Ca $= 5$ are typical there (the solar ratio is 150); but the absolute concentration of that ^{42}Ca is 100 times less than in the oxygen-burning core. This feature may give evidence of the condensing matter within presolar supernova grains. When the isotopic composition of the entire spectrum of massive supernova explosions is computed, they show that all isotopes of Ca except ^{48}Ca are created by those supernova explosions, which are their source.

Anomalous isotopic abundance

Calcium–aluminum-rich inclusions   After systematic linear variations with isotopic mass (mass fractionation) of the measured isotopic abundances has been removed by defining the ratio ^{44}Ca/^{40}Ca as normal, anomalies in the ^{42}Ca/^{40}Ca ratio are found. Both excesses and deficits of ^{42}Ca have been recognized in this way, but no larger than about 0.2%. Based on a few CAIs the ^{42}Ca excess (or deficiency) seems to correlate with whether ^{48}Ca is in excess or deficient.

Presolar grains   Supernova condensates, called SUNOCONs, are presolar grains that condensed from the expanding supernova interior gas. The ^{42}Ca/^{40}Ca ratio is approximately solar in the two such grains studied to date that show evidence of extinct ^{44}Ti as proof of supernova origin. This could have been much different, because the ^{42}Ca/^{40}Ca ratio varies greatly among distinct supernova nuclear shells. But the ^{42}Ca/^{40}Ca ratio is significantly greater than solar in the more than 20 graphite grains from supernovae in which ^{44}Ti is not so abundant. This is roughly understandable from the radial structure of supernova composition.

Mainstream presolar SiC grains are believed to have condensed in the atmospheres of AGB carbon stars; and calcium seems consistent with that expectation. Bulk samples show measurable excess of ^{42}Ca/^{40}Ca ratio, as expected from s-process nucleosynthesis.

^{43}Ca | $Z = 20, N = 23$
Spin, parity: $J, \pi = 7/2-$;
Mass excess: $\Delta = -38.408$ MeV; $\qquad S_n = 7.93$ MeV

Abundance

From the isotopic decomposition of normal calcium one finds that the mass-43 isotope, ^{43}Ca, is second least abundant of the stable Ca isotopes; 0.135% of all calcium. On the scale where one million silicon atoms is taken as the standard for solar-system matter, this isotope has

solar abundance of ^{43}Ca = 82.5 per million silicon atoms.

Nucleosynthesis origin

Like ^{42}Ca, ^{43}Ca exists primarily owing to the oxygen-burning process (both hydrostatic and explosive) of stellar nucleosynthesis and to the s process. Within the supernova structure, therefore, the ^{43}Ca concentration leaps to 100 times greater than its concentration within the initial star, and it does so in the central zones that are burning oxygen at the time the star explodes. The problem is that the calculated ratio ^{43}Ca/^{42}Ca tends to be about five times less than solar almost everywhere. This may reflect a small problem in nucleosynthesis theory and is under study today.

Although the oxygen-burning process in the core accounts for the bulk of ^{43}Ca production in stars, a larger isotopic ratio ^{43}Ca/^{40}Ca is created in the star's mantle by neutrons liberated during helium burning (the s process, see **Glossary**). The neutrons deplete ^{40}Ca by turning it in part into ^{42}Ca and ^{43}Ca. Inverse ratios as low as ^{40}Ca/^{43}Ca = 14 are typical there (the solar ratio is 718); but the absolute concentration of that ^{43}Ca is 30 times less than in the oxygen-burning core, which also contains large ^{43}Ca/^{40}Ca ratios.

Anomalous isotopic abundance

Calcium–aluminum-rich inclusions   After systematic linear variation with isotopic mass (mass fractionation) of the measured isotopic abundances has been removed by defining the ratio ^{44}Ca/^{40}Ca as normal, anomalies in the ^{43}Ca/^{40}Ca ratio are found. Both excesses and deficits of ^{43}Ca have been recognized in this way, but no larger than about 0.2%. Based on a few CAIs the ^{43}Ca excess (or deficiency) seems to correlate with whether ^{48}Ca is in excess or deficient. These correlations are probably a *cosmic chemical memory* (see **Glossary**) of the prior interstellar matter which carried them in larger degree.

Presolar grains Supernova condensates, called SUNOCONs, are presolar grains that condensed from the expanding supernova interior gas. The ^{43}Ca/^{40}Ca ratio is greater than solar in the two such grains studied to date that show evidence of extinct ^{44}Ti as proof of supernova origin, one graphite grain and one SiC grain. This could have gone either way, because the ^{43}Ca/^{40}Ca ratio varies greatly among distinct supernova nuclear shells. But the ^{43}Ca/^{40}Ca ratio is much greater (two to three times) than solar in the more than 20 graphite grains from supernovae that did not contain abundant ^{44}Ti. This is not hard to understand in one sense, because the oxygen-rich zones contain much larger ^{43}Ca/^{40}Ca ratios than the zones bearing ^{44}Ti.

　　Mainstream presolar SiC grains are believed to have condensed in the atmospheres of AGB carbon stars; and calcium seems consistent with that expectation. Bulk samples show measurable excess of ^{43}Ca/^{40}Ca ratio, as expected from s-process nucleosynthesis.

^{44}Ca | $Z = 20$, $N = 24$
Spin, parity: J, $\pi = 0+$;
Mass excess: $\Delta = -41.469$ MeV;　　　$S_n = 11.14$ MeV

Next to ^{36}S, ^{44}Ca is the lightest nucleus of all nuclei containing four more neutrons than protons that is stable against beta decay. It is, however, not created as itself in supernovae, but as a radioactive ^{44}Ti parent, which has $Z = N$.

Abundance

From the isotopic decomposition of normal calcium one finds that the mass-44 isotope, ^{44}Ca, is second most abundant of the stable Ca isotopes; 2.086% of all calcium. On the scale where one million silicon atoms is taken as the standard for solar-system matter, this isotope has

　　　solar abundance of ^{44}Ca = 1275 per million silicon atoms,

about 50% greater than the abundance of the element fluorine.

Nucleosynthesis origin

^{44}Ca exists primarily owing to the decay of radioactive ^{44}Ti (60-yr halflife), which is synthesized along with ^{40}Ca during oxygen-burning and silicon-burning processes (both hydrostatic and explosive) of stellar nucleosynthesis and which mostly decays during the first century following each explosion. Put simply, during oxygen burning two ^{16}O nuclei collide, causing the reaction $^{16}O + {}^{16}O \rightarrow {}^{28}Si + {}^4He$. Capture of four alpha particles produces ^{44}Ti. For this reason ^{44}Ca is, like ^{28}Si, ^{32}S, ^{36}Ar, and ^{40}Ca, considered an "alpha nucleus," meaning assembled from alpha particles as ^{44}Ti, which truly can be thought of as eleven alpha particles. Within the supernova structure, therefore, the ^{44}Ti concentration leaps to 100 times greater than the concentration of ^{44}Ca within the initial star, and it does so in the central zones that are burning oxygen at the time the star explodes.

Although the oxygen-burning process in the core accounts for some of the ^{44}Ca production in stars via the ^{44}Ti progenitor, an even larger amount of ^{44}Ti arises from the *alpha-rich freezeout* (see **Glossary**) nearer to the supernova center. Matter there was broken down into alpha particles by the high temperature of the supernova shock, and then quickly reassembled during the expansion into nuclei having a large concentration of ^{44}Ti, 1000 times greater than the concentration of ^{44}Ca within the initial star. This is actually nature's major source of ^{44}Ca and of the gamma rays emitted when the ^{44}Ti decays in the young supernova (see **^{44}Ti**).

Some ^{44}Ca is created as itself in the massive star's mantle by neutrons liberated during helium burning (the *s* process, see **Glossary**). The neutrons deplete ^{40}Ca by turning it in part into ^{44}Ca. Ratios as high as $^{44}Ca/^{42}Ca = 1.7$ are typical there (the solar ratio is 0.34); but the absolute concentration of that ^{44}Ca is several hundred times less than in the central alpha-rich freezeout, and it is not an important bulk source of ^{44}Ca. It could, however, appear in presolar grains condensed from the s-process zones. Similarly, the *AGB stars* (see **Glossary**), which produce s-process elements, make some ^{44}Ca.

Anomalous isotopic abundance
Calcium–aluminum-rich inclusions   The CAIs reveal a lot of isotopic action in calcium. But identification of $^{44}Ca/^{40}Ca$ variations is obscured by a common normalization procedure. ^{40}Ca itself is taken as a reference for the other Ca isotopes because it is the most abundant by far. What is measured is the ratio of the other isotope abundances to that of ^{40}Ca. Because of systematic linear variations with isotopic mass of the measured isotopic abundances, moreover, it is common to remove that mass-fractionation effect by defining the ratio $^{44}Ca/^{40}Ca$ as normal. Doing so hides its variations, although a cogent argument can reverse the interpretation. That some such variations probably exist within the *cosmic chemical memory* (see **Glossary**) is revealed by the huge excess of ^{44}Ca found in presolar supernova grains, which probably leave an isotopic fingerprint in some early solar-system solids.

Presolar grains   Supernova condensates, called SUNOCONs, are presolar grains that condensed from the expanding supernova interior gas. The $^{44}Ca/^{40}Ca$ ratio is much greater than solar in several such grain types studied to date, graphite and SiC. In the most extreme example in a graphite grain the ^{44}Ca/^{40}Ca ratio is measured to be 138 times greater than solar! That makes ^{44}Ca 89% as abundant as ^{40}Ca in that grain. Nowhere within a young supernova is so much ^{44}Ca found; but the explanation had been predicted twenty years earlier by the writer. The supernova creates abundant ^{44}Ti (60-yr halflife), which condenses in a year or two into the grain as titanium, but within a few centuries the ^{44}Ti decays to ^{44}Ca. Both graphite and SiC are more chemically fond of Ti than Ca, so their condensed Ti/Ca ratio is larger than solar, with resultant emphasis after the ^{44}Ti decays leaving a large fractional excess in the element Ca. But the ^{44}Ca/^{40}Ca ratio is only modestly greater than solar in the more than 20 graphite grains from supernovae in which ^{44}Ti is not so abundant.

^{46}Ca | $Z = 20, N = 26$
Spin, parity: $J, \pi = 0+$;
Mass excess: $\Delta = -43.135$ MeV; $\qquad S_n = 10.40$ MeV

^{46}Ca is the lightest stable nucleus to possess six excess neutrons. It is also extremely rare.

Abundance

From the isotopic decomposition of normal calcium one finds that the mass-46 isotope, ^{46}Ca, is least abundant of the stable Ca isotopes. It is but 0.004% of all calcium isotopes. On the scale where one million silicon atoms is taken as the standard for solar-system matter, this isotope has

solar abundance of ^{46}Ca = 2.4 per million silicon atoms.

It falls to a low 99th place in the order of isotopic abundances. Among lighter isotopes, only ^{9}Be is less abundant.

Nucleosynthesis origin

^{46}Ca is not created in the oxygen-burning and silicon-burning processes (both hydrostatic and explosive) that are so prolific for the lighter isotopes of Ca; it is too difficult to direct so many excess neutrons into a Ca nucleus at thermonuclear flow conditions. Instead, a solar-like isotopic ratio ^{46}Ca/^{40}Ca is created in the star's interior by neutrons liberated during helium burning (the s process, see **Glossary**); but it has to be a high-neutron-flux capture chain, so that radioactive ^{45}Ca is as likely to capture a free neutron as it is to beta decay with its 162-day halflife. The neutron flux is sufficiently intense to achieve this in the hottest s-process regions. The neutrons deplete ^{40}Ca by turning it in part into ^{42}Ca. Ratios as large as ^{46}Ca/^{42}Ca = 0.4 (the solar ratio is 0.0067) are achieved in a thin shell of material that explosively ignites a fraction of carbon when the supernova shock wave passes. When the isotopic composition of the entire spectrum of massive supernova explosions is computed, they show that all isotopes of Ca except ^{48}Ca are created by those supernova explosions, which are their source.

Anomalous isotopic abundance

The ^{46}Ca/^{40}Ca ratio has not yet been measurable in presolar grains owing to the small number of ^{46}Ca atoms.

^{48}Ca | $Z = 20, N = 28$
Spin, parity: $J, \pi = 0+$;
Mass excess: $\Delta = -44.215$ MeV; $\qquad S_n = 9.95$ MeV

*^{48}Ca is such an extraordinarily neutron-rich isotope that it drives nuclear physicists to poetry. Beams of it are used in nuclear collisions in the attempt to create neutron-rich nuclei in the laboratory. It is the only stable nucleus below and including iron to have eight more neutrons than protons. Its excess neutrons count one-sixth of all its nucleons, such a large fraction that one must look up the chart of nuclides to ^{82}Se to exceed that ratio. With 20 protons and 28 neutrons, ^{48}Ca is one of three stable heavy nuclei possessing a "magic" number of both. This makes it stable against beta decay. It also accounts for the neutron separation energy being almost 10 MeV despite having eight excess neutrons. ^{48}Ca is one of five isotopes (of five different elements) having $N = 28$, a multiplicity that does not recur until also-magic $N = 50$ neutrons. Its total binding energy per nucleon is greater than that of all lighter nuclei because of its doubly-magic numbers (20 and 28)! A stable nucleus having eight excess neutrons does not occur again until ^{64}Ni. Its neutron-richness is more like those of r-process isotopes (see **Glossary**) than of nuclei formed in thermonuclear processes. And yet it is not an r-process isotope. Its true source surprises in its subtlety and beauty.*

Abundance

The high abundance of ^{48}Ca has been one of the great puzzles of nucleosynthesis. From the isotopic decomposition of normal calcium one finds that the mass-48 isotope, ^{48}Ca, is third most abundant of the six stable Ca isotopes, despite having a whopping eight excess neutrons. It is 0.187% of all calcium isotopes, making it 47.5 times more abundant than ^{46}Ca, and even more abundant than ^{43}Ca, which has the advantage of being a main-line product of oxygen burning, which ^{48}Ca is not. On the scale where one million silicon atoms is taken as the standard for solar-system matter, this isotope has

solar abundance of ^{44}Ca $= 114$ per million silicon atoms,

about 50% of the abundance of ^{41}K, another main-line isotope from oxygen burning.

Nucleosynthesis origin

The ^{48}Ca natural abundance can not be a simple consequence of a string of neutron captures because the ratio ^{48}Ca/^{46}Ca $= 47.5$ is too large. The r process is inadequate for similar reasons. For decades it has been recognized that a successful explanation must involve its large neutron binding energy and large total binding energy, leading to ideas of thermal nuclear equilibrium, albeit in very neutron-rich gas to favor neutron-rich isotopes. The problem was to understand how this could happen and where the appropriate site might be found. The entire gas must have an excess of one neutron for every six total nucleons to efficiently produce ^{48}Ca as a near-equilibrium product. A severe problem is met in the even larger overproduction of ^{66}Zn and ^{82}Se in the most obvious hypothesized settings, however; and, furthermore, supernova expansions do not easily lead to the right thermal conditions. The core collapse of a Type II supernova was long thought to be a promising site, but the entropy of matter is too great there, causing the ample ^{48}Ca formed early when conditions in the neutron-rich alpha-rich freezeout are near *thermal equilibrium* (see **Glossary**) to disappear during the expansion and cooling phases. Technically, the near-equilibrium turns into a *quasiequilibrium* (see **Glossary**) having *too few* assembled heavy nuclei to allow ^{48}Ca to survive the expansion. Recent studies show that the quasiequilibrium must instead be one of very low entropy, which requires the site to be the center of an exploding white-dwarf star, called a Type Ia supernova. But the common models of Type Ia supernovae do not have enough excess neutrons for the ^{48}Ca to be a natural winner in the fierce competition that exists there. These restrictions illuminate what nature must have done; namely, there exists a rare class of Type Ia explosions in which the star's mass was near the Chandrasekhar mass, in which case electron-captures after ignition render the matter sufficiently neutron-rich for ^{48}Ca dominance and also for low-entropy in order that the quasiequilibrium have *too many* nuclei rather than too few to be a thermal equilibrium, but just right to allow ^{48}Ca to not only survive the expansion but to be an abundance pillar to which the quasiequilibrium is attached. That quasiequilibrium buffer preserves ^{48}Ca during the expansion. Although this abstract conclusion stumps even astrophysicists, save for a few specialists, nature presents a clear requirement in high ^{48}Ca abundance of the ecological struggle that nuclei face for existence in such circumstances; namely, a small fraction of Type Ia supernovae evolve to explosion with low entropy and high neutron excess, so that ^{48}Ca defeats all other nuclei in the local competition. As in life forms, ^{48}Ca's great stabilty is not enough to account for its high abundance; it also needs the proper environment to prosper and survive.

Anomalous isotopic abundance

Presolar grains   Supernova condensates, called SUNOCONs, are presolar grains that condensed from the expanding supernova interior gas. The ^{48}Ca/^{40}Ca ratio has not yet been measureable within them, however, owing to the small number of ^{48}Ca atoms. When it is, the firm prediction is an absence of ^{48}Ca, which is not created in Type II supernovae. If SUNOCONs from Type Ia supernovae exist, however, some should bear huge ^{48}Ca excess.

Calcium–aluminum-rich inclusions   After systematic linear variation with isotopic mass (mass fractionation) of the measured isotopic abundances has been removed by defining the ratio ^{44}Ca/^{40}Ca as normal, sizeable anomalies in the ^{48}Ca/^{40}Ca ratio are found in CAIs. Both excesses and deficits of ^{48}Ca have been recognized in this way, some larger than 1%. Based on a few CAIs the ^{48}Ca excess (or deficiency) seems to correlate with whether ^{42}Ca is in excess or deficient. Excitingly large ^{48}Ca variations, up to $\pm 10\%$, have been found in *hibonite* (CaAl$_{12}$O$_{19}$) crystals. These also appear to be tiny Al-rich rocks made in the solar system by the processes that made the more complex CAIs. The setting for this and the meaning of these very large isotopic anomalies has not been deciphered, but seems likely to be related to the *cosmic-chemical-memory* (see **Glossary**) interpretation of isotopic variations in solar-system rocks; namely, because the nucleosynthesis source for ^{48}Ca is unique among Ca isotopes, it probably resides in the interstellar medium in different chemical sites than do the other Ca isotopes, an interpretation devised by the writer.

21 | Scandium (Sc)

Scandium nuclei have charge $+21$ electronic units, so that 21 electrons orbit the nucleus of the neutral atom. Its electronic configuration can be abbreviated as an inner core of inert argon (a noble gas) plus three more electrons: $(Ar)3d^1 4s^2$, which locates Sc atop Group IIIB of the chemical periodic table, above yttrium (Y) and lanthanum (La). Scandium has oxidation state $+3$, because its three valence electrons are given up almost equally. Its oxide is Sc_2O_3 and its chloride $ScCl_3$. It thus bears some similarity to aluminum and the IIIA group; but it really heads a group of relatively rare elements called "the rare earths." Rare earths are metals having very similar chemistry, so that they are usually found in nature as mixes of several rare-earth elements. Procedures for separating them chemically require care.

Although Sc was difficult to isolate, its existence and several of its chemical properties were predicted in 1871 by Dmitri Mendeleyev, who called it *ekaboron* because of its expected similarity to boron. Lars Fredrik Nilson first isolated its oxide (*scandia*) while persuing the oxide of a different rare earth.

Scandium is one of the odd-Z elements that has only a single stable isotope, ^{45}Sc, similar to the elements F, Na, Al, and P. Many odd-Z elements have two stable isotopes, *e.g.* Cl and K; but none have three. The most interesting thing about the abundance of Sc is that it is small in comparison with its immediate neighbor elements, Ca and Ti, which demands nuclear interpretation. At one time scandium was thought to be the rarest of elements, but it is actually the 32nd most abundant element in the universe, after gallium and before strontium. But it is certainly very rare for its mass, being the most rare element between carbon and germanium. An interesting difference from the other rare earths is revealed by their abundances in the Earth's crust. Sc is the eighth most abundant rare earth in the crust, but the more abundant rare earths there are all actually much less abundant than Sc in the cosmos. This requires that a much higher fraction of Sc has settled to the Earth's core than occurred for other rare earths, requiring Sc to have a fundamental geochemical difference from them.

Metallic Sc is of importance as an alloying addition, especially in the aerospace industry, owing to its light weight (2.99 g cm^{-3}), resistance to corrosion, and high melting point ($1541\ ^\circ$C). It is silver white and is fairly soft. It tarnishes slightly in air, taking a yellow-pink hue, but remains fairly resistant to further corrosion owing to the thin layer of oxide.

Scandium compounds are not heavily studied. Its oxide, Sc_2O_3 is used in the manufacture of high-intensity electric lamps; and its iodide, ScI_3, is used in lamps to produce light having color similar to sunlight.

Because Sc has but a single stable isotope it sheds no light on isotopic anomalies within presolar material. Sc nucleosynthesis derives almost entirely from a single main-line source, oxygen burning, which occurs in massive stars that will become Type II supernovae.

Natural isotopes of scandium and their solar abundances

A	Solar percent	Solar abundance per 10^6 Si atoms
45	100	34.2

^{45}Sc | $Z = 21, N = 24$
Spin, parity: $J, \pi = 7/2-$;
Mass excess: $\Delta = -41.069$ MeV; $S_n = 11.32$ MeV

Scandium is the lightest odd-Z element to not have a stable isotope at $N = Z + 1$.

Abundance

^{45}Sc is the only stable isotope of scandium; therefore all elemental observations of its elemental abundance are of the abundance of ^{45}Sc. This isotope has

solar abundance of ^{45}Sc $= 34.2$ per million silicon atoms.

^{45}Sc is the 69th most abundant nucleus in the universe and 32nd most abundant element.

Nucleosynthesis origin

In the mix of interstellar atoms from which the solar system formed, ^{45}Sc exists primarily owing to two source zones in Type II supernovae: the *oxygen-burning* process of stellar nucleosynthesis; and the *alpha-rich freezeout* reassembly of alpha particles near the collapsed core – the same zone responsible for ^{44}Ti, which differs only by a single proton from ^{45}Sc and by a single neutron from ^{45}Ti. Most ^{45}Sc is ejected as radioactive ^{45}Ti (halflife 184 minutes) but a significant fraction as ^{45}Sc itself. Type Ia supernovae are responsible for only a small fraction of ^{45}Sc.

Astronomical observations

Scandium abundances have been measured in stars by spectroscopic interpretation of the 5526 and 5657 Å lines of the Sc$^+$ ion, which is isoelectronically similar to neutral calcium. The abundance ratio Sc/Fe is roughly constant in stars of widely differing metallicity. However, owing to the ^{45}Sc production by very different amounts from Type II supernovae of differing mass and to the Fe nucleosynthesis being primarily in Type Ia supernovae, more accurate interpretation of the Sc/Fe ratio may one day be possible and lead to interesting time-dependent trends of that ratio.

22 | Titanium (Ti)

All titanium nuclei have charge +22 electronic units, so that 22 electrons orbit the nucleus of the neutral atom. Its electronic configuration can be abbreviated as an inner core of inert argon (a noble gas containing eighteen inner electrons) surrounded by four more electrons: $(Ar)3d^2 4s^2$, which locates titanium at the top of Group IVB elements of the periodic table. These IVB "titanium metals" include zirconium and hafnium. Titanium has valences +2, +3 and +4 owing to its two 4s electrons and two 3d electrons, and combines readily with oxygen (-2) atoms, doing so with +2 valence as TiO, with +3 valence as Ti_2O_3 and with +4 valence as TiO_2. Its good binding within refractory minerals formed at high temperature enables Ti to be abundant in many presolar grains that had such an origin. For example, TiN and TiC subcrystals appear to be significant in presolar supernova grains, and those diatomic molecules are firmly bound.

Astronomical observations in red-giant stars of molecular bands in the molecules TiO and TiC have provided a sensitive test of the transition from an O-rich star to a C-rich star ("carbon star"). If the star's atmosphere contains more oxygen than carbon, free oxygen remains after chemically forming as much of the tightly bound CO molecule as is possible. Such stars reveal bands of TiO in their spectra. Conversely, in carbon stars, excess C remains after the formation of the maximum amount of CO molecules. These stars reveal bands of TiC in their spectra. And it is from the atmospheres of such stars that the "mainstream SiC" presolar grains are recovered in meteorites.

Pure titanium is a shiny gray metal. It does not corrode easily despite having a strong bond with oxygen. Because of its high strength-to-weight ratio it is used as a base metal for structural alloys. These find such diverse uses as aircraft, golf clubs, and surgical pins because Ti does not react with flesh and bone. The pure metal is not a very good conductor of electricity; and, as a paramagnetic, it is not very responsive to magnetic fields.

Titanium is the 21st most abundant element in the universe, following chlorine and before cobalt; thus it is a significant datum for nucleosynthesis theories. It is even more significant in terrestrial abundances, where its high reactivity incorporates it into many common minerals, making it the ninth most abundant element in Earth's rocky crust. Its most abundant ore is rutile (TiO_2), with an abundance almost 1% that of common sand. Titanium is highly depleted from interstellar gas by having been condensed in interstellar grains, many of which were formed when the atoms first left the stars. So it has been a very important element in stardust. And the Ti element abundance has been measured in stars of various metallicities.

Titanium is composed of five stable isotopes, and one radioactive one of high significance. Its stable masses are 46, 47, 48, 49, and 50; and the radioactive isotope ^{44}Ti (60-yr halflife) is demonstrably present in the Cassiopeia A supernova that exploded somewhat more than three centuries ago. The many isotopes and a confluence of interesting processes render titanium perhaps the richest of all elements for the development of nucleosynthesis theory. Almost simultaneous formation and thermal decomposition of titanium nuclei occur in the cores of massive stars. Formation is mostly during silicon burning and during alpha-rich reassembly during the exhaustion of ^{28}Si.

Titanium isotopic anomalies in presolar grains and in *calcium–aluminum-rich inclusions* (CAIs) from meteorites are common. For the former it is fortunate that Ti condenses so efficiently in graphite and in SiC grains. One of the largest anomalies and the most stunning is the variable ^{50}Ti that is so common not only in presolar grains but also in CAIs. Initially unexpected for rocks made in the early solar system, these isotopic anomalies are probably to be understood within the *cosmic-chemical-memory* (see **Glossary**) interpretation of isotopic variations inherited from preexisting interstellar material.

Natural isotopes of titanium and their solar abundances

A	Solar percent	Solar abundance per 10^6 Si atoms
44	gamma rays in supernovae; extinct radioactivity in presolar grains	
46	8.0	192
47	7.3	175
48	73.8	1771
49	5.5	132
50	5.4	130

---

^{44}Ti | $Z = 22$, $N = 22$ $t_{1/2} = 60$ years
Spin, parity: $J, \pi = 0+$;   **Mass excess:** $\Delta = -37.548$ MeV

Measuring the halflife of ^{44}Ti required a series of difficult experiments; its value is important for analysis of isotopes in presolar grains and for gamma-ray astronomy. The large solar abundance of ^{44}Ca, daughter of ^{44}Ti, reflects the nuclear stability of parent ^{44}Ti, which can be assembled from eleven alpha particles.

Extinct radioactivity

This isotope of titanium has great importance for the science of *extinct radioactivity* (see **Glossary**). Owing to its short halflife it did not have significant abundance in the early solar system, but is instead of the type of extinct radioactivity that shows itself within promptly condensing presolar grains. ^{44}Ti created in explosive events, helium novae or supernovae, can remain alive for the time needed to grow grains within the expanding and cooling gas of the explosions. Its halflife near 60 yr exceeds the time (weeks for a nova and one year for a supernova) required for the dust to grow.

Extinct radioactive ^{44}Ti has been recognized experimentally within low-density graphite grains found within carbonaceous meteorites by the excess of daughter ^{44}Ca that those grains contain. The calcium within some is almost pure ^{44}Ca, arguing strongly that the condensation of chemically reactive titanium is its reason for being there. This identification is secure because nature provides no other natural source of isotopically enriched ^{44}Ca. From the ratio of the number of ^{44}Ca atoms in these graphite and SiC grains today to the number of stable ^{48}Ti atoms in the same grains one infers that the initial ratio when the supernova grains condensed was ^{44}Ti/^{48}Ti = 0.001−0.6. These are very high and interesting values for supernova theory, because the explosive He-burning origin creates ratios near unity, whereas the Si-burning origin (see below) creates ratios near 0.001. Most portions of supernova ejecta have ratios near zero. Thus the exact ratio in each grain appears to be saying something about the mixtures of atoms from these two adjacent production sites. The full implications for supernovae have yet to be drawn, but they should be very significant.

Gamma-ray astronomy

Live ^{44}Ti has been detected by its gamma-ray lines emitted following its beta decay, first to ^{44}Sc and then to ^{44}Ca. Observable gamma rays of 68 keV and 78 keV are emitted following the ^{44}Ti decay to ^{44}Sc; and a higher-energy gamma-ray line at 1.16 MeV is emitted following the subsequent decay of ^{44}Sc to ^{44}Ca. The writer predicted this possibility in 1969 in work with Stirling Colgate and Jerry Fishman, and it provided additional incentive for the construction and launching of energy-sensitive spectrometers into space. This and other goals culminated with NASA's *Compton Gamma Ray Observatory*, the second of NASA's planned sequence of Great Space Observatories, which made the first detections of ^{44}Ti. The three-century-old supernova remnant Cassiopeia A, probably detected optically by the Astronomer Royal, Flamsteed, was found to emit barely detectable 1.16 MeV photons by the COMPTEL (COMPton TELescope) instrument; however, the OSSE (Oriented Scintillation Spectroscopy Experiment) instrument was unable to confirm detection at the flux measured by COMPTEL. Statistical agreement between these two instruments aboard the *Compton Gamma Ray Observatory* suggests that about 0.0001 solar masses of radioactive ^{44}Ti was ejected by that event. Fortunately, observations made by the Italian satellite *BeppoSAX* confirm the decay rate of ^{44}Ti in the Cas A supernova remnant by detection of the 68 keV and 78 keV lines emitted following the ^{44}Ti decay; and the results agree quantitatively. Since the supernova explosion was seen about five ^{44}Ti halflives in the past, only 1/32 of the initial amount persists today in the Cas A remnant. The larger question is why a larger number of recent supernova remnants are not visible by the same means, inasmuch as theoretical expectations were that half a dozen or so events would be detectable that occurred primarily during the past two centuries. Instead one has seen only a single remnant, one more than three centuries old and that can therefore only barely be seen, and only because it is so close to the Earth. The absence of other detectable supernova remnants first became apparent via NASA's *Solar Maximum Mission* Gamma Ray Spectrometer, and it is now a significant astrophysical puzzle. Either our Milky Way Galaxy has not hosted supernovae during the past two centuries, despite statistics from more distant galaxies that we experience three per century, or most supernovae eject much less ^{44}Ti than did the Cas A supernova.

Nucleosynthesis origin

^{44}Ti is primarily the result of the explosive fusion of helium into heavy elements. Two venues for this exist: the *alpha-rich freezeout* (see **Glossary**) of shock-decomposed nuclei near the core of a core-collapse supernova (Type II); and explosive helium burning within a cap of helium atop a low-mass electron-degenerate CO core as part of one variation of Type I supernovae. In addition, some ^{44}Ti is also synthesized in the process by which silicon transforms to nickel (within both types of supernovae). This process is called *silicon burning* (see **Glossary**) by astrophysicists. The nuclei having $Z = N = A/2$ in this mass region, ^{28}Si, ^{32}S, ^{36}Ar, and ^{40}Ca, are spoken of as "alpha nuclei," assembled from alpha particles; and ^{44}Ti logically is one of them.

^{46}Ti | $Z = 22, N = 24$
Spin, parity: $J, \pi = 0+$;
Mass excess: $\Delta = -44.125$ MeV; $\qquad S_n = 13.19$ MeV

The neutron separation energy is greater than for any lighter neutron-rich isotope, giving ^{46}Ti an advantage in the fight for neutrons at the high temperature of explosive oxygen burning.

Abundance

From the isotopic decomposition of normal titanium one finds that the mass-46 isotope, ^{46}Ti, is one of the four lesser abundant of the stable titanium isotopes; 8.0% of all Ti. On the scale where one million silicon atoms is taken as the standard for solar-system matter, this isotope has

solar abundance of ^{46}Ti $= 192$ per million silicon atoms;

i.e about half the elemental abundance of copper.

Nucleosynthesis origin

^{46}Ti exists primarily owing to the oxygen-burning and silicon-burning processes (both hydrostatic and explosive) of stellar nucleosynthesis in massive stars. The second follows the first quite quickly in the evolution of massive stars. Put simply, during oxygen burning two ^{16}O nuclei collide, causing the reaction ^{16}O $+$ ^{16}O $\rightarrow$ ^{28}Si $+$ ^{4}He. Subsequent alpha-captures produce ^{44}Ti, and two free neutrons added make ^{46}Ti. The high neutron binding hinders removal of a neutron from ^{46}Ti by thermal gamma rays at high temperature. This may equally well be viewed as the addition of a final alpha particle to ^{42}Ca; therefore ^{46}Ti becomes abundant in the same oxygen-burning zone that synthesizes ^{42}Ca. Their productions are linked. This occurs already during the latter stages of exhaustion of oxygen, when the local concentration of ^{46}Ti leaps to a value 1000 times greater than the initial solar concentration. If the burning proceeds into silicon burning, the ^{46}Ti (and ^{42}Ca) erodes away quickly. Although ^{46}Ti contains excess neutrons, it cannot strictly be labeled a "secondary nucleosynthesis" product because of the positive beta decays that occur during oxygen burning, which render the bulk neutron-rich to some extent.

Anomalous isotopic abundance

Presolar grains and CAIs   After removing the progressive mass-dependent fractionation that occurs in the measuring process and in the formation process for the samples by choosing the ^{46}Ti/^{48}Ti ratio to be normal, the isotopes of Ti provide a rich spectrum of isotopic deviations from the solar ratios. This procedure in CAIs defines the ^{46}Ti/^{48}Ti ratio as normal; so isotopic anomalies for ^{46}Ti (and ^{48}Ti) are not often reported in CAIs. In presolar grains the ^{46}Ti/^{48}Ti ratio is measured by SIMS (see **Glossary**), and it is usually quite near the solar value, but sometimes 5–10% in excess of it in supernova grains.

Mainstream SiC grains   These, which form in AGB stars, typically have small excesses in ^{46}Ti abundance. This fits the interpretation of dredged up s-process matter, which is enriched slightly in ^{46}Ti by neutron capture by ^{45}Sc and ^{44}Ca. The measured ^{46}Ti/^{48}Ti ratio seems to correlate with the measured ^{29}Si/^{28}Si ratio, implicating the initial compositions of the stars since these ratios increase together in the *chemical evolution of the Galaxy* (see **Glossary**).

^{47}**Ti** ⁞ $Z = 22, N = 25$
⁞ **Spin, parity:** $J, \pi = 5/2-$;
⁞ **Mass excess:** $\Delta = -44.932$ MeV; $S_n = 8.88$ MeV

Abundance

From the isotopic decomposition of normal titanium one finds that the mass-47 isotope, ^{47}Ti, is one of the four lesser abundant of the stable titanium isotopes; 7.3% of all Ti. On the scale where one million silicon atoms is taken as the standard for solar-system matter, this isotope has

solar abundance of ^{47}Ti = 175 per million silicon atoms;

i.e. about half the abundance of copper.

Nucleosynthesis origin

^{47}Ti exists primarily owing to the oxygen-burning and silicon-burning processes (both hydrostatic and explosive) of stellar nucleosynthesis in massive stars. The second follows the first quite quickly in the evolution of massive stars. Put simply, during oxygen burning two ^{16}O nuclei collide, causing the reaction ^{16}O + ^{16}O → ^{28}Si + ^{4}He. Subsequent alpha-captures produce ^{44}Ti, and three free neutrons added make ^{47}Ti. This may equally well be viewed as the addition of a final neutron to ^{46}Ti; therefore ^{47}Ti becomes abundant in the same oxygen-burning zone that synthesizes ^{46}Ti. Their productions in that zone are linked. This occurs already during the latter stages of exhaustion of oxygen, when the local concentration of ^{47}Ti leaps to a value 100 times greater than the initial solar concentration. If the burning proceeds into silicon burning, the ^{47}Ti remains equally abundant, unlike ^{46}Ti which disappears in Si burning. In that zone, adding a proton to ^{47}Ti approaches *quasiequilibrium* (see **Glossary**) with the removal of a proton from ^{48}Cr, the parent of ^{48}Ti . Additional ^{47}Ti exists in the quasiequilibrium within the *alpha-rich freezeout* (see **Glossary**) near the mass cut with the remnant neutron star. Some models of Type Ia supernovae also contribute significantly to interstellar ^{47}Ti. Despite all this, a theoretical problem is that the calculated supernova models of both types do not make enough ^{47}Ti. Some unidentified error is causing a calculated underabundance for ^{47}Ti, or some efficient process for making it is omitted in current thinking. Because its neutron-capture cross section is larger than for other Ti isotopes, making ^{47}Ti by s-process neutron irradiation is not the solution.

Anomalous isotopic abundance

Presolar grains   Supernova condensates, called SUNOCONs, are presolar grains that condensed from the expanding supernova interior gas. Graphite and SiC SUNOCONs reveal compatible evidence in Ti isotopes. Variations in ^{47}Ti abundance are the rule, an excess of 50% in the most extreme example. It is hard to interpret that excess as grains condensed with nuclei from the s-process burning shells, because the large neutron-capture cross section tends to deplete ^{47}Ti. The several cases with depleted ^{47}Ti may be from that cause, however. Thus ^{47}Ti is mysterious in nucleosynthesis and in presolar grains.

Mainstream SiC grains   These, which form in AGB stars, typically have very small variations in ^{47}Ti abundance. This fits the interpretation of dredged up s-process matter, which is depleted slightly in ^{47}Ti by neutron capture, but is mixed with the original ^{47}Ti-bearing atmosphere before condensation, making the variation small.

Calcium–aluminum-rich inclusions (CAIs) After removing the progressive mass-dependent fractionation that occurs in the measuring process and in the formation process for the samples by choosing the $^{46}Ti/^{48}Ti$ ratio to be normal, isotopic anomalies for ^{47}Ti are observed in the FUN subgroup of CAIs (see **Glossary**, *Calcium–aluminum-rich inclusions*). The anomalies occur sometimes as an excess and sometimes as a deficit of ^{47}Ti. These are also seen in smaller *hibonite* crystals (Ca–Al oxides) that are probably genetically related to the same process that created the CAIs.

^{48}Ti | $Z = 22, N = 26$
Spin, parity: $J, \pi = 0+$;
Mass excess: $\Delta = -48.487$ MeV; $\qquad S_n = 11.63$ MeV

^{48}Ti is synthesized as a radioactive progenitor, ^{48}Cr, in explosive burning. Although it contains four excess neutrons, these are established after the nucleosynthesis of ^{48}Cr, which, having no excess neutrons, experiences two successive radioactive positron decays, destroying two protons and creating two neutrons within the nucleus. This is analogous to the synthesis of ^{44}Ca, which is synthesized as a radioactive progenitor, ^{44}Ti, in explosive burning.

Abundance

From the isotopic decomposition of normal titanium one finds that the mass-48 isotope, ^{48}Ti, is, by a factor near 10, most abundant of the stable titanium isotopes; 73.8% of all Ti. On the scale where one million silicon atoms is taken as the standard for solar-system matter, this isotope has

solar abundance of $^{48}Ti = 1771$ per million silicon atoms;

i.e. about half the elemental abundance of chlorine. Of all stable isotopes, ^{48}Ti is the 41st most abundant, immediately after cobalt.

Nucleosynthesis origin

^{48}Ti exists primarily owing to nucleosynthesis of ^{48}Cr in stellar explosions. Two beta decays after ejection lead to daughter ^{48}Ti. This occurs primarily in explosive silicon burning and during fusion of helium into heavy elements. Two venues for the latter exist: the alpha-rich freezeout of shock-decomposed nuclei near the core of a core-collapse supernova (Type II); and explosive helium burning within a cap of helium atop a low-mass electron-degenerate CO core as part of one model of Type I supernovae. The concentration of radioactive ^{48}Cr rises to 500 times the initial solar concentration of ^{48}Ti in the process by which silicon transforms to nickel (within both types of supernovae). This process is called *silicon burning* (see **Glossary**) by astrophysicists. An equally high concentration is established in the *alpha-rich freezeout* (see **Glossary**) just outside the residual neutron-star core. The nuclei having $Z = N = A/2$ in this mass region, ^{28}Si, ^{32}S, ^{36}Ar, and ^{40}Ca, are spoken of as "alpha nuclei," assembled from alpha particles; and ^{48}Cr logically is one of them. Although ^{48}Ti is more tightly bound than is ^{48}Cr ($\Delta = -42.82$ MeV), the near equality of neutrons and protons ($Z = N$) in the explosive gas favors the mass-48 nucleus also having that property, namely ^{48}Cr, because of the *quasiequilibrium* (see **Glossary**) nature of the silicon-burning process.

Astronomical measurements

The Ti abundance has been studied in stars. Because ^{48}Ti dominates the spectra of Ti in stars, and especially because ^{48}Ti is a primary "alpha nucleus," the galactic evolution of ^{48}Ti is the same as the observed galactic evolution of the titanium elemental abundance. As nucleosynthesis and chemical evolution progressed in time, the galactic Ti/H increased by a factor of 10^3 in step with the corresponding increase of the Fe/H abundance ratio. The behavior is very similar to that of oxygen, and indeed to all alpha nuclei that can be created from fusion of He, C and O. These include the elements Ne, Mg, Si, S, Ar, Ca, and Ti. The ratio to iron, Ti/Fe, declines, however, from about three-times solar in metal-poor stars to solar in young stars. This is understood as the rapid coproduction of the alpha nuclei, along with O, in massive Type II supernovae. These produce less iron; but the later onset of Type Ia supernova explosions dramatically increases Fe nucleosynthesis and causes the interstellar Ti/Fe ratio to decline to its solar value.

Anomalous isotopic abundance

Since the number of other Ti isotopes in a measurement is expressed in ratio to the number of ^{48}Ti atoms, the ^{48}Ti normalization does not logically admit of ^{48}Ti anomalies.

^{49}Ti | $Z = 22$, $N = 27$
Spin, parity: J, $\pi = 7/2-$;
Mass excess: $\Delta = -48.558$ MeV; $S_n = 8.15$ MeV

^{49}Ti is the lightest stable nucleus having five excess neutrons. It also has the largest alpha-particle binding energy of any odd-mass isotope above ^{29}Si, further enhancing its abundance in quasiequilibrium events having excess neutrons. In quasiequilibrium events having few excess neutrons, however, the mass-49 nucleosynthesis occurs as the ^{49}Cr radioactive progenitor. This has high significance for presolar-grain isotopes because its activity lives almost a year.

Abundance

From the isotopic decomposition of normal titanium one finds that the mass-49 isotope, ^{49}Ti, is one of the four lesser abundant of the stable titanium isotopes; 5.5%

of all Ti. On the scale where one million silicon atoms is taken as the standard for solar-system matter, this isotope has

solar abundance of ^{49}Ti $= 132$ per million silicon atoms;

i.e. about one-third the elemental abundance of copper.

Nucleosynthesis origin

^{49}Ti exists primarily owing to nucleosynthesis of radioactive ^{49}Cr in stellar explosions. The isotope ^{49}Cr is primarily the result of the explosive fusion of helium into heavy elements. Two venues for this exist: the alpha-rich freezeout of shock-decomposed nuclei near the core of a core-collapse supernova (Type II); and explosive helium burning within a cap of helium atop a low-mass electron-degenerate CO core as part of one model of Type I supernovae. In addition, ^{49}Cr is also synthesized in the "silicon burning" which transforms silicon to nickel (within both types of supernovae).

Of potential importance to both gamma-ray-line astronomy and to presolar supernova grains, the ^{49}V to which ^{49}Cr decays has a halflife of 330 days, allowing escape of its gamma rays and allowing much of the supernova grain condensation of mass-49 to be as the refractory element, vanadium. Effects of the latter have been seen as ^{49}Ti isotopic anomalies.

Anomalous isotopic abundance

The isotopes of Ti provide a rich spectrum of isotopic deviations from the solar ratios in presolar materials and in solar materials prepared retaining a *cosmic chemical memory* (see **Glossary**) of presolar materials.

Presolar supernova grains   Supernova condensates, called SUNOCONs, are presolar grains that condensed within the gas of the expanding supernova interior. Graphite and SiC SUNOCONs reveal compatible evidence in Ti isotopes. Large excesses in ^{49}Ti abundance are the rule, more than a factor of 2 in one example. That excess is probably because the grains condensed with nuclei from the s-process burning shells, which are enriched in ^{49}Ti owing to neutron capture by ^{48}Ti, or because they condensed from material near the supernova center, where enhanced condensation of radioactive ^{49}V (see above) has been shown to exaggerate the ^{49}Ti excess.

Mainstream SiC grains   These, formed in AGB stars, typically have large (about 10%) excess in ^{49}Ti abundance. This fits the interpretation of dredged up s-process matter, which is enriched in ^{49}Ti from neutron capture by ^{48}Ti.

Calcium–aluminum-rich inclusions (CAIs)   Isotopic anomalies for ^{49}Ti are observed in certain types of CAI ("FUN" inclusions). These anomalies occur as either an excess or as a deficit of ^{49}Ti. These are also seen in smaller *hibonite* crystals (Ca–Al oxides) that are probably genetically related to the same process that created the CAIs. The ^{49}Ti anomaly usually has the same sign as the larger ^{50}Ti anomaly, which occurs either as an excess or as a deficit. These are probably a cosmic chemical memory of ^{49}Ti-rich presolar dust.

^{50}Ti | $Z = 22, N = 28$
Spin, parity: $J, \pi = 0+$;
Mass excess: $\Delta = -51.426$ MeV; $\qquad S_n = 10.94$ MeV

> ^{50}Ti is the second lightest stable nucleus to have a magic $N = 28$ neutrons; and, like ^{48}Ca, much of its nucleosynthesis reflects that tight neutron binding. It also has the largest alpha-particle binding energy of any Ti isotope, further enhancing its abundance in quasiequilibrium events having excess neutrons.

Abundance

From the isotopic decomposition of normal titanium one finds that the mass-50 isotope, ^{50}Ti, is one of the four lesser abundant of the stable titanium isotopes; 5.4% of all Ti. On the scale where one million silicon atoms is taken as the standard for solar-system matter, this isotope has

solar abundance of ^{50}Ti $= 130$ per million silicon atoms;

i.e. about one-third the abundance of copper.

Nucleosynthesis origin

^{50}Ti is produced primarily in those Type Ia supernovae having masses close to the Chandrasekhar limiting mass, so that electron capture during the thermonuclear explosion turns the composition neutron-rich. Its quasiequilibrium nucleosynthesis is therefore correlated with those of ^{48}Ca, ^{54}Cr, ^{58}Fe, ^{62,64}Ni, and ^{66,70}Zn. These special Type Ia supernovae must be rather rare (see ^{48}**Ca**). But unlike the case of ^{48}**Ca**, some ^{50}Ti is also made by s-process neutron captures in the burning shells of presupernova massive stars and in AGB stars. Those ejecta may play a role in isotopically abnormal presolar grains. But the explosive burning within Type II supernovae does not make significant ^{50}Ti.

Anomalous isotopic abundance

The isotopes of Ti provide a rich spectrum of isotopic deviations from the solar ratios in presolar materials and in solar materials prepared retaining a cosmic chemical memory of presolar materials.

Presolar grains Supernova condensates, called SUNOCONs, are presolar grains that condensed from the expanding supernova interior gas. Graphite and SiC

SUNOCONs reveal compatible evidence in Ti isotopes. Large excesses in ^{50}Ti abundance are the rule, more than a factor of 2 in one example. That excess may be because the grains condensed with nuclei from the s-process burning shells, which are enriched in ^{50}Ti owing to neutron capture by ^{49}Ti. It is not known if any grains survive from the rare Type Ia supernovae that make abundant ^{50}Ti.

Mainstream SiC grains These, formed in AGB stars, typically have large (about 20%) excess in ^{50}Ti abundance. This fits the interpretation of dredged up s-process matter, which is enriched in ^{50}Ti from neutron capture by ^{48}Ti and ^{49}Ti.

Calcium–aluminum-rich inclusions (CAIs) After removing the progressive mass-dependent fractionation that occurs in the measuring process and in the formation process for the samples by choosing the ^{46}Ti/^{48}Ti ratio to be normal, large isotopic anomalies for ^{50}Ti are observed in certain types of CAI ("FUN" inclusions). These anomalies occur as both excess and as deficit ^{50}Ti atoms, up to four parts per thousand, which is quite large for a solar-system rock. The anomalies are much larger in smaller *hibonite* crystals (Ca–Al oxides) that are probably genetically related to the same process that created the CAIs. Excesses up to a factor of 2 and deficits of 70% have been observed in individual hibonite crystals. These are enormous for solar rocks, suggesting prior existence of material reservoirs of gas and dust that are enriched or depleted in the presolar-dust carriers of ^{50}Ti. Some form of *cosmic chemical memory* (see **Glossary**) is involved. One hypothesis utilizes differing sizes of the presolar grains in different reservoirs. If ^{50}Ti resides primarily in the finest sizes, for example, a hibonite crystal made from them will have large excess ^{50}Ti; whereas, a hibonite crystal made from predominantly larger presolar grain sizes will be deficient in ^{50}Ti. The same chemical heating processes make the hibonite grains from the presolar materials in both reservoirs, but the ^{50}Ti inventory is "remembered."

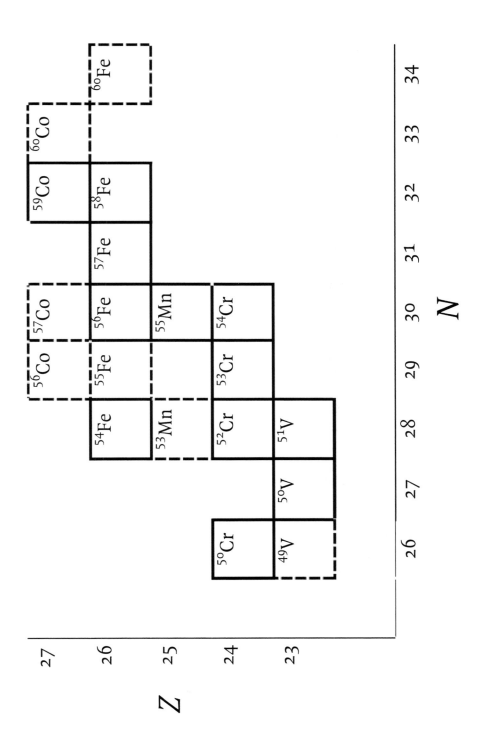

23 | Vanadium (V)

Vanadium nuclei have charge $+23$ electronic units, so that 23 electrons orbit the nucleus of the neutral atom. Its electronic configuration can be abbreviated as an inner core of inert argon (a noble gas) plus five more electrons: $(Ar)3d^3 4s^2$, which locates V atop Group VB of the chemical periodic table, above niobium (Nb) and tantalum (Ta). Vanadium uses four oxidation states, $+2$, $+3$, $+4$ and $+5$. Its four oxides are respectively VO, V_2O_3, VO_2 and V_2O_5. Four different halides assume the same set of valences. This reflects the great versatility of having three 3d electrons and two 4s electrons. Not surprisingly, vanadium forms a large variety of compounds because it combines with nearly all nonmetals. Many of its compounds are colorful. Vanadium participates in the crystal structure of at least 56 different minerals. The element was difficult to isolate, however, so that pure samples were not obtained until 1867.

Vanadium is one of the odd-Z elements that has only a single stable isotope, ^{51}V; but it has another very rare isotope, ^{50}V, that is so long lived that it is essentially stable, and a second radioactive isotope for which evidence exists in presolar supernova grains. Vanadium is about six times more abundant than the rare scandium, but it is underabundant relative to its immediate even-Z neighbors Ti and Cr. It is the 26th most abundant element.

Much recovered vanadium is used in producing an alloy with iron that is especially strong and corrosion resistant. It is used in cutting tools.

Because V has but a single abundant isotope, its isotopic composition is not measurable within presolar material. However, extinct ^{49}V with a halflife of 330 days has been shown to have been alive in presolar grains from supernovae at the time of their condensation (see ^{49}Ti).

Natural isotopes of vanadium and their solar abundances

A	Solar percent	Solar abundance per 10^6 Si atoms
49	extinct radioactive parent of ^{49}Ti in presolar supernova grains	
50	0.25	0.73
51	99.75	292

^{49}V | $Z = 23, N = 26$ $t_{1/2} = 330$ days
Spin, parity: $J, \pi = 7/2-$;
Mass excess: $\Delta = -47.956$ MeV; $S_n = 11.56$ MeV

Nucleosynthesis origin

^{49}V exists in young supernovae primarily owing to nucleosynthesis of radioactive ^{49}Cr in their explosions. ^{49}Cr is primarily the result of the explosive fusion of helium into heavy elements. Two venues for this exist: the *alpha-rich freezeout* (see **Glossary**) of shock-decomposed nuclei near the core of a core-collapse supernova (Type II); and explosive helium burning within a cap of helium atop a low-mass electron-degenerate CO core as part of one model of Type I supernovae. In addition, ^{49}Cr is also synthesized in the *silicon burning* (see **Glossary**) which transforms silicon to nickel (within both types of supernovae).

Of potential importance to both gamma-ray-line astronomy and to presolar supernova grains, the ^{49}V to which ^{49}Cr decays has a halflife of 330 days, allowing escape of its gamma rays and allowing much of the supernova grain condensation of mass-49 to be as the refractory element, vanadium. Effects of the latter have been seen as ^{49}Ti isotopic anomalies.

Anomalous isotopic abundance

Presolar supernova grains   Supernova condensates, called SUNOCONs, are presolar grains that condensed within the gas of the expanding supernova interior. Graphite and SiC SUNOCONs reveal compatible evidence of ^{49}V. Large excesses in ^{49}Ti abundance are the rule, more than a factor of 2 in one example. Such an excess could either exist because the grains condensed with nuclei from the s-process burning shells, which are enriched in ^{49}Ti owing to neutron capture by ^{48}Ti, or because they condensed from material near the supernova center, where enhanced condensation of radioactive ^{49}V (see above) has been shown to exaggerate the ^{49}Ti excess. Measurements on a suite of Type X SiC grains have recently shown that the ^{49}Ti/^{48}Ti ratio in the SUNOCONs is larger in those having larger element abundance ratios V/Ti. This correlation points to the existence of the radioactive ^{49}V in the grains at the time they condensed. This seems reasonable, because the SUNOCONs condensed within the first couple of years, when much ^{49}V was still alive.

⁵⁰V | $Z = 23, N = 27$ $t_{1/2} = 1.4 \times 10^{17}$ years
Spin, parity: $J, \pi = 6+$;
Mass excess: $\Delta = -49.218$ MeV; $S_n = 9.34$ MeV

> *⁵⁰V may have the longest known radioactive halflife, ten million times the age of the universe! This is a consequence of the large spin change $\Delta J = 6$ of its beta decay and of its low energy. It is but the third lightest of the odd-Z odd-N isotopes that exist on Earth owing to nucleosynthesis in stars; and it is the only one having four more neutrons than protons, a situation that is very common among even-Z elements.*

Abundance

⁵⁰V is much the lesser abundant isotope of vanadium; it is but $\frac{1}{4}$% of all V on Earth. Therefore, all observations of vanadium's elemental abundance are of the abundance of ⁵¹V. This isotope has

solar abundance of ⁵⁰V = 0.73 per million silicon atoms.

⁵⁰V is a very rare isotope, especially for its mass range. It is the 118th most abundant nucleus in the universe, comparable only to heavy elements such as iodine or to the rare light isotopes such as ⁹Be.

Nucleosynthesis origin

In the mix of interstellar atoms from which the solar system formed, ⁵⁰V exists primarily owing to the explosive *oxygen-burning* (see **Glossary**) process of stellar nucleosynthesis and to the neon burning which immediately precedes oxygen burning. ⁵⁰V is made as itself only, because both neighboring isobars (⁵⁰Ti and ⁵⁰Cr) are stable. The stability of ⁵⁰Cr severely limits ⁵⁰V production in explosive oxygen burning, preventing the ⁵⁰V abundance that might otherwise have been made there; and the stability of ⁵⁰Ti prevents production of ⁵⁰V by the s process. Type Ia supernovae probably contribute more ⁵⁰V than do the core-collapse Type II supernovae.

Astronomical observations

See ⁵¹V, which dominates V abundances.

Isotopic anomalies

Although variations of the isotopic abundance ratio ⁵⁰V/⁵¹V are expected in presolar grains, they are not likely to be measured in the near future owing to the great rarity of ⁵⁰V. Astronomical measurements are even less likely.

^{51}V | $Z = 23$, $N = 28$
Spin, parity: J, $\pi = 7/2-$;
Mass excess: $\Delta = -52.198$ MeV; $S_n = 11.05$ MeV

> *^{51}V has a larger neutron binding energy than any heavier odd-A nucleus, a consequence of its magic number $N = 28$ neutrons. It is one of five elements (Ca, Ti, V, Cr, and Fe) having a stable isotope with 28 neutrons, a record exceeded only by $N = 82$ neutrons.*

Abundance

^{51}V is much the more abundant isotope of vanadium; 99.75 % of all V on Earth. Therefore all observations of the elemental abundance are of the abundance of ^{51}V. This isotope has

> solar abundance of ^{51}V $= 292$ per million silicon atoms.

^{51}V is a moderately rare isotope for its mass range. It is the 55th most abundant nucleus in the universe, comparable to ^{54}Cr or ^{41}K.

Nucleosynthesis origin

In the mix of interstellar atoms from which the solar system formed, ^{51}V exists primarily owing to the explosive *oxygen-burning* process of stellar nucleosynthesis and to *silicon burning*, into which oxygen burning merges. ^{51}V is made primarily as radioactive progenitors, either ^{51}Cr or ^{51}Mn, which decays quickly to ^{51}Cr. Although the halflife of ^{51}Cr is 27.7 days, no observable consequence of that fact seems likely. Type Ia supernovae probably contribute more ^{51}V than do the core-collapse Type II supernovae.

Astronomical measurements

The neutral V atom has a large number of observable lines in stellar spectra. The abundance ratio V/Fe remains essentially constant in observed galactic stars ranging in Fe/H from 10^{-3} of solar to solar. This is unsurprising considering the close link between the nucleosynthesis of V and that of Fe.

Isotopic anomalies

Although variations of the isotopic abundance ratio ^{50}V/^{51}V are expected in presolar grains, they are not likely to be measured in the near future owing to the great rarity of ^{50}V. Astronomical measurements are even less likely.

24 | Chromium (Cr)

All chromium nuclei have charge $+24$ electronic units, so that 24 electrons orbit the nucleus of the neutral atom. Its electronic configuration is $1s^2 2s^2 2p^6 3s^2 3p^6 3d^5 4s^1$. This can be abbreviated as an inner core of inert argon (a noble gas) plus six more electrons: $(Ar)3d^5 4s^1$, which locates it atop Group VIB of the periodic table. This electron structure gives chromium three oxidation states designated by valence $+2$, $+3$ and $+6$. Its $+3$ oxide, Cr_2O_3, is the ninth most abundant mineral in the Earth's crust. Cr is found in many chromates, such as *lead chromate* ($PbCrO_4$), used in "chrome yellow" pigment. Indeed the name derives from the Greek *chroma*, "color," because of its many pigment compounds – "Siberian red lead" and "chrome oxide green." Its readiness to combine chemically means that it is seldom found as pure (free) chromium. It is usually produced from the natural mineral *chromite* ($FeCr_2O_4$) by heating it in the presence of a more oxygen-retaining element (Al or charcoal, for example). The pure metal is rather hard and brittle, and corrosion-resisitant at normal temperature. It melts at $1857\,°C$ and boils at $2672\,°C$.

The discovery of Cr began with a curious red lead found in Siberia. In 1797 it was shown that the mineral contained not only lead but also another unknown element, which was isolated a year later. Today chromium is used in various steels – spring, nickel–chromium, and stainless, and in heat-resistant nickel-based alloys. My own generation of automobiles was adorned with "chrome" (chromium-plated) bumpers and fender and window trims – an excess that staggers the mind attuned to today's automobile manufacturing.

Chromium is the 16th most abundant element (of 81), just behind nickel and just ahead of phosphorus. It is therefore an abundant element. It possesses four stable isotopes, masses 50, 52, 53 and 54. Their abundances are useful diagnostics of several key processes of nucleosynthesis. Indeed in Hoyle's original nucleosynthesis theory, Cr was to be produced along with Fe, Mn, Co and Ni in statistical equilibrium in evolved massive stars. That theory had a problem with the low abundance of Cr (which was not then well known) and predicts a Cr/Fe ratio that is much larger than the ratio observed in the Sun, in which Cr is but 1.5% of Fe. A decisive theoretical change of picture occurred in 1968 when it was shown that both elements are synthesized in supernovae as radioactive Fe and Ni parents of Cr and Fe, respectively. In that modern picture, which the writer played a key role in formulating, the radioactive progenitors are synthesized as Fe/Ni in the observed Cr/Fe abundance ratio. The modern theory was shown to be correct by observations of gamma-ray lines from the radioactive progenitors of Fe as predicted by that theory (see ^{56}Fe, **Nucleosynthesis origin**).

Chromium spectral lines are observed in stars. They show that the abundance ratio Cr/Fe has remained constant, even while each has increased in its ratio to H. Although the spectral lines do not require it, this observed Cr abundance is surely dominated by the ^{52}Cr isotope. It is quite expected that in the surfaces of some stars and in the nucleosynthesis ejecta of certain stars, Cr isotopes will have nonsolar proportions. However, Cr isotopic compositions have not been measurable by astronomers owing to their low abundance relative to ^{52}Cr and to the small differences in the wavelengths of the radiation emitted by each isotope.

Chromium exhibits a rich display of isotopic anomalies. The clearly documented cases include an unknown presolar component rich in ^{54}Cr, and variability in the ^{53}Cr/^{52}Cr ratio owing to various effects from radioactive ^{53}Mn. Cosmochemists rejoice at the richness of the Cr isotopic data but remain perplexed about the detailed causes, which derive from *cosmic chemical memory* (see **Glossary**) rather than from a supernova admixture into the forming solar system.

Natural isotopes of chromium and their solar abundances

A	Solar percent	Solar abundance per 10^6 Si atoms
50	4.35	587
52	83.8	11 310
53	9.50	1283
54	2.365	319

^{50}Cr | $Z = 24$, $N = 26$
Spin, parity: $J, \pi = 0+$;
Mass excess: $\Delta = -50.255$ MeV; $S_n = 12.93$ MeV

Abundance

From the isotopic decomposition of normal chromium one finds that the mass-50 isotope, ^{50}Cr, is the second least abundant of all the Cr isotopes; 4.35% of all Cr. Using the total abundance of elemental Cr $= 1.35 \times 10^4$ per million silicon atoms in solar-system matter, this isotope has

solar abundance of ^{50}Cr $= 587$ per million silicon atoms.

This is almost as abundant as the halide fluorine. It ranks 48th in abundance among nuclei, what might be called moderately abundant.

Nucleosynthesis origin

In the mix of interstellar atoms from which the solar system formed, ^{50}Cr exists primarily owing to quasiequilibrium during explosive oxygen and silicon burning in supernovae. Its nucleosynthesis is closely allied with that of ^{46}Ti. The synthesis of ^{52,53}Cr (as ^{52,53}Fe) occurs just slightly more centerward in the explosive object than that of ^{50}Cr. *Quasiequilibrium* (see **Glossary**) applies when the different heavy nuclei are in nuclear equilibrium with each other, but their total number is not correct to be in equilibrium with the numbers of light nuclei (protons, neutrons, and helium).

Anomalous isotopic abundance

Wide variations of ^{50}Cr/^{54}Cr are expected within presolar grains, and perhaps also in the ratio ^{50}Cr/^{52}Cr; but the small chromium abundance within the grains has rendered the measurement impractical so far. Also, the ^{50}Cr/^{52}Cr ratio is often assumed to be normal in order that corrections for mass-dependent fractionation can be made. Such normalization suppresses manifest variability in ^{50}Cr; but of course it can be recovered from the data if one selects some other assumption about the mass fractionation.

---

^{52}**Cr** | $Z = 24$, $N = 28$
Spin, parity: J, $\pi = 0+$;
Mass excess: $\Delta = -55.413$ MeV; $\qquad S_n = 12.03$ MeV

The mass excess, $\Delta = -55.413$ MeV, shared by 52 nucleons is -1.066 MeV per nucleon, which is the most negative mass per nucleon of any lighter nucleus, exceeded four mass units later by the -1.082 MeV for ^{56}Fe. This means that the ^{52}Cr nucleons are more tightly bound by about 1.07 MeV per nucleon than are the twelve nucleons in ^{12}C, the standard for which $\Delta = 0$. These numbers played a big role in Hoyle's initial formulation, but were later seen to be not relevant to the actual radioactive progenitors, which were ^{52}Fe and ^{56}Ni respectively.

Abundance

From the isotopic decomposition of normal chromium one finds that the mass-52 isotope, ^{52}Cr, is the most abundant of the Cr isotopes; 83.8% of all Cr. Using the total abundance of elemental Cr $= 1.35 \times 10^4$ per million silicon atoms in solar-system matter, this isotope has

solar abundance of ^{52}Cr $= 1.13 \times 10^4$ per million silicon atoms.

This is about the abundance of the common element phosphorus. ^{52}Cr ranks 31st in abundance among nuclei, and must be reckoned as an abundant isotope.

Nucleosynthesis origin

In the mix of interstellar atoms from which the solar system formed, ^{52}Cr exists primarily owing to the quasiequilibrium process of stellar nucleosynthesis in explosive silicon burning. Almost all of this is produced as radioactive ^{52}Fe, but its decay to ^{52}Cr is complete by the end of the first day in supernova matter that is still too hot for chemistry to occur.

The nucleosynthesis of chromium is fascinating and historically significant. The ^{56}Fe abundance stands like a mountain peak above the abundance plain of the other metal isotopes between mass numbers $A = 45$ and $A = 65$. It was this striking dominance that inspired Hoyle's formulation of stellar nucleosynthesis theory in 1946. He first demonstrated that the centers of evolved massive stars reach such high temperatures and densities that the nuclei reach an equilibrium, and that that equilibrium could have ^{56}Fe as its most abundant nucleus if the gas has enough excess neutrons. In their review paper in 1957, B^{2}FH called this *statistical equilibrium* "the e process," with the e standing for equilibrium. One flaw was that it predicted a ratio ^{52}Cr/^{56}Fe much larger than the observed value 0.015. Although Hoyle's e process was and is regarded as one of the triumphs and lynchpins of nucleosynthesis theory, its details were other than initially envisioned. By 1967 it had become apparent that most of the matter ejected from stars in near-equilibrium would not have this desired ratio of neutrons to protons; instead it would be characterized by almost equal numbers of protons and neutrons within the total number. This meant that ^{52}Cr and ^{56}Fe would have to be synthesized as the radioactive isotopes ^{52}Fe and ^{56}Ni, respectively. This nucleosynthesis utilizes explosive quasiequilibria established during ejection from the supernova.

Although the core-collapse supernovae, called Type II, occur about five times more frequently than the explosions of white-dwarf stars, called Type I supernovae, the quantities of ^{52}Cr ejected over a period of time by each of these two sources is comparable.

Astronomical measurements

Chromium spectral lines are visible in stars. They show that the abundance ratio Cr/Fe has, through most of galactic history, remained constant, even while each has increased in proportion to H. A puzzle exists only in the most metal-poor stars, where stunning recent observations show that the Cr/Fe ratio is almost five times less than solar when Fe/H is but 1/10 000th of that in the Sun, and that the ratio climbs to solar by the galactic epoch when Fe/H has climbed to 1/300th of that in the Sun. Cr/Fe then remains at the solar ratio during subsequent galactic chemical evolution. The first Fe produced was surely from massive supernovae because they evolved fastest after the first stars were born. But why they ejected larger Fe/Cr ratios is unexplained today, although it surely reveals much about the detailed structure of the first supernovae. It suggests that *alpha-rich freezeout* (see **Glossary**) dominated in the first supernovae, since that quasiequilibrium synthesizes Cr/Fe ratios less than solar. Although the spectral lines do not require it, this observed Cr abundance is surely dominated by the ^{52}Cr isotope. Those oldest stars formed almost 10 billion years ago, whereas the Sun's birth 4.6 billion years ago was from interstellar gas that had by that time been much enriched in Cr. The constancy of the Cr/Fe ratio during the mature growth of galactic metallicity represents a compromise between ^{52}Cr nucleosynthesis in both massive Type II supernovae and in Type Ia supernovae, and ^{56}Fe mostly in Type Ia.

Anomalous isotopic abundance

Wide variations of ^{52}Cr/^{54}Cr are expected within presolar grains, and perhaps also in the ratio ^{52}Cr/^{50}Cr; but the small chromium abundance within them has rendered the measurement impractical so far. However, ^{52}Cr itself is almost always used as a normalization, so that the abundances of the other isotopes are expressed as their ratio to ^{52}Cr.

^{53}**Cr** | $Z = 24, N = 29$
Spin, parity: $J, \pi = 3/2-$;
Mass excess: $\Delta = -55.281$ MeV; $\qquad S_n = 7.94$ MeV

The proton binding energy 11.13 MeV is the largest of any odd-A nucleus since ^{29}Si.

Abundance

From the isotopic decomposition of normal chromium one finds that the mass-53 isotope, ^{53}Cr, is second most abundant of all Cr isotopes; 9.50% of all Cr. Using the total abundance of elemental Cr $= 1.35 \times 10^4$ per million silicon atoms in solar-system matter, this isotope has

> solar abundance of ^{53}Cr $= 1283$ per million silicon atoms.

This is about as abundant as the halide chlorine. It ranks 43rd in abundance among nuclei, what might be called moderately abundant.

Nucleosynthesis origin

In the mix of interstellar atoms from which the solar system formed, ^{53}Cr exists primarily owing to the quasiequilibrium process of stellar nucleosynthesis in explosive silicon burning. Most of this is produced as radioactive ^{53}Fe, whose decay to ^{53}Mn is rapid. The decay of ^{53}Mn to ^{53}Cr, however, has a halflife of 3.7 Myr. Thus ^{53}Mn nuclei exist outside of the supernova that produced them for several million years. This leads to an observational opportunity for X-ray astronomy (see ^{53}**Mn, *X-ray-line astronomy***). It also leads to many of the ^{53}Mn nuclei decaying in whatever interstellar grains accreted them prior to their decay. Lastly, ^{53}Mn lives long enough that measurable amounts of it were present in the early solar system, where it is today one of the *extinct radioactivities* (see ^{53}**Mn**, **Extinct radioactivity**). The eventual decay of ^{53}Mn to ^{53}Cr, however, is the cosmic source of ^{53}Cr and makes it one of the most radiogenic of all stable nuclei.

Anomalous isotopic abundance

Wide variations of ^{53}Cr/^{52}Cr are expected within *presolar grains* (see **Glossary**); but the small chromium abundance within them has rendered the measurement impractical so far. Undiscovered presolar grains of some unknown type must, theoretically, carry widely varying concentrations of ^{53}Cr within the Cr because of the variable ^{53}Mn radioactive abundances in diverse presolar grains. It is a challenge for the future to find these grains and measure their small Cr isotopic abundances. Disentangling nucleosynthesis variations in ^{53}Cr will be difficult because of the variations from ^{53}Mn decay (see ^{53}**Mn**).

The *meteorite* (see **Glossary**) Orgueil (that reveals such widespread ^{54}Cr variations) contains much smaller variations in ^{53}Cr. This suggests that the *cosmic chemical memory* (see **Glossary**) selecting ^{54}Cr carries no signal at ^{53}Cr and/or that the ^{53}Cr variations are radiogenic from ^{53}Mn.

Calcium–aluminum-rich inclusions (see **Glossary**) also contain ^{53}Cr variations that are smaller than those of ^{54}Cr. These may be ^{53}Mn-derived or intrinsic isotopic patterns.

^{54}**Cr** | $Z = 24$, $N = 30$
Spin, parity: $J, \pi = 0+$;
Mass excess: $\Delta = -56.929$ MeV; $\qquad S_n = 9.72$ MeV

Abundance

From the isotopic decomposition of normal chromium one finds that the mass-54 isotope, ^{54}Cr, is least abundant of all Cr isotopes; 2.37% of all Cr. Using the total abundance of elemental Cr $= 1.35 \times 10^4$ per million silicon atoms in solar-system matter, this isotope has

solar abundance of ^{54}Cr $= 319$ per million silicon atoms.

This is comparable in abundance to the element vanadium. It ranks 54th in abundance among nuclei, what might be called moderately abundant.

Nucleosynthesis origin

In the mix of interstellar atoms from which the solar system formed, ^{54}Cr exists primarily owing to neutron-rich quasiequilibrium regions ejected from the neutron-rich cores of some Type Ia supernovae. A smaller portion is produced from preexisting ^{53}Cr by s-process neutron capture in the carbon-burning shells of massive stars. The neutron-rich version of the equilibrium processes of stellar nucleosynthesis in Type Ia occurs more rarely but is responsible for the closely correlated nucleosynthesis of the neutron-rich stable isotopes ^{50}Ti, ^{54}Cr, and ^{48}Ca. It is noteworthy that the major source for ^{54}Cr produces none of the other Cr isotopes, so that the special variability of ^{54}Cr in meteorites may one day be amenable to a clear interpretation.

Anomalous isotopic abundance

Wide variations of ^{52}Cr/^{54}Cr are expected within *presolar grains* (see **Glossary**); but their small sizes and the low numbers of chromium atoms have rendered the measurement impractical so far.

It is in the *meteorites* (see **Glossary**) themselves that very interesting variations of ^{54}Cr abundance have been documented. The meteorites assembled in the solar system and suffered many forms of chemical rearrangements owing to either heating of the meteorites or alteration by liquid water within them, or both. The carbonaceous meteorite Orgueil has come under extensive study; and it seems to contain no isotopically normal Cr at all! When differing components of the meteorites are chemically isolated by dissolving in differing series of acids, most of the dissolved Cr has a shortage of about 0.6 parts per thousand of ^{54}Cr. Although this variation (with respect to normal solar Cr isotopes) is small, it is measured with more than adequate precision to demonstrate this reproduceable effect. But subsequent dissolving in strong hydrochloric acid reveals a component having an excess of ^{54}Cr by as much as 2.1%. This is a very big excess within minerals formed in the solar system. The explanation probably will utilize the *cosmic chemical memory* (see **Glossary**), the idea that the meteorite minerals remember the special presolar components from which they were constructed by chemistry in the solar system. The question is: how does ^{54}Cr chemically remember the Type Ia explosions that produced that isotope alone? The answers to such questions are the details by which the *cosmic chemical memory* (see **Glossary**) works. Two good answers have been suggested: (1) that a special grain rich in chromium condenses within those Type Ia explosions; (2) that the Type Ia ejecta emerge gaseous so that subsequently the ^{54}Cr atoms adhere to the surfaces of the very abundant tiny (0.01μm) grains that are known by astronomers to fill the *interstellar medium* (see **Glossary**).

In the *calcium–aluminum-rich inclusions* (see **Glossary**) within meteorites both excesses and deficits of ^{54}Cr have been measured. These include a large 0.5% excess of ^{54}Cr in a FUN (showing Fractionation and Unknown Nuclear anomalies) CAI named EK1-4-1 and a large deficit of -0.2% in FUN inclusion C1. In the non-FUN CAIs the ^{54}Cr is usually in excess, but smaller in value. The FUN values are probably the result of the same cosmic chemical memory that appears in most of the acid residues extracted from the Orgueil meteorite throughout. The sign of the large ^{54}Cr effects correlates with the sign of large ^{50}Ti effects.

25 | Manganese (Mn)

All manganese nuclei have charge +25 electronic units, so that 25 electrons orbit the nucleus of the neutral atom. Its electronic configuration can be abbreviated as an inner core of inert argon (a noble gas) plus seven more electrons: $(Ar)3d^5 4s^2$, which makes it the first of Group VIIB of the periodic table. Manganese manifests several oxidation states designated by valence +2, +3, +4 or +7, so that its rich coterie of oxides are MnO, Mn_2O_3, MnO_2 and Mn_2O_7. *Manganosite* (MnO) is the tenth most abundant compound in the Earth's crust. Mn halides use only the +2 or +3 oxidation states. The metal, which is hard, brittle, and gray-white, resembles iron and, like iron, it oxidizes in air. It melts at 1244 °C and boils at 1962 °C. Manganese is best known as an alloying addition to facilitate the forming and hot-working of steel.

The existence of the element Mn was famously predicted along with chlorine, oxygen and barium in a dissertation by Carl Wilhelm Scheele in 1774. It was purified for the first time later in that same year by J. G. Gahn, who fired a mineral that was later shown to be MnO_2 in charcoal, thereby removing the oxygen. As for most metals, the bulk of Mn in the Earth lies in its molten core. But in the universe it is quite abundant, being the 18th most abundant element and comprising an important part of the so-called "iron abundance peak." The theory of nucleosynthesis in stars was introduced in 1946 by Fred Hoyle specifically to account for that mountainous peak in the abundances of the elements when they are plotted against atomic number.

Manganese has but one stable isotope. Within nuclear astrophysics, a second naturally occurring isotope of Mn is radioactive. It was present in the early solar system and may be detectable by its X-ray emission following supernova explosions.

Natural isotopes of manganese and their solar abundances

A	Solar percent	Solar abundance per 10^6 Si atoms
53	extinct radioactivity in meteorites; X-ray emitter in ISM	
55	100	9550

^{53}Mn | $Z = 25$, $N = 28$ $t_{1/2} = 3.74 \times 10^6$ years
Spin, parity: J, $\pi = 7/2-$;   **Mass excess:** $\Delta = -54.684$ MeV

Extinct radioactivity

^{53}Mn is the decay product of radioactive ^{53}Fe during the supernova explosion, the main synthesized isotope at $A = 53$. It, in turn, decays to stable ^{53}Cr after a few million years, and is the main nucleosynthesis avenue to ^{53}Cr. ^{53}Mn carries high significance for the science of *extinct radioactivity*. It is the only extinct radioactivity having its origin restricted to the explosive silicon-burning inner shells of supernovae that also gives demonstrable evidence of having existed in the early solar-system cloud. Other extinct radioactivities can be produced in the envelopes of supernovae; but ^{53}Mn can not. Its presence within the presolar molecular cloud therefore requires that at least one supernova within the previous 20 Myr or so had been partially incorporated into that molecular cloud. It appears to have had small but measurable abundance in the early solar system, about $(8-30) \times 10^{-6}$ of stable manganese, with conflicting evidence about the exact value. Extinct ^{53}Mn in the early solar system was the result of one of four phenomena: it was either produced by a supernova unrelated to solar birth near the molecular cloud that gave birth to the Sun, or by a supernova that triggered the Sun's formation within the molecular cloud, or by young cosmic rays within the molecular cloud, or later by solar energetic-particle collisions in the planetary disk around the emerging young Sun.

Presolar grains   Extinct radioactive ^{53}Mn has not yet been recognized as having once been present within presolar grains because grains rich in manganese are not common. Over galactic history much ^{53}Mn will undoubtedly have decayed within interstellar grains, so that many of them, especially the abundant smallest interstellar grains, will carry excess ^{53}Cr on that account. But getting the evidence has been hard because the only presolar supernova grains discovered to date are refractory condensates (SUNOCONs) that thermally condensed during the expansions.

Solar-system rocks   Extinct radioactive ^{53}Mn is recognized as having once been present within rocks formed by chemical processes on protoplanetary bodies in the early solar system. These rock types include the Ca–Al-rich inclusions (CAIs), the chondrules found in meteorites, and larger rocks from meteorites that suffered early melting and differentiation among their constitutive elements within the cooling of the melt. These last "differentiated rocks" are from cores of small planets (asteroids), and are much larger than presolar grains. They contain the smallest observed concentration of ^{53}Mn because many millions of years had to pass before those asteroidal cores cooled enough to solidify, and the remaining ^{53}Mn abundance was then smaller than in the early-forming CAIs or in chondrules. Live ^{53}Mn within the solidifying rocks of all three types is attested to by the excess of daughter ^{53}Cr that those rocks contain. It is necessary to demonstrate that the rocks containing more manganese contain more ^{53}Cr in order to pin this down. The melting and chemical separations (called "differentiation") that occur within molten asteroids provide Mn-rich portions wherein the effect of intial ^{53}Mn can be isolated. The types of differentiated meteorites are known as "eucrites" and "angrites," two of which have yielded convincing ^{53}Cr excesses in the portions of highest Mn/Cr element ratio. The main problem here is knowing how old these meteorites were when they cooled enough to form the minerals being studied. This is believed to require 10–20 Myr; and allowance must be made for the amount of ^{53}Mn that will already have decayed by that time. Since 10 Myr is about two halflives for ^{53}Mn, the live ^{53}Mn in the rock after it cooled and solidified is no more than one-quarter of that that was initially present in the molten meteorite. This correction for a period of free decay in these meteorites causes uncertainty over the initial ^{53}Mn/^{55}Mn ratio.

Calcium–aluminum inclusions (CAIs) A measured ratio Δ^{53}Cr/^{55}Mn $= 4 \times 10^{-5}$ in CAIs, where Δ^{53}Cr is the number of excess ^{53}Cr atoms in comparison with normal solar Cr isotope ratios, appears to be about fivefold greater than the initial ^{53}Mn/^{55}Mn value that has been inferred from the above data. But one must ascertain that the excess Δ^{53}Cr has truly resulted from the decay of live ^{53}Mn rather than being a *cosmic chemical memory* (see **Glossary**) of *fossil* ^{53}Cr in the presolar materials. (For an explanation of the term "fossil" see **Glossary**, *Extinct radioactivity*.) If this is indeed the correct intial ^{53}Mn/^{55}Mn ratio in the solar system, the time to cool the differentiated asteroids (above) is perhaps 20 Myr.

Chondrules A ratio ^{53}Mn/^{55}Mn $= 9.4 \times 10^{-6}$ has been inferred from chondrules in the primitive Chainpur meteorite. In this case chondrules having larger Mn/Cr abundance ratio have been shown to also have larger Δ^{53}Cr/^{55}Mn ratio. If these numbers could all be taken with certainty, one could infer that the chondrules formed a few million years after the CAIs, which would be a very important conclusion.

These and other vagaries plague the sureness of chronologies based on *in-situ* decay of live ^{53}Mn. This age confusion unsettles astrophysicists who naturally wish for a simple and clear statement of the correct value for the initial ^{53}Mn/^{55}Mn ratio in the solar cloud. The reader may concur; but the truth is that a proper temporal picture of the events in the early solar system is needed to ascertain the correct initial isotopic ratio. The isotopic ratio ^{53}Mn/^{55}Mn $= 10^{-5}$ is not far from the ratio that might be expected on average within molecular clouds. These clouds take 50 Myr to form from dilute interstellar gas, so continuous mixing of dilute gas into molecular clouds at a rate needed to replenish them in 50 Myr maintains live ^{53}Mn within them and rather naturally accounts for the inferred ^{53}Mn abundance. This has an important consequence; namely, the extinct ^{53}Mn in the early solar system need not have been injected from the postulated simultaneous supernova trigger that may have caused the solar-system cloud to collapse. It may be the residue of several supernovae that occurred earlier than the trigger. So analysis of the causes for the extinct radioactivities can be somewhat ambivalent about ^{53}Mn. (For comparison, see ^{26}Al or ^{41}Ca.)

Planets   Three bulk planetary bodies at different distances from the Sun reveal differing amounts of excess ^{53}Cr. This has been interpreted as evidence of differing initial ^{53}Mn/^{55}Mn ratios within those planets. These three are the Earth at 1.0 AU from the Sun, Mars at 1.5 AU from the Sun, and the eucrite parent planet at 2.4 AU from the Sun. This could mean either a variation with distance from the Sun of the local ^{53}Mn/^{55}Mn ratio or a change of that ratio with time if these planetary bodies did not accrete their surface matter at the same time. Alternatively, in a *cosmic-chemical-memory* interpretation, these bulk variations may reflect differing proportions of grain sizes that came together to make these planets. The observed effect would then come about because the ^{53}Mn/^{55}Mn ratios are greater in the smaller grains (an expected consequence of cosmic chemical memory in astrophysics). What's more, the putative ^{53}Mn might instead be fossil ^{53}Cr in that picture. Interpreting excess ^{53}Cr in planets of all types seems sure to remain a significant science activity for many decades.

Cosmogenic radioactivity ^{53}Mn is created continuously as a nuclear collision fragment in meteorites when cosmic rays strike the meteoritic iron nuclei during the meteorite's journey to the Earth. This radioactivity is counted in the lab, alive today in the meteorite, and gives information on the transit time of the meteorite from parent body to Earth.

Nucleosynthesis origin

^{53}Mn is synthesized primarily in the process by which silicon transforms to nickel (within both types of supernovae). This process is called *silicon burning* (see **Glossary**) by astrophysicists. It is produced by the same processes that build the abundances in the iron-group abundance peak. Both Type II core-collapse supernovae and Type I exploding-white-dwarf supernovae create it in substantial quantities – probably about equally responsible for its natural abundance. Models predict that Type II eject about $(2-4) \times 10^{-4}$ solar masses of ^{53}Mn and that Type Ia eject 2–3 times as much, but in rarer events. During the core collapse of Type II supernovae lesser quantities of ^{53}Mn are also created by the reassembly near the supernova core of nuclei that were broken down into alpha particles during the supernova shock wave passage and then reassemble during the early expansion and cooling (the *alpha-rich freezeout*, see **Glossary**). Each of these processes primarily make ^{53}Fe rather than ^{53}Mn, but ^{53}Fe beta decays to ^{53}Mn with an 8.5-min halflife. It must be appreciated that ^{53}Fe and ^{53}Mn production is the major source of stable daughter ^{53}Cr in nature. This makes ^{53}Cr one of the most radiogenic of all natural abundances.

X-ray-line astronomy

Although the decay of ^{53}Mn is not blessed with a gamma-ray line (which would be easily detectable), it does emit X-ray lines. It is hoped that as ^{53}Mn accumulates in the interstellar medium, its abundance in regions of active star formation may give rise to a detectable emission. This would make ^{53}Mn similar to ^{26}Al and to ^{60}Fe in its ability to trace star formation and nucleosynthesis activity in the Milky Way.

^{55}**Mn** | $Z = 25$, $N = 30$
Spin, parity: $J, \pi = 5/2-$;
Mass excess: $\Delta = -57.708$ MeV; $\qquad S_n = 10.22$ MeV

Abundance

The only stable isotope of natural Mn has

solar abundance of ^{55}Mn = 9550 per million silicon atoms,

making it the 18th most abundant element and 33rd most abundant of all stable isotopes. The solar abundance of Mn is but 72% of its abundance inferred from meteorites, and the reason for this mild discrepancy is not known. A lighter radioactive isotope appears in nature as an extinct radioactive nucleus in the early solar system (see ^{53}**Mn**).

Nucleosynthesis origin

In the mix of interstellar atoms from which the solar system formed, ^{55}Mn exists primarily owing to the *quasiequilibrium* (see **Glossary**) processes of stellar nucleosynthesis. It is synthesized during explosive silicon burning and during *alpha-rich freezeout* (see **Glossary**) not as itself but as a radioactive progenitor, ^{55}Co, having more nearly equal numbers of protons and neutrons. Unfortunately, the ^{55}Co decay has no observable effects that have yet been measurable; but X-ray lines from the decay of ^{55}Fe (halflife 2.7 yr) to ^{55}Mn may be detectable in supernovae. The decay occurs during ejection from the supernovae, wherein this synthesis occurred. Of the two types of quasiequilibrium process (which really means processes whose final abundances are well simulated by a quasiequilibrium at a higher freezeout temperature), the ^{55}Co/^{56}Ni ratio created is very much larger in silicon burning than in the alpha-rich freezeout. This means that the ^{55}Co/^{56}Ni ratio ejected from Type II core-collapse supernovae depends in theory upon the relative masses ejected from these two types of quasiequilibrium processes. Those relative masses depend in the Type II supernovae upon the mass cut dividing material that gets out of the core from material that falls back onto the remnant. The greater the mass that falls back, the greater is the mass cut, and the greater is the ratio of the ^{55}Co/^{56}Ni ejected. Supernova theory can yield the solar abundance ratio easily; however, that theory cannot really "predict" the mass ratios ejected. The ^{55}Mn/^{56}Fe ratio has the value 1/191 inferred from the solar spectrum, but the value 1/95 measured in meteorites. The cause of this twofold discrepancy is unknown, but astrophysical arguments depend upon the correct choice because the nucleosynthesis of these two iron-peak nuclei is tightly coupled in the theory.

Type Ia supernovae (the explosions of white dwarfs) also create both ^{55}Mn and ^{56}Fe; they contribute a significant proportion, perhaps even most, of these nuclei and all iron-group nuclei. Calculations of differing physical models (deflagration models or delayed detonation models) of these explosions produce ^{55}Mn/^{56}Fe ratios varying between 1/29 and 1/120, further complicating the goal of an exact decomposition of the solar abundances of these nuclei into contributions from at least these three classes of supernova. The hope remains to use the solar ^{55}Mn/^{56}Fe ratio to determine the correct mass cuts within Type II supernovae, the true explosion model for Type I supernovae, and the relative frequencies of these two classes with the aid of the observed abundances of the iron-group isotopes. But that goal remains elusive.

Astronomical observations

The atomic lines of the Mn atom and ion are measurable in the spectra of stars, where iron is also observable and is usually taken as a standard measure of the abundance of heavy elements within stars. Often the ratio Fe/H, the ratio of the abundances of iron to hydrogen, is taken as a measure of the "metallicity" of the star.

The abundance ratio Mn/Fe has increased during most of the abundance evolution of the Galaxy. Both Mn and Fe are primary products of nucleosynthesis, so this evolving ratio is surprising and unexplained. The most dramatic puzzle exists in the most metal-poor stars, where stunning recent observations show that the Mn/Fe ratio is ten times less than solar when Fe/H is but 1/10 000th of that in the Sun, and that the ratio climbs to 1/3 solar by the galactic epoch when Fe/H has climbed to 1/1000th of that in the Sun. The ratio Mn/Fe then remains at 1/3 solar during subsequent galactic chemical evolution until Fe/H has grown to 1/10th of solar, after which it begins a climb toward the solar Mn/Cr ratio. The first Mn and Fe produced was surely from massive Type II supernovae because they evolved fastest after the first stars were born. This suggests that alpha-rich freezeout dominated the first supernova ejecta, because that quasiequilibrium can produce Fe/Mn ratios greater than solar. But precisely why the first supernovae did this is unexplained today, although it surely reveals much about the detailed structure of the first supernovae. So although Mn/H increases in the interstellar gas in rough proportion to the increases of Fe/H, the very early metal-poor stars provide exciting demonstration that not all elements grew together at all times in their galactic abundance ratio.

Anomalous isotopic abundance

Because Mn has but a single stable isotope, there exists no isotopic ratio to study, save that with ^{53}Mn, which is interpreted as varying amounts of that radioactive isotope.

26 | Iron (Fe)

All iron nuclei have charge $+26$ electronic units, so that 26 electrons orbit the nucleus of the neutral atom. Its electronic configuration is $1s^2 2s^2 2p^6 3s^2 3p^6 3d^6 4s^2$. This can be abbreviated as an inner core of inert argon (a noble gas) plus eight more electrons: $(Ar)3d^6 4s^2$, which locates it in Group VIII of the periodic table. Iron commonly has two oxidation states designated by valence $+3$ (*ferric*) or $+2$ (*ferrous*) and combines readily with other elements. Its readiness to combine chemically means that it is seldom found as pure (free) iron. It melts at 1535 °C and boils at 2750 °C.

Iron was one of the nine elements known to the ancient world. Its symbol (Fe) derives from its Latin name, *ferrum*. It is rather rare in its free (chemically pure element) state, being found as pure metal only in meteorites and in certain basaltic rocks from Greenland. Discovery of smelting methods allowed iron to be obtained from chemical iron ores, so that the Iron Age began (about 1200 BC); weapons and tools were made of iron. Most iron on Earth settled gravitationally into its molten core during the early hot epochs of the Earth's history; but it is still the second most abundant metal (after aluminum) in the Earth's crust. Its alloys are mostly oxides, but sulfides and carbonates are also common. Mined ores are roughly half iron, and smelting is designed to remove the other elements. Its wonderful alloy with carbon (steel) was ideal for the industrial revolution. Carbon steel is made possible because molten iron is able to dissolve carbon into the liquid and into the solid crystalline structure. Commercial steel has about 1% C dissolved in its structure. Unquestionably iron is one of the important elements for the history of mankind.

From the point of view of physics, iron is important for its magnetic properties. It is the most magnetic of all elements. The spin pairing of the valence electrons make each atom a tiny magnet, and when the atoms in a piece of iron are all aligned by a magnetic field, that piece becomes "magnetized." Therefore iron is strongly attracted by a magnet. Iron is magnetically soft, meaning that its magnetization closely follows the strength of the magnetizing field. This makes iron industrially useful in electric motors, relays and transformers. Iron can be made magnetically "hard" by alloying with nickel and cobalt. In this form it finds many uses as permanent magnets.

The spectral lines of the Fe atom and ions are very prominent in the spectra of stars. For this reason, coupled with its high abundance, iron has been a standard measure of the abundance of heavy elements within stars. Often the ratio Fe/H, the ratio of the abundance of iron to that of hydrogen, is taken as a measure of the "metallicity" of the star. It provided one of the earliest and best indications that very old stars within the Milky Way have smaller Fe/H; that is, they are more nearly the pure Big-Bang residues of H and He with which star formation began in the universe. Indeed, the most metal-poor stars of our Galaxy have ratios Fe/H that are only 1/10 000th of that in the Sun! Those oldest stars formed almost 10 billion years ago, whereas the Sun's birth 4.6 billion years ago was from interstellar gas that had by that time been much enriched in the element iron. It is quite expected that in the surfaces of some stars and in the nucleosynthesis ejecta of certain stars, Fe isotopes will have nonsolar proportions. However, its isotopic composition has not been measurable by astronomers owing to the small abundance of the minor Fe isotopes and to the small differences in the wavelengths of radiation emitted by each isotope of this element.

Iron is also one of the most important elements (arguably the most important element) for the history of nucleosynthesis in stars. Iron is hugely more abundant than its immediate neighbors. This feature was for British cosmologist Fred Hoyle the springboard to the theory of nucleosynthesis in stars. Iron possesses four stable isotopes (A = 54, 56, 57, 58), each of which has a unique and interesting consequence for understanding the origin of the abundances. It has one short-lived radioactive isotope, ^{55}Fe (2.7-yr halflife), that is synthesized abundantly in supernovae as the source of ^{55}Mn (see **^{55}Mn**), and which may be detectable in young supernovae with X-ray telescopes. It also possesses one long-lived radioactive isotope (A = 60) having measurable abundance in certain circumstances and having profound consequences for the origin of nuclei and of the solar system. To study the origin of iron isotopes is to study the history of nucleosynthesis in stars.

Natural isotopes of iron and their solar abundances

A	Solar percent	Solar abundance per 10^6 Si atoms
54	5.8	5.22×10^4
55	radioactive: X-ray emitter in supernovae	
56	91.72	8.25×10^5
57	2.2	1.98×10^4
58	0.28	2520
60	extinct radioactivity in meteorites	

^{54}Fe | $Z = 26, N = 28$
Spin, parity: $J, \pi = 0+$;
Mass excess: $\Delta = -56.249$ MeV; $\qquad S_n = 13.62$ MeV

During silicon burning in supernovae, the number of excess neutrons per nucleon determines the abundance of ^{54}Fe, which has two excess neutrons and which becomes the host for essentially all of the excess neutrons. As a result, the abundance of ^{54}Fe during silicon burning is half the abundance of excess neutrons. This is a very high abundance, producing locally a ^{54}Fe/^{56}Fe ratio that is much larger than the solar ratio.

Abundance

From the isotopic decomposition of normal iron one finds that the mass-54 isotope, ^{54}Fe, is the second most abundant of all the iron isotopes; 5.8% of all Fe. Using the vast total abundance of elemental Fe $= 9 \times 10^5$ per million silicon atoms in solar-system matter, this isotope has

solar abundance of ^{54}Fe $= 5.22 \times 10^4$ per million silicon atoms.

That is, it is 5.22% as abundant as the element silicon, making ^{54}Fe one of the abundant nuclear species. It ranks 21st in abundance among nuclei.

Nucleosynthesis origin

In the mix of interstellar atoms from which the solar system formed, ^{54}Fe exists primarily owing to the equilibrium process of stellar nucleosynthesis. This is technically a rather special equilibrium sometimes called *quasiequilibrium* (see **Glossary**). This term applies when the different heavy nuclei are in nuclear equilibrium with each other, but their total number is not correct to be in equilibrium with the numbers of light nuclei (protons, neutrons, and helium). In the quasiequilibrium established during silicon burning, which supernovae seem to create and eject, the ratio of abundances, ^{54}Fe/^{56}Fe, is considerably greater than the solar ratio (6.3%). But in the quasiequilibrium established during *alpha-rich freezeout* (see **Glossary**), reassembly of matter broken down by the core-rebound shock wave into alpha particles, the abundance ratio ^{54}Fe/^{56}Fe compensates by being less than the solar value. True nuclear equilibria can produce a ratio ^{54}Fe/^{56}Fe near the solar value 6.3%; but much of the Fe actually arises from the two quasiequilibria, which scatter in their production to both sides of the solar ratio, and by considerable margins. Explosive silicon burning in Type Ia supernova explosions is where most ^{54}Fe production occurs, and in this setting the ^{54}Fe/^{56}Fe ratio is much greater than solar, even becoming greater than unity in select regions.

Anomalous isotopic abundance

The isotopes of Fe are not expected to occur in all natural samples in their usual proportions. Wide variations are expected within presolar grains; but the small iron abundance within them has hindered measurements.

Cosmic rays   Cosmic rays constitute another sample of cosmic matter. The ^{54}Fe/^{56}Fe abundance ratio in the cosmic rays arriving at the solar system and recorded by NASA space missions raises dramatic questions. Its value in the cosmic rays is 9.3%, with an uncertainty of about 0.6%, which is almost half again as large as the solar ratio 6.3%. This very large difference demands good explanation; but the correct explanation depends upon knowing just what sample the cosmic rays represent. Two basic pictures dominate thinking about the origin of cosmic rays: (1) they represent local atoms of interstellar matter today that have been accelerated to their high energies (typically several hundred MeV per atomic mass unit) by interstellar processes driven by shock waves from supernovae; (2) they represent atoms from the supernova ejecta that have been accelerated by its shock waves. In picture (1) one is forced to suppose that the interstellar-medium (ISM) ^{54}Fe/^{56}Fe ratio near the Sun today is greater than when the Sun formed 4.6 billion years ago. Plausible as this sounds, it faces difficulty under tight inspection. The problem is that the Sun is a metal-rich object for its age, and theory holds that the ^{54}Fe/^{56}Fe ratio in the ISM should be near the Sun's value because the Fe/H abundance ratio is. Picture (2) is growing in favor, because the average ejecta from supernovae today are expected on theoretical grounds to carry a larger ^{54}Fe/^{56}Fe ratio than in the matter from which the Sun formed 4.6 billion years ago (the solar ratio). This theoretical expectation for supernova ejecta today is tricky, however. One needs to be sure of two competing effects, both of which may be involved. Recent supernovae formed from gas intially richer in carbon and oxygen, the seeds from which a portion of the supernova ^{54}Fe must be produced, suggesting that the current ^{54}Fe/^{56}Fe ejecta have higher values than that from the average, ancient, galactic supernovae that are responsible for the solar system's ^{54}Fe/^{56}Fe ratio. On top of that, the rate of Type I supernovae relative to the number of Type II supernovae is greater today than in the galactic history of the matter that went to produce the Sun 4.6 billion years ago. Some recent calculations indicate that the ^{54}Fe/^{56}Fe production ratio is greater in Type I supernovae than in Type II. But both pictures have their difficult points. What can be said is that the meaurements of this ratio in the cosmic rays have produced a complicated and rich astrophysical question.

Presolar grains   The first measurements of ^{54}Fe/^{56}Fe ratios in presolar grains were published in 2002. The concentration of iron in SiC grains shows that Fe is abundant enough in the grains for a measurement of all its isotopic ratios. The three Type X SiC grains measured originated in supernova interiors. In the material that had been exposed to the *s process* (see **Glossary**) in the presupernova, the ratio ^{54}Fe/^{56}Fe is predicted to be much less than solar. But in material from the silicon burning it is greater than solar. Both answers are found among these first such measurements. So the measured ^{54}Fe/^{56}Fe ratio will help considerably in the determination of the matter from which the SiC grains condensed.

^{55}Fe | Z = 26, N = 29 $t_{1/2}$ = 2.73 years
Spin, parity: J, π = 3/2–;   **Mass excess:** Δ = −57.476 MeV

> ^{55}Fe has a mass excess only 230 keV greater than its daughter ^{55}Mn, less than half the mass of a single electron. It is therefore stable against positron emission. Its decay requires that the ^{55}Fe nucleus capture one of the K-shell 1s electrons of the ^{55}Fe atom. The resulting vacancy in the ^{55}Mn K shell is what creates a 5.89-keV X-ray emission when it is refilled.

X-ray-line astronomy

^{55}Fe will be detectable in young supernovae by future observations of the X-ray lines emitted following its radioactive decay there to ^{55}Mn. About 0.001 solar masses of ^{55}Fe is ejected. When this ^{55}Fe decays to ^{55}Mn over the following few years, X-rays are emitted from the excited daughter ^{55}Mn atom. About one-quarter of such decays result in the 5.89-keV X-ray emission that is familiar in terrestrial labs as a calibration source for X-ray spectrometers. It is the flux of X-rays of that energy that will be detectable by high-resolution X-ray spectrometer experiments in space.

Nucleosynthesis origin

^{55}Fe is synthesized by quasiequilibrium processes during supernova nuclear burning. Both explosive *silicon burning* and *alpha-rich freezeout* (see **Glossary**) contribute to its production. Most is actually synthesized as ^{55}Co, which decays with a 17.5-hr halflife to ^{55}Fe. About 0.001 solar masses of ^{55}Fe is ejected.

^{56}Fe | Z = 26, N = 30
Spin, parity: J, π = 0+;
Mass excess: Δ = −60.601 MeV; S_n = 11.20 MeV

> The binding energy per nucleon of this isotope is greater than any other. This led to an intellectual detour in nucleosynthesis theory in which it was presumed that this high binding was the source of the very high abundance of ^{56}Fe. But that dramatic fact was but a coincidence, showing that causal relations in science must be demonstrated rather than assumed on the grounds of plausibility.

Abundance

From the isotopic decomposition of normal iron one finds that the mass-56 isotope, ^{56}Fe, is the most abundant of all iron isotopes; 91.72% of all Fe. Using the huge abundance of elemental Fe = 9×10^5 per million silicon atoms in solar-system matter, this isotope has

solar abundance of ^{56}Fe = 8.25×10^5 per million silicon atoms.

That is 82.5% as abundant as the element silicon, making ^{56}Fe one of the hugely abundant nuclear species. In the universe it is the 10th most abundant nucleus. On the crust of the Earth, however, iron is nowhere near as abundant as silicon. Sand (silica) is much more abundant than the metal iron and its ores because much of the iron settled to the center of the Earth during its early molten phase.

Nucleosynthesis origin

In the mix of interstellar atoms from which the solar system formed, ^{56}Fe exists primarily owing to the equilibrium process of stellar nucleosynthesis.

The nucleosynthesis of iron is the most fascinating and historically significant of all the isotopes. Its only competitors for top billing would be the Big-Bang light nuclei, which reveal so much of the early universe, and the ^{12}C nucleus, which provided a turning point in the path of stellar nucleosynthesis. The abundance of ^{56}Fe stands like a mountain peak above the abundance plain of the other metal isotopes between mass numbers $A = 45$ and $A = 65$. It was this striking dominance that inspired Hoyle's 1946 formulation of stellar nucleosynthesis theory. He first demonstrated that the centers of evolved massive stars reach such high temperatures and densities that the nuclei reach an equilibrium, and that that equilibrium could have ^{56}Fe as its most abundant nucleus. It was shown capable of comprising more than 90% of all nuclei in the equilibrium. In their review paper in 1957, B^{2}FH called this "the e process," with the e standing for equilibrium.

The nature of an e process can be imagined in everyday terms as follows. Many chips of ice, some at the freezing point, 0 °C, but some initially supercooled to less than 0 °C, are dropped into a glass of water. The glass is then placed in a freezer. The sequence of events as that system approaches equilibrium can be described as follows. The water first brings each ice chip up to the 0 °C freezing point of water, the water cooling slightly; the water then melts each ice chip, the smallest first and finally the largest, the water cooling considerably; then the loss of heat to the freezer turns all of the water into one large ice cube. That one large ice cube is analogous to ^{56}Fe. The analogy is that the binding energy, on average, of all nucleons is greatest if they are assembled into ^{56}Fe rather than into other less tightly bound nuclei. Quantitatively, the atomic mass excess of ^{56}Fe, $\Delta = -60.60$ MeV, represents a mass deficit of -1.082 MeV per each one of the 56 nucleons in comparison with the standard atomic mass unit, which itself corresponds to the average nucleon mass as it is bound in the ^{12}C atom. The nucleons are on average less massive in the ^{56}Fe nucleus than in ^{12}C owing to their reduced energy within ^{56}Fe, according to Einstein's famous energy–mass equation, $E = mc^2$. The nucleons are on average more bound, by 1.082 MeV, in ^{56}Fe than in ^{12}C.

Although Hoyle's e process was and is regarded as one of the triumphs and lynchpins of nucleosynthesis theory, it is now actually quite different than initially envisioned. For two decades, during the construction of most of nucleosynthesis theory by its pioneers, it was maintained that the equilibrium possessed about 15% more neutrons than protons in the total equilibrium bath. This ratio allowed the ^{56}Fe to be synthesized as itself within the equilibrium. By 1967 it had become apparent that most of the matter ejected from stars in near-equilibrium would not have this desired n/p ratio; instead, it would be characterized by almost equal numbers of protons and neutrons. This meant that ^{56}Fe would have to be synthesized as a radioactive isotope of nickel, ^{56}Ni. Although the halflife of ^{56}Ni is only 6.1 days, that is easily long enough for the ejected matter to cease its nuclear transmutations and freeze as ^{56}Ni.

Although the core-collapse supernovae, called Type II, occur about five times more frequently than the explosions of white-dwarf stars, called Type I supernovae, comparable quantities of ^{56}Ni are ejected over time from each of these two sources when their relative rates of occurrence are taken into account. The Type I supernova is a prolific event for ^{56}Ni nucleosynthesis. Each one ejects almost half of a solar mass of ^{56}Ni. The Type II supernova, on the other hand, ejects about 0.07 solar masses of ^{56}Ni. But both do so by variants of the equilibrium nucleosynthesis processes. In the Type I it is the e process much as Hoyle envisioned it except that the number of excess neutrons is small, so that ^{56}Ni is the parent for ^{56}Fe. The ratio ^{54}Fe/^{56}Fe from Type I supernovae is greater than the solar ratio (6.3%). In Type II the entropy in the exploding matter is much greater than in Type I. This means that at the equilibrium temperature, near five billion degrees, the density of the Type II is much smaller than that of the Type I. Therefore the ^{56}Ni in Type II are assembled in what is now called an *alpha-rich freezeout* (see **Glossary**). This is a near-equilibrium, except that the low density has prevented all He from reassembling into ^{56}Ni, so that excess He nuclei exist in the gas making the ^{56}Ni nuclei. The equilibrium-like distribution of abundances is actually a *quasiequilibrium* (see **Glossary**). It is one of the important concepts in nucleosynthesis that was not seen in the earliest works, but was introduced by the writer in 1967 and has gained many clarifications in subsequent decades. These high entropies arise when a shock wave impacts the silicon and oxygen shells overlying the core of iron. That core of iron will not itself emerge in the ejecta, but will remain behind, compressed into the most compact of all stars, the neutron star. In the alpha-rich freezeout the ^{54}Fe abundance is small relative to that of ^{56}Ni, so the final ^{54}Fe/^{56}Fe ratio from Type II supernovae is probably less than the solar ratio (6.3%). This expectation enters in the interpretation of cosmic-ray isotopes.

Astronomical measurements

Iron is also involved in the demonstration that not all elements grow together in their galactic abundance. The best measure of this has been the ratio of oxygen to iron. Both are very abundant in stars. Both are *primary products* of nucleosynthesis. But observations of old stars show convincingly that the O/Fe was larger in the gas from which those old stars formed than in that giving birth to later-forming stars. The earliest values of O/Fe are about three times greater than those in the middle-aged Galaxy. The reason is that only massive core-collapse supernovae can be born and evolve fast, and in their ejecta the O/Fe ratio is several times the ratio in solar abundances. But after 100 million years or so, the white-dwarf star population has been established, so that their later explosions as Type I supernovae can eject large quantities of iron. These cause the O/Fe ratio to decline to a value that seems to hold steady thereafter. Evidence of different nuclei synthesized in different types of supernovae is established by these and similar data.

Gamma rays   The equilibrium process leading to ^{56}Ni nucleosynthesis allowed a unique test of nucleosynthesis theory. In 1969 it was realized that ^{56}Co, to which ^{56}Ni decays in about a week, would live longer before its own decay. The 111-day-halflife decay of that radioactive ^{56}Co into the final ^{56}Fe nucleus emits many gamma rays of predictable energies. By that time the explosive ejecta would be sufficiently thin for the gamma rays to get out. These made it possible to confirm nucleosynthesis theory by a new form of astronomy, the spectroscopy of gamma radiation. This prediction was confirmed when a Type II supernova occurred in the Large Magellanic Cloud, a satellite galaxy to the Milky Way, in February 1987. It was called supernova 1987A, because it was the first supernova of 1987. Because the Magellanic Clouds are so much nearer than other galaxies, this supernova was very bright, the first visible to the naked eye in over four centuries. It is one of the most studied astronomical objects in the sky. Gamma-ray telescopes, which must fly above the Earth's atmosphere because the gamma rays cannot penetrate to the ground, were launched to observe supernova 1987A. They successfully detected the 847-keV and 1264-keV gamma rays that had been predicted in 1969. From these and the great heating that they produced, it is known that supernova 1987A ejected 0.075 solar masses of radioactive ^{56}Ni. This amounts to a mass of radioactive ^{56}Ni that is 20 000 times the mass of the Earth! It may be satirized as the largest verified nuclear accident of all time. This was the first time in history when the radioactivity produced in a specific stellar explosion had been measured.

NASA's *Compton Gamma Ray Observatory*, the second in NASA's sequence of great space observatories, was motivated in part by the hope of such astronomy. The writer was a champion of that astronomy and its potential for measuring nucleosynthesis theory, as well as a Co-Investigator on that mission. It was disappointing that its launch in April 1991, more than four years after supernova 1987A, came too late for detectable levels of 111-day-halflife ^{56}Co to remain within its ejecta. But it was still possible to detect the less abundant but longer lived ^{57}Co by its 122-keV gamma-ray line, making it the second radioactive nucleus to be detected from a specific supernova. The ratio of ^{57}Co/^{56}Co in the supernova 1987A ejecta was about 1.5 times the known ratio ^{57}Fe/^{56}Fe of their daughters on Earth. This confirmed a prediction from 1974 that ^{57}Fe is also synthesized as a detectable radioactive parent.

Cosmic rays   One of the interesting aspects of the cosmic rays arriving in the solar system is that they are rich in iron. To be more specific, the elemental Fe/C ratio is almost three times the solar value, and the ratio Fe/O is more than four times the solar ratio. Whether this be described as "an excess of Fe" or a deficiency of C and O is a matter of choice. A sensible choice depends on placing these into the spectrum of all element ratios in the cosmic rays. When this is done, Fe appears in its normal solar ratio to most heavy elements, so that it appears more sensible to regard the cosmic rays as deficient in C and O, rather than overabundant in Fe.

Anomalous isotopic abundance

The isotopes of Fe are not expected to occur in all natural samples in their usual proportions. Wide variations are expected within presolar grains; but the small iron abundance within them has hindered measurements.

Cosmic rays   Cosmic rays constitute another sample of cosmic matter. The ^{54}Fe/^{56}Fe abundance ratio in the cosmic rays arriving at the solar system and recorded by NASA space missions raise dramatic questions. Its value in the cosmic rays is 9.3%, with an uncertainty of about 0.6%, which is almost half again as large as the solar ratio 6.3%. This very large difference demands good explanation; but the correct explanation depends upon knowing just what sample the cosmic rays represent. Because it is a ratio of abundances it is difficult to say whether the anomaly belongs to ^{56}Fe or ^{54}Fe. That ambiguity is characteristic of all isotopic ratios unless measurements of yet other isotopic ratios enable one to conclude which isotope is actually anomalous. It is probably more logical and satisfying to think of these variations in ^{54}Fe/^{56}Fe as reflecting ^{54}Fe variations in the cosmic rays (see **^{54}Fe**).

^{57}**Fe** | $Z = 26$, $N = 31$
Spin, parity: $J, \pi = 1/2-$;
Mass excess: $\Delta = -60.176$ MeV; $S_n = 7.64$ MeV

Abundance

From the isotopic decomposition of normal iron one finds that the mass-57 isotope, ^{57}Fe, is the third most abundant of the iron isotopes; 2.2% of all Fe. Using the total abundance of elemental Fe $= 9 \times 10^5$ per million silicon atoms in solar-system matter, this isotope has

solar abundance of ^{57}Fe $= 1.98 \times 10^4$ per million silicon atoms,

making it 1.98% as abundant as the element silicon and in 27th rank of all nuclear species.

Nucleosynthesis origin

In the mix of interstellar atoms from which the solar system formed, ^{57}Fe exists primarily owing to the equilibrium process of stellar nucleosynthesis (see **^{56}Fe** for discussion of this in the context of iron's major isotope). Like ^{56}Fe, ^{57}Fe is synthesized as an isotope of radioactive nickel. Similar too is the distribution between Type II supernova core-collapse events and the Type I white-dwarf explosion events. They probably synthesize ^{57}Fe at comparable rates in the Galaxy, although each Type I event produces a greater quantity of ^{57}Fe than the more frequent Type II event. ^{57}Fe-rich iron can, to a limited extent, be created by exposing iron to free neutrons (the *s process*, see **Glossary**); although that process may make observable Fe isotopic anomalies in presolar grains, it is not a significant source of bulk ^{57}Fe.

Astronomical measurements

Although iron is detectable in may stars, and is a diagnostic of the chemical enrichment of the Galaxy during its ageing, that spectroscopy has not been able to distinguish the isotopes of Fe from the dominant ^{56}Fe. The isotopic shift in wavelength is too small, especially when coupled to the much lower abundance of ^{57}Fe. This may one day be possible, however, using molecules containing iron.

Gamma rays   Gamma-ray astronomy has the great virtue of being able to detect specific isotopes, with the added virtue of knowing, because of the halflife, that those isotopes are newly synthesized. The ^{57}Co parent of ^{57}Fe was one of a handful of nuclei predicted to form the basis of that new astronomy. NASA's *Compton Gamma Ray Observatory*, the second of NASA's sequence of great space observatories, was motivated in part by the hope of gamma-ray-line astronomy. It was disappointing that its launch in April 1991, more than four years after supernova 1987A, came too late for detectable levels of the 77-day-halflife ^{56}Co to remain within its ejecta. But it was still possible to detect the less abundant but longer lived (halflife 272 days) ^{57}Co by its 122-keV gamma-ray line, making it the second radioactive nucleus to be detected from a specific supernova. The ratio ^{57}Co/^{56}Co in the supernova 1987A ejecta was about 1.5 times the known ratio ^{57}Fe/^{56}Fe of their daughters on Earth. This confirmed a prediction from 1974 that ^{57}Fe is also synthesized as a detectable radioactive parent. The supernovae create ^{57}Ni, which decays to ^{57}Co, which in turn decays to stable ^{57}Fe. To appreciate how supernova 1987A made an iron isotope ratio 1.5 times greater than solar abundances, it is necessary to take into account that most of the Ni production in this supernova occurred by the *alpha-rich freezeout* (see **Glossary**) version of the equilibrium processes. But even so, mysteries remain concerning this modest deviation from normal iron ratios.

These gamma-ray detections of ^{56}Co and ^{57}Co in supernova 1987A are among the most important observations in the history of astronomy. This gave birth to a new type of astronomy. They confirmed that new nuclei are created in supernova explosions. They confirmed the continuous nature of stellar nucleosynthesis, that it is ongoing even today. They confirmed the equilibrium processes to be those that produce iron as radioactive nickel rather than as iron itself. They confirmed that the amount produced per supernova is correct for creating the entire galactic abundance from about 10 billion years of supernova explosions, occurring at the rate of about 5 per century. The factor 1.5 in the isotope ratio reveals yet unsolved aspects of the mechanics of the core-collapse event.

Anomalous isotopic abundance

The isotopes of Fe are not expected to occur in all natural samples in their usual proportions. Wide variations are expected within presolar grains; but the small iron abundance within them has hindered the measurements.

Cosmic rays   Cosmic rays constitute another sample of cosmic matter. The ^{57}Fe/^{56}Fe abundance ratio in the cosmic rays arriving at the solar system and recorded by NASA space missions raise dramatic questions. Its value in the cosmic rays is 3.7%, with an uncertainty of about 0.4%, which is almost half again as large as the solar ratio 2.4%. This very large difference demands good explanation; but the correct explanation depends upon knowing just what sample the cosmic rays represent. Are they accelerated interstellar atoms, or are they accelerated ejecta from supernovae? (See **^{54}Fe**, where the situation is analogous.)

Presolar grains   The ^{57}Fe/^{56}Fe ratio has not been published in presolar grains at the time of writing. But measurements of the concentration of iron in SiC grains show that Fe is abundant enough in the grains for a measurement of its isotopic ratios. In the material that had been exposed to the *s process* (see **Glossary**), the ratio ^{57}Fe/^{56}Fe is predicted to be several times greater than solar. But in material from the silicon burning, it is predicted to be less than solar. In material from the *alpha-rich freezeout* (see **Glossary**), which is the source of most of the radioactive Ni parents of ^{57}Fe and ^{56}Fe in Type II supernovae, the ^{57}Fe/^{56}Fe ratio is expected to be slightly in excess of solar. So the measured ^{57}Fe/^{56}Fe ratio in presolar grains will help considerably in the determination of the matter from which the SiC grains condensed. Values both greater than the solar ratio and less than that ratio are to be expected. The ^{57}Fe/^{54}Fe ratio may be even more diagnostic, because it is much greater than solar in the alpha-rich freezeout but less than solar in silicon burning. It seems likely that the presolar supernova grains will provide exciting new information about supernova explosions. The usefulness of iron as a diagnostic in presolar grains was predicted by the writer in 1975.

^{58}**Fe** | Z = 26, N = 32
Spin, parity: J, π = 0+;
Mass excess: $\Delta = -62.149$ MeV; $\qquad$ $S_n = 10.04$ MeV

Abundance

From the isotopic decomposition of normal iron one finds that the mass-58 isotope, ^{58}Fe, is the least abundant of the iron isotopes; 0.28% of all Fe. Using the total abundance of elemental Fe = 9×10^5 per million silicon atoms in solar-system matter, this isotope has

solar abundance of ^{58}Fe = 2520 per million silicon atoms.

^{58}Fe is one of the moderately abundant nuclear species, ranking 39th among all nuclei.

Nucleosynthesis origin

In the mix of interstellar atoms from which the solar system formed, ^{58}Fe exists primarily owing to s-process neutron capture in stars by the ^{56}Fe and ^{57}Fe isotopes which had been made in earlier stars by the equilibrium process of stellar nucleosynthesis. Substantial portions of ^{58}Fe can also be synthesized by the equilibrium process of stellar nucleosynthesis; but adequate yield requires a neutron-rich version of the equilibrium processes of stellar nucleosynthesis, which occurs more rarely and is responsible for much of the most neutron-rich stable isotopes of Ti, Cr, Fe, and Ni. At this time the relative importance of neutron capture and of the equilibrium process of stellar nucleosynthesis to ^{58}Fe nucleosynthesis is hard to state with precision; but perhaps two-thirds by neutron capture is not far from the mark. Although ^{58}Fe holds the distinction of being the most abundant of all s-process products, its yield is undetermined owing to the uncertainty over the total amount from n-rich Type Ia supernovae.

Anomalous isotopic abundance

The isotopes of Fe are not expected to occur in all natural samples in their usual proportions. Wide variations are expected within presolar grains; but the small iron abundance within them has hindered the measurements.

Cosmic rays $\quad$ Cosmic rays constitute a sample of cosmic matter. The ^{58}Fe/^{56}Fe abundance ratio in the cosmic rays arriving at the solar system and recorded by NASA space missions raise dramatic questions. Its value in the cosmic rays is 0.2% (not very accurately measurable yet, with an uncertainty of about 0.1%), which is about 2/3 of the solar ratio 0.31%. This very large difference demands good explanation; but the correct explanation depends upon knowing just what sample the cosmic rays represent. Are they accelerated interstellar atoms, or are they accelerated ejecta from supernovae? (See **^{54}Fe**, where the situation is analogous.)

Presolar grains   The first measurements of ^{58}Fe/^{56}Fe ratios in presolar grains were published in 2002. The concentration of iron in SiC grains shows that Fe is abundant enough in the grains for a measurement of all its isotopic ratios. In the material that had been exposed to the s process, the ratio ^{58}Fe/^{56}Fe is predicted to be about 50 times greater than solar. But in material from the silicon burning, where ^{58}Fe is virtually absent, it is much less than solar. So the measured ^{58}Fe/^{56}Fe ratio will help considerably in the determination of the matter from which the SiC grains condensed. Values both greater than the solar ratio and less than that ratio are to be expected; and both have been found. The usefulness of iron as a diagnostic in presolar grains was predicted by the writer in 1975.

^{60}Fe | $Z = 26$, $N = 34$ $t_{1/2} = 1.49$ million years
Spin, parity: J, $\pi = 0+$;
Mass excess: $\Delta = -61.407$ MeV; $S_n = 8.92$ MeV

Abundance

The radioactive ^{60}Fe nucleus is one of the most complex and interesting isotopes to occur in nature. Its 1.49-Myr halflife is, though long in terms of human civilization, short in terms of the age of the Earth. After 100 Myr for example, when the Earth's surface was still undergoing chemical formation, the number of initial atoms of ^{60}Fe on the Earth would have already halved 66 times. Each gram of ^{60}Fe on Earth, initially containing 10^{22} atoms of ^{60}Fe, would by that time have only 100 atoms remaining; and those 100 atoms will virtually be gone after an additional seven halflives. So, ^{60}Fe has long been extinct on Earth. It is an *extinct radioactivity* (see **Glossary**). But evidence of live ^{60}Fe has been obtained in three ways.

Excess ^{60}Ni in meteorites   Solids that formed during the earliest phases of solar-system history would have contained the amount of ^{60}Fe that was initially present in the interstellar cloud from which the solar system formed. The time required for the interstellar cloud to collapse and the solar planetary disk to form is estimated to be of order one million years; so a significant fraction of ^{60}Fe might have remained alive in solids formed on small planetesimals; such as the meteorites. A certain class of meteorites, the "angrites," underwent heating and chemical crystallization within the first few million years of solar-system history. Within their iron-containing minerals (especially) the ^{60}Fe nuclei would decay to ^{60}Ni nuclei, leaving a small excess of those ^{60}Ni nuclei in that mineral. The larger the chemical Fe/Ni abundance ratio that was dissolved intially in that mineral, the greater will be the number of daughter ^{60}Ni atoms per initial stable Ni atom; i.e. the greater will be the percentage excess in the ^{60}Ni isotopic abundance. Such studies have shown that the initial isotopic ratio in the solar system was near ^{60}Fe/^{56}Fe $= 1.5 \times 10^{-6}$, if the time of 3 Myr required for those meteorites to crystallize has been well reckoned. Even this small abundance ratio poses severe astrophysical problems. Where did the ^{60}Fe atoms within that initial ^{60}Fe/^{56}Fe $=$ 1.5×10^{-6} come from? What created them so recently prior to the Earth's formation that they were not all gone? These are the classic questions of *extinct radioactivity* (see **Glossary**).

The proposed solutions to that problem have featured one of four possibilities. (1) The first involves the average amount of live ^{60}Fe that can be expected to be present in all molecular clouds where stars form. This average expectation is, however, too small because the lifetime of ^{60}Fe is too short for that picture. (2) The second is that the live ^{60}Fe was created by one dying star in the cloud near to the place where the Sun was about to form within a dense pocket of cloud gas. The dying star could be either a supernova explosion, or the wind from a presupernova massive Wolf–Rayet star, or an AGB star transitioning to a planetary nebula. It has proved difficult, however, to understand how the ejecta from one star can mix within the gas of another forming star and do so within a few million years. (3) The third solution involves creating the live ^{60}Fe within the forming solar system rather than entraining it in the presolar cloud. This creation would be achieved by the nuclear collision of two heavy atoms, atoms that had been locally accelerated to low-cosmic-ray energy. This picture presents a satisfactory joint solution for the known cohort of short-lived radioactivities in their observed amounts; but for ^{60}Fe the appropriate nuclear collisions fail. That ^{60}Fe is not effectively created by nuclear collisions sets it apart from all other extinct radioactivities, and endows the abundance of ^{60}Fe in the early solar system with unique importance. (4) The fourth solution questions whether the ^{60}Fe was actually alive when the meteorites formed. It utilizes the large ^{60}Fe abundance initially present in supernova grains of Fe. It envisions the excess abundances of the daughters as an inherited effect; the excess daughter abundances (^{60}Ni in this case) are carried chemically into larger solids by their robust prior presence in the presolar iron grains from which the larger solids are subsequently fused. This solution has not been accepted by chemists who have not been able to envision chemical pathways for such chemical memory. Although these four classes of solutions are attempted for each of the extinct radioactivities, the excess ^{60}Ni from extinct ^{60}Fe speaks most troublingly but most powerfully to the correct solution, what ever that may prove to be. The majority of scientists supports the second solution above.

Gamma rays During every million-year period the Milky Way stars are believed to have produced some 30 000 supernova explosions, explosions creating some ^{60}Fe nuclei. The first discussion of the astrophysical importance of ^{60}Fe (in 1971) advanced the idea that the associated ^{60}Fe radioactivity within interstellar gas could be detected by telescopes designed to record the characteristic gamma rays that are emitted following the decays of ^{60}Fe nuclei. About 10^{-5} solar masses of ^{60}Fe (about 3.4 Earth masses) are expected to be ejected from each individual supernova. The specific gamma-ray energies that identify the ^{60}Fe nucleus have energies 1.17 MeV and 1.33 MeV. A very interesting upper limit to the amount of interstellar ^{60}Fe was established by NASA's *Solar Maximum Mission*. Detecting them has more recently been attempted, unsuccessfully, by NASA's *Compton Gamma Ray Observatory*. The expected flux of gamma rays, based on supernova calculations of the relative amounts of ^{60}Fe and of ^{26}Al produced, is about 15% of the flux of detected gamma rays from ^{26}Al, rendering it near the limit attainable by the COMPTEL telescope. Although that telescope has successfully detected gamma rays from ^{26}Al, the rate from ^{60}Fe was too small to be seen. The next-generation gamma-ray-line telescope, one capable of a tenfold improvement in sensitivity, will achieve powerful astronomical results based on its measurements of these gamma-ray lines from the Milky Way. The first chance comes from the launch in 2002 of the European INTEGRAL mission.

Accretion by the Earth   The Earth and its associated solar-wind bubble moves through the space among the stars, and as it does so it encounters dilute gas and dust particles. Isolated Fe atoms would not penetrate the outflowing solar wind, whose magnetic field would sweep up the Fe atoms after they are ionized by the Sun's radiation and carry them back out of the solar system. But dust particles can penetrate through the solar wind and reach the Earth. The dust detector on the *Ulysses* spacecraft has recorded such particles entering the solar system. These are tiny, too small to see with the human unaided eye, containing a few billion atoms from interstellar space. Only about 10^{-13} grams of mass, some of them may contain live ^{60}Fe within them if those grains formed from atoms created in part by recent nearby supernovae (say within the last few million years).

Only since 1998 has the quest to detect ^{60}Fe atoms captured by the Earth been technically feasible. German scientists reported live ^{60}Fe in ocean sediments. They took a crust sample from a depth of 1300 meters in the Pacific Ocean, and within it were live atoms (live today!) of ^{60}Fe in numbers sufficient to be detected and counted. The measured isotopic ratios were tiny, near ^{60}Fe/^{56}Fe $= 2 \times 10^{-15}$ (requiring almost 10^{15} Fe atoms within a sample to contain a single ^{60}Fe!). The ages of these oceanic crust samples are known from other techniques to be about 3 Myr old. The interpretation is that a supernova exploded several Myr ago at a nearby distance of about 30 parsecs (100 light-years), and that its ^{60}Fe-containing ejecta created some dust grains within its outflowing gas that were able, owing to their motion, to encounter the solar system and be deposited in this sedimentary layer in the South Pacific. The dust particles would be vaporized on atmospheric entry, but the released ^{60}Fe atoms could drift down to the oceans and settle on the bottom as part of the sedimentary growth. The total rate of entry of ^{60}Fe atoms into the Earth's atmosphere needs to be near 7 per cm^2 per year. This would have been a very bright event to emerging life on Earth, millions of times brighter than the stars that are visible to the naked eye.

Nucleosynthesis origin

In an average mix of interstellar atoms, ^{60}Fe is deposited by the ejecta from dying stars, within which it was created. In those ejecta, ^{60}Fe exists primarily owing to neutron capture within those stars by the lighter and more abundant Fe isotopes. Those abundant isotopes already existed in the stellar matter, because they had been produced earlier by stellar nucleosynthesis. Substantial portions of ^{60}Fe may also be synthesized by the equilibrium process of stellar nucleosynthesis; but it must be a neutron-rich version of the equilibrium processes of stellar nucleosynthesis, which necessarily occurs more rarely and is responsible for significant portions of the most neutron-rich stable isotopes of Ti, Cr, Fe, and Ni. Presently, the relative importance of neutron capture and of the equilibrium process of stellar nucleosynthesis to ^{60}Fe nucleosynthesis in the Galaxy is unknown. If the ^{60}Fe was indeed alive in the early solar system, as meteoritic evidence seems to indicate, it is most probable that it was ejected from the carbon core or from the He-burning shell of a nearby supernova that may have caused the solar cloud to collapse to form the Sun. When the supernova shock from core rebound reaches the carbon-burning shell or the He-burning shell of the presupernova, the sudden increase in temperature produces a neutron flux sufficiently strong for radioactive ^{59}Fe, before it decays, to capture a neutron, thus creating ^{60}Fe. At the base of the He-burning shell the shock liberates a burst of free neutrons that create, in a narrow shell, such a high concentration of ^{60}Fe that is even greater than those of the stable iron isotopes. It is produced in both locations when stable iron captures free neutrons caused by the primary neutron-liberating reaction, ^{22}Ne(alpha, n)^{25}Mg. Although supernova injection into the solar cloud is a plausible scenario, to avoid simultaneously injecting too much extinct ^{53}Mn radioactivity, the oxygen-and-silicon burning core of the supernova ejecta must not be admixed into the solar cloud. Reasons for this sharp division are under lively debate. It is also possible that the ^{60}Fe was inherited by the solar-birth cloud from a local AGB star.

When attention turns to alternative causes for the shorter-lived extinct radioactivities, local production of ^{60}Fe within the solar system by nuclear collisions of fast nuclei, the major impediment to such ^{60}Fe nucleosynthesis quickly reveals itself. The nucleus ^{60}Fe is so neutron-rich (34 neutrons versus 26 protons) that collisions between common nuclei cannot produce it. The reactions that can produce ^{60}Fe (such as that between ^{48}Ca and ^{18}O) involve nuclei of such low natural abundance that their collisions are unfavorably rare. This obstacle to ^{60}Fe nucleosynthesis by cosmic-ray collisons endows ^{60}Fe with unique importance for resolving the extinct radioactivity question.

27 | Cobalt (Co)

All cobalt nuclei have charge +27 electronic units, so that 27 electrons orbit the nucleus of the neutral atom. Its electronic configuration is $1s^2 2s^2 2p^6 3s^2 3p^6 3d^7 4s^2$. This can be abbreviated as an inner core of inert argon (a noble gas) plus nine more electrons: $(Ar)3d^7 4s^2$, which locates it in Group VIII of the periodic table. Cobalt commonly has two oxidation states designated by valence +2 or +3, so that its oxides are CoO and Co_2O_3. The metal, which is hard, ductile, and lustrous blue-gray, melts at $1857\,^\circ C$ and boils at $2672\,^\circ C$.

In the 16th century it was known that certain minerals could turn glass deep blue. Cobalt, first discovered in 1739 was shown to be the element contained within them that caused this – hence the term "cobalt blue". The most important ores are *cobaltite* ($CoAsS$) and *erythrite* ($Co_3(AsO_4)_2$).

From the point of view of physics, cobalt is not nearly as good a conductor as copper, but it is important for its magnetic properties. Cobalt metal must be heated to higher temperature than any other metal ($1121\,^\circ C$, its "Curie point") in order to destroy its magnetization. Iron can be magnetically "hardened" by alloying it with nickel and cobalt. In this form Co finds many uses in permanent magnets, whose preparation consumes a quarter of cobalt production. Cobalt is recovered from operations used for the refining of nickel, copper, and iron. As for those elements, most of Co in the Earth lies in its molten core.

Within nuclear astrophysics, the two most important isotopes of Co are radioactive! Gamma-ray lines from supernovae reveal shortly after the explosions detectable concentrations of ^{56}Co ($t_{1/2} = 77.3$ days) and ^{57}Co ($t_{1/2} = 271.8$ days) within the ejecta. They are present by virtue of being the daughters of radioactive ^{56}Ni and ^{57}Ni, both of which were predicted 24 years ago to be created directly by the equilibrium and quasiequilibrium nucleosynthesis and to provide nature's origin for stable ^{56}Fe and ^{57}Fe (see $^{56}\mathbf{Ni}$ and $^{57}\mathbf{Ni}$ for more information). Amazingly, a third radioactive isotope, ^{60}Co ($t_{1/2} = 1925$ days), also offers great hopes for gamma-ray detection by a more sensitive instrument than those designed for the *Compton Gamma Ray Observatory*; but that requires the good fortune of a galactic supernova to fulfil this promise. For diagnostic purposes in quality testing in manufacturing and in medicine, ^{60}Co finds practical applications owing to its X-ray emission accompanying the radioactive decay process. And cobalt's only stable isotope is synthesized in supernovae as radioactive ^{59}Ni, which may be detectable there with X-ray telescopes. Cobalt isotopes are replete with ramifications for cosmic radioactivity.

Natural isotopes of cobalt and their solar abundances

A	Solar percent	Solar abundance per 10^6 Si atoms
56	radioactive: gamma-ray emitter in supernovae	
57	radioactive: gamma-ray emitter in supernovae	
59	100	2250
60	radioactive: gamma-ray emitter in supernovae	

^{56}Co | $Z = 27$, $N = 29$ $t_{1/2} = 77.3$ days
Spin, parity: $J, \pi = 4+$;   **Mass excess:** $\Delta = -56.035$ MeV

Abundance

^{56}Co has been seen by observations of the gamma-ray lines emitted from supernova 1987A following its radioactive decay there. According to theory, ^{56}Co would not be present there except for the abundant nucleosynthesis of ^{56}Ni in supernovae. The observability of gamma-ray lines at 847 keV and 1238 keV were predicted in 1969 and provided strong motivation for gamma-ray-line telescopes. Supernova 1987A did not expand rapidly enough to become transparent to all interesting gamma-ray lines, but a significant fraction were detected. Type Ia supernovae expand so much faster and make so much ^{56}Ni that ^{56}Co is the prime target for the future of gamma-ray-line astronomy (see ^{56}**Ni** for more information). For a time the light curves of many supernovae decline at the same rate as the declining abundance of ^{56}Co (halflife 77 d) because the heating of the remnant by that radioactive decay provides the source for the optical and infrared photons that are emitted. This explained the tracking of exponential decay by the luminosity radiated optically, an observation from the 1950s that had inspired the "californium hypothesis." Right idea, wrong radioactivity.

Nucleosynthesis origin

See ^{56}**Fe** (to which ^{56}Co decays) for a discussion of the nucleosynthesis of ^{56}Ni (parent of ^{56}Co) by equilibrium and quasiequilibrium processes during supernova nuclear burning.

^{57}Co | $Z = 27$, $N = 30$ $t_{1/2} = 272$ days
Spin, parity: $J, \pi = 7/2-$;   **Mass excess:** $\Delta = -59.340$ MeV

Abundance

^{57}Co has been seen by observations of the gamma-ray lines emitted from supernova 1987A following its radioactive decay there. According to theory, ^{57}Co would not be present except for the abundant nucleosynthesis of ^{57}Ni in supernovae.

NASA's *Compton Gamma Ray Observatory*, the second in NASA's sequence of great space observatories, was motivated in part by the hope of gamma-ray-line astronomy. It was disappointing that its launch in April 1991, more than four years after supernova 1987A, came too late for detectable the levels of 77-day-halflife ^{56}Co to remain within its ejecta. Launch occurred 19 halflives after that event. But it was still possible to detect the less abundant but longer lived ^{57}Co daughter of ^{57}Ni by its 122-keV gamma-ray line, making it the second radioactive nucleus to be detected from a specific supernova. It had decayed for only slightly more than five halflives. The initial ratio of ^{57}Co to ^{56}Co in the supernova 1987A ejecta, was about 1.5 times the known ratio on Earth of their daughters, ^{57}Fe/^{56}Fe.

An unexpected consequence of this detection was its demonstration that the mass of ^{57}Ni initially synthesized was not great enough (by a factor near 3) to account for the total radiation emerging after three years from supernova 1987A. It had been thought that the energy of the ^{57}Co decay would provide the heat needed to maintain the total light output of the remnant after the faster decay rate of ^{56}Co had declined to a level too small to accomplish that heating. But it was not so. The discoverers therefore advanced the idea that it was delayed recombination of free electrons that was providing the power. This was itself an exciting advance. These free electrons had been created by earlier radioactive decays rather than by concurrent radioactivity. This step forward in the theory of young supernova remnants had been inspired by the gamma-ray measurement showing the amount of ^{57}Co to be inadequate.

Nucleosynthesis origin

See ^{57}**Fe** for a discussion of the nucleosynthesis of ^{57}Ni (parent of ^{57}Co) by equilibrium and quasiequilibrium processes during supernova nuclear burning. ^{57}Ni and ^{57}Ni are synthesized together in both explosive *silicon burning* (see **Glossary**) and by the *alpha-rich freezeout* (see **Glossary**).

^{59}**Co** | Z = 27, N = 32
Spin, parity: J, $\pi = 7/2-$;
Mass excess: $\Delta = -62.226$ MeV; $S_n = 10.47$ MeV

Abundance

The mass-59 isotope is the only stable isotope of natural cobalt. It has

solar abundance of ^{59}Co = 2250 per million silicon atoms,

which is substantial, rendering it the 40th most abundant of all nuclei. Solar and meteoritic abundances are in good agreement. Two lighter radioactive isotopes, ^{56}Co and ^{57}Co, appear naturally as radioactive nuclei in supernova. They are measured via the gamma-ray lines emitted when they decay to Fe (see ^{56}**Fe**).

Nucleosynthesis origin

In the mix of interstellar atoms from which the solar system formed, ^{59}Co exists owing to both the *quasiequilibrium* (see **Glossary**) processes of stellar nucleosynthesis and to the *s process* (see **Glossary**). In the former it is synthesized during explosive silicon burning and during the alpha-rich freezeout, not as itself but as a radioactive progenitor, ^{59}Cu, having more nearly equal numbers of protons and neutrons. Unfortunately, the ^{59}Cu decay has no observable effects that have yet been measurable; but the decay of its daughter, ^{59}Ni (halflife 75 000 yr), is associated with X-rays of 6.93 keV that will in the future be detectable in young supernovae and in regions of active star formation. The ^{59}Cu decay occurs during its ejection from supernovae. Of the two types of quasiequilibrium process, the ^{59}Cu/^{56}Ni ratio created is very much larger in the *alpha-rich freezeout* (see **Glossary**). During silicon burning the excess neutrons just do not effectively populate ^{59}Co in the competition of that quasiequilibrium. This means that the ^{59}Cu/^{56}Ni ratio ejected from Type II core-collapse supernovae depends in theory upon the relative masses ejected from these two types of quasiequilibrium processes. Those relative masses depend upon the mass cut dividing material that gets out of the supernova core from material that falls back onto the remnant. The greater the mass that falls back, the greater is the mass cut, and the smaller is the ratio of the ^{59}Cu/^{56}Ni ejected because the portion rich in ^{59}Cu has fallen back.

In massive stars a comparable amount of ^{59}Co is also synthesized as itself in the s process, as it occurs in the thermonuclear shells of pre-core-collapse presupernova stars. The weak s process occurs in helium burning as it establishes the large (C,O) core of the presupernova star, and also during the C burning in the presupernova structure.

Type I supernovae (the explosions of white dwarfs) also create both ^{59}Co and ^{56}Fe; and they contribute a significant number, perhaps even most, of these nuclei and all iron-group nuclei. Calculations of differing physical models (deflagration models or delayed detonation models) of these explosions produce ^{59}Co/^{56}Fe ratios varying between 1/600 and 1/750, somewhat smaller than the solar ratio. The hope remains to use the solar ^{59}Co/^{56}Fe ratio to determine the correct mass cuts within Type II supernovae, the true explosion model for Type I supernovae, and the relative frequencies of these two classes with the aid of the observed abundances of the iron-group isotopes.

Astronomical measurements

The strengths of the spectral lines of the cobalt atom and ion are measurable in composite spectra from stars, where iron is also observable and is usually taken as a standard measure of the abundance of heavy elements within stars. Observations show that the abundance ratio Co/Fe has, through most of galactic history, remained constant, even while each has increased in its proportion to H. A puzzle exists only in the most metal-poor stars, where stunning recent observations reveal a Co/Fe ratio that is almost five times greater than solar when Fe/H is near 1/10 000th of that in the Sun, and that ratio falls to solar by the galactic epoch when Fe/H has climbed to 1/500th of that in the Sun. The ratio Co/Fe then remains at its solar value during subsequent *chemical evolution of the Galaxy* (see **Glossary**). The first Fe produced was surely from massive supernovae because they evolved fastest after the first stars were born. But why they ejected larger Co/Fe ratios is unexplained today, although it surely reveals much about the detailed structure of the first supernovae. Those oldest stars formed almost 10 billion years ago, whereas the Sun's birth 4.6 billion years ago was from interstellar gas that had by that time settled to the solar Co/Fe ratio. The early values seem to indicate the dominance of the alpha-rich freezeout, which produces larger Co/Fe ratios. The constancy of the Co/Fe ratio during the mature growth of galactic metallicity represents a compromise utilizing the full spectrum of both massive Type II supernovae and Type Ia supernovae.

X-ray-line astronomy

Core-collapse supernovae (Type II) eject between 10^{-4} and 10^{-3} solar masses of radioactive ^{59}Ni (halflife 75 000 yr), which is the parent of ^{59}Co. About 30% of these decays are accompanied by a 6.93-keV K-shell X-ray which X-ray telescopes will, in the future, be able to detect both from young supernova remnants and from regions of active star formation, wherein many unobserved supernovae may have occurred during the past 100 000 yr. Important consequences for human understanding of supernovae can result. Type Ia supernova explosions are even more prolific in this regard; but they are more infrequent and usually farther away, making their detection less likely.

^{60}**Co** | $Z = 27$, $N = 33$ $t_{1/2} = 5.27$ yr
Spin, parity: $J, \pi = 5+$;   **Mass excess:** $\Delta = -61.645$ MeV

Gamma-ray lines from ^{60}Co decay are expected to be significant for astronomy not only by virtue of ^{60}Co being the daughter of the ^{60}Fe decay but also by the direct production of ^{60}Co itself in supernovae. Following either mode of production, the large angular momentum of the ^{60}Co nucleus causes its decay to be to excited states of ^{60}Ni, which then emit the sought-for gamma-ray lines.

Astronomical measurement

During every million-year period the Milky Way stars are believed to have produced some 30 000 supernova explosions, explosions creating some ^{60}Co nuclei in close association with their ^{60}Fe production. The yield of ^{60}Co is about twice that of ^{60}Fe, but they are detectable in differing situations owing to their very different halflives. The first discussion of the astrophysical importance of ^{60}Fe for gamma-ray astronomy (in 1971) advanced the idea that the associated ^{60}Fe radioactivity within interstellar gas could be detected by telescopes designed to record the characteristic gamma rays that are emitted following the decays of ^{60}Fe nuclei. About 10^{-5} solar masses of both ^{60}Fe and of ^{60}Co (about 3.4 Earth masses of each) are expected to be ejected from each individual supernova. The ^{60}Co gamma-ray line from a young individual supernova remnant can be detected for up to a decade after its explosion because its shorter halflife gives a much larger emission rate; but the ^{60}Fe with a million-year halflife must be seen as the sum of all supernovae in the field of view during the past million years. The specific gamma-ray energies that identify the ^{60}Co nucleus have values 1.17 MeV and 1.33 MeV, the same that the ^{60}Fe decay will produce by virtue of its first decaying to ^{60}Co, which then decays to ^{60}Ni. Detecting these gamma-ray lines has been attempted unsuccessfully by NASA's *Compton Gamma Ray Observatory*. The next-generation gamma-ray-line telescope, one capable of a tenfold improvement in sensitivity, will achieve powerful astronomical results from these gamma-ray lines if a supernova occurs within our Galaxy. The first chance will follow the launch in 2002 of the European *INTEGRAL* mission.

Nucleosynthesis origin

In an average mix of interstellar atoms, ^{60}Co and ^{60}Fe are deposited by the ejecta from dying stars, within which they were created. In those ejecta, ^{60}Co exists primarily owing to neutron capture within those stars by the lighter stable ^{59}Co isotope. Its parent ^{59}Co already existed in the stellar matter, not only by virtue of its initial composition but even overabundant because it had been enriched by earlier s-process stellar nucleosynthesis derived from isotopes of iron. The masses of ^{60}Co and ^{60}Fe ejected are comparable and come from the same s-process zones, although ^{60}Co can be made by a small neutron flux whereas ^{60}Fe requires one large enough to penetrate through radioactive ^{59}Fe. Because ^{60}Fe can not be detected from individual supernovae owing to its slow decay rate, the best measure of ^{60}Fe yield will be through the detection of young ^{60}Co.

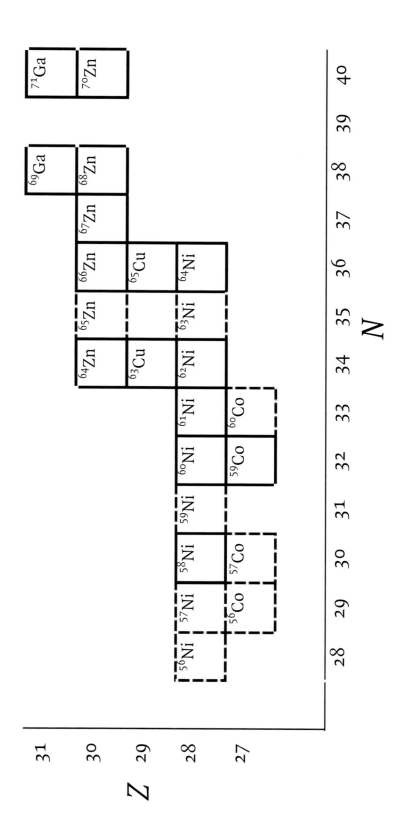

28 | Nickel (Ni)

All nickel nuclei have charge $+28$ electronic units, so that 28 electrons orbit the nucleus of the neutral atom. Its electronic configuration is $1s^2 2s^2 2p^6 3s^2 3p^6 3d^8 4s^2$. This can be abbreviated as an inner core of inert argon (a noble gas) plus ten more electrons: $(Ar)3d^8 4s^2$, which locates it in Group VIII of the periodic table. Like iron, nickel has two oxidation states designated by valence $+3$ or $+2$, but $+2$ is almost all that one finds in practice. Its only oxide is thus NiO. Nickel melts at 1453 °C and boils at 2732 °C.

A reddish brown mineral containing Ni was known to the glassmakers of the ancient world for its ability to make green glass, just as similar cobalt can turn glass blue. Miners in the 1700s called it *kupfernickel*, or "false copper" since it accompanies copper ores. In 1751 Cronstedt obtained a white metal from a new mineral named *niccolite*, which gave Ni its name. The primary commercial ore is the mineral *pentlandite* $(NiS(FeS)_2)$, mostly from Ontario. Nickel is not as abundant on the Earth's surface as in the universe, because the bulk of terrestrial Ni settled early into our molten metal core.

The purified metal is hard but malleable and ductile. Its silvery white appearance can be polished to a lustrous finish, and provides about 1/4th of the American 5-cent piece. Its corrosion resistance makes it good for coinage. The ancient Chinese produced an alloy called "Paktong" by smelting ores containing copper, nickel, and zinc that is now known as nickel silver and is the base for high-grade silver-plated ware.

From the point of view of physics, nickel is not nearly as good a conductor as copper, but it is important for its magnetic properties. Iron can be magnetically "hardened" by alloying it with nickel and cobalt. In this form it finds many uses in permanent magnets. Nickel has many other interesting properties when alloyed with iron or copper, and these contribute to a wide number of technological uses.

Astronomical observations of Ni in stars are not as copious as those of Fe; but the results are interesting. Nickel spectral lines show that the abundance ratio Ni/Fe has remained almost constant, even while each has increased in its ratio to H during the metal enrichment of galactic history. Although the spectral lines do not require it, this observed Ni abundance is dominated by the ^{58,60}Ni isotopes. In detail the ratio Ni/Fe seems to be a factor 2–3 greater than solar in the most metal-poor stars, then evolve with time to a minimum 20% less than solar, then increase very gradually up to the present. This reflects the transition from Type II supernova domination of Ni nucleosynthesis in the early Galaxy to Type Ia domination later.

Nickel is an important element for nucleosynthesis. It possesses five stable isotopes (A = 58, 60, 61, 62 and 64), each of which has interesting consequences for understanding the origin of the abundances. One important radioactive isotope, ^{59}Ni (halflife 75 000 yr), is produced abundantly in supernovae as the parent of stable cobalt (see ^{59}Co, *Nucleosynthesis origin*). Its radioactive decay will be detectable with X-ray telescopes (see ^{59}Co, *X-ray-line astronomy*) and provide unique information about supernovae and their rate of occurrence in star-forming regions.

Natural isotopes of nickel and their solar abundances

A	Solar percent	Solar abundance per 10^6 Si atoms
56	radioactive: gamma-ray emitter in supernovae	
57	radioactive: gamma-ray emitter in supernovae	
58	68.3	33 700
60	26.1	12 900
61	1.13	557
62	3.59	1770
64	0.91	449

^{56}Ni | $Z = 28$, $N = 28$ $\qquad t_{1/2} = 6.077$ days
Spin, parity: $J, \pi = 0 +$;
Mass excess: $\Delta = -53.90$ MeV; $\qquad S_n = 16.64$ MeV

Abundance

^{56}Ni has not yet been seen in nature; but its presence in supernovae has been indirectly confirmed by observations of the gamma-ray lines emitted from supernova 1987A following the radioactive decay there of its daughter, ^{56}Co. According to theory, ^{56}Co would not be present except for the abundant nucleosynthesis of ^{56}Ni in supernovae. Hopes are high for the detection of the gamma-ray lines following the decay of ^{56}Ni itself. These were predicted in 1969 and provided strong motivation for gamma-ray-line telescopes. Supernova 1987A did not expand rapidly enough to become transparent to the ^{56}Ni lines before the ^{56}Ni had all decayed; but Type Ia supernovae expand so much faster and make so much ^{56}Ni that it is a prime target for the future of gamma-ray astronomy. Measurements can delineate the structure of the Type Ia explosion mechanism.

When supernova 1987A occurred in February 1987, gamma-ray telescopes, which must fly above the Earth's atmosphere because the gamma rays cannot penetrate to the ground, successfully detected the 847-keV and 1264-keV gamma rays that had been predicted in 1969 to follow the radioactive decay there of the daughter of ^{56}Ni, ^{56}Co. From these and the great heating that they produced, it is known that supernova 1987A ejected 0.075 solar masses of radioactive ^{56}Ni. This amounts to a mass of radioactive ^{56}Ni that is 20 000 times the mass of the Earth! This was the first time in history when the radioactivity produced in a specific stellar explosion had been measured. After the explosions the light curves of many supernovae decline at the same rate as the declining abundance of ^{56}Co (halflife 77 d) because the heating of the remnant by that radioactive decay provides the source for the optical and infrared photons that are emitted. This radioactive ^{56}Ni $\rightarrow$ ^{56}Co $\rightarrow$ ^{56}Fe light source replaced the californium hypothesis of the 1950s.

Nucleosynthesis origin

See ^{56}**Fe** (to which ^{56}Ni decays) for discussion of the nucleosynthesis of ^{56}Ni by equilibrium and quasiequilibrium processes during supernova nuclear burning. Roughly 0.5 to 0.8 solar masses of ^{56}Ni is created by the thermonuclear explosions of white dwarfs (Type Ia supernovae). Only about 1/10th as much is created by each core-collapse supernova (Type II), but they are about 4 to 5 times more frequent than Type Ia explosions. This means that about two-thirds of the galactic total of ^{56}Ni is synthesized by Ia explosions and about one-third by core-collapse Type II.

^{57}**Ni** | $Z = 28, N = 29$ $t_{1/2} = 35.6$ hours
Spin, parity: $J, \pi = 3/2-$;
Mass excess: $\Delta = -56.076$ MeV; $S_n = 10.25$ MeV

Abundance

^{57}Ni has not yet been seen in nature; but its presence in supernovae has been indirectly confirmed by observations of the gamma-ray lines emitted from supernova 1987A following the radioactive decay there of its daughter, ^{57}Co. According to theory, ^{57}Co would not be present except for the abundant nucleosynthesis of ^{57}Ni in supernovae.

NASA's *Compton Gamma Ray Observatory*, the second in NASA's sequence of great space observatories, was motivated in part by the hope of gamma-ray-line astronomy. The writer championed that astronomy and its potential for measuring nucleosynthesis theory, and became Co-Investigator on that mission. It was disappointing that its launch in April 1991, more than four years after supernova 1987A, came too late for detectable levels of the 111-day-halflife ^{56}Co to remain within its ejecta. But it was still possible to detect the less abundant but longer lived ^{57}Co daughter of ^{57}Ni by its 122-keV gamma-ray line, making it the second radioactive nucleus to be detected from a specific supernova. The ratio of ^{57}Co to ^{56}Co in the supernova 1987A ejecta was about 1.5 times the known ratio of their daughters, ^{57}Fe/^{56}Fe, on Earth. This confirmed a prediction that ^{57}Fe is also synthesized as a detectable radioactive parent.

An unexpected consequence of this detection was its demonstration that the mass of ^{57}Ni initially synthesized was not great enough (by a factor near 3) to account for the total radiation emerging from supernova 1987A. It had been thought that the energy of that decay would provide the heat needed to maintain the total light output of the remnant after the faster decay rate of ^{56}Co had declined to a level too small to accomplish that heating. But it was not so. The discoverers therefore advanced the idea that it was delayed recombination of free electrons that was providing the power. These free electrons had been created by the previous radioactivity rather than by the current radioactivity.

Nucleosynthesis origin

See ^{57}**Fe** for a discussion of the nucleosynthesis of ^{57}Ni by equilibrium and quasiequilibrium processes during supernova nuclear burning. ^{57}Ni and ^{56}Ni are synthesized together in both explosive *silicon burning* (see **Glossary**) and by the *alpha-rich freezeout* (see **Glossary**). Roughly 0.01–0.02 solar masses of ^{57}Ni is created by the thermonuclear explosions of white dwarfs (Type Ia supernovae). Only about 1/10th as much is created by each core-collapse supernova (Type II), but they are about 4 to 5 times more frequent than Type Ia explosions. This means that about two-thirds of the galactic total of ^{57}Ni is synthesized by Ia explosions and about one-third by core-collapse Type II, just as for ^{56}Ni.

---

^{58}**Ni** | $Z = 28, N = 30$
Spin, parity: $J, \pi = 0+$;
Mass excess: $\Delta = -60.223$ MeV; $\qquad S_n = 12.20$ MeV

Nickel is the first even-Z element whose most abundant stable isotope, ^{58}Ni, is synthesized as a neutron-rich nucleus.

Abundance

From the isotopic decomposition of normal nickel one finds that the mass-58 isotope, ^{58}Ni, is the most abundant of all nickel isotopes; 68.3% of all Ni. Using the total abundance of elemental Ni $= 4.93 \times 10^4$ per million silicon atoms in solar-system matter, this isotope has

solar abundance of ^{58}Ni $= 3.37 \times 10^4$ per million silicon atoms.

That is, it is 3.37% as abundant as the element silicon, revealing ^{58}Ni as one of the abundant nuclear species. It is 24th most abundant of all nuclei.

Nucleosynthesis origin

Along with ^{56}Ni and ^{57}Ni, which are radioactive sources for ^{56}Fe, and ^{57}Fe, respectively, stable ^{58}Ni is also synthesized by equilibrium and quasiequilibrium processes during supernova nuclear burning. They are synthesized together in both explosive silicon burning and by the alpha-rich freezeout. ^{58}Ni is anticorrelated with ^{54}Fe in these explosive burnings in the sense that silicon burning makes little ^{58}Ni and much ^{54}Fe, whereas the capture of free alpha particles during the alpha-rich freezeout shifts the ^{54}Fe upward to ^{58}Ni by the capture of another alpha particle. This shift should not be thought of as a freezeout effect, however, but as a quasiequilibrium consequence of a larger alpha-particle density. Roughly 0.01–0.04 solar masses of ^{58}Ni is created by the thermonuclear explosions of white dwarfs (Type Ia supernovae). Only about 1/5th as much is created by the alpha-rich freezeout of each core-collapse supernova (Type II), but they are about 4 to 5 times more frequent than Type Ia explosions. This means that perhaps half of the galactic total of ^{58}Ni is synthesized by Ia explosions and about half by core-collapse Type II.

^{60}**Ni** | $Z = 28, N = 32$
Spin, parity: $J, \pi = 0+$;
Mass excess: $\Delta = -64.468$ MeV; $\qquad S_n = 11.38$ MeV

Abundance

From the isotopic decomposition of normal nickel one finds that the mass-60 isotope, ^{60}Ni, is the second most abundant of the nickel isotopes; 26.1% of all Ni. Using the total abundance of elemental Ni $= 4.93 \times 10^4$ per million silicon atoms in solar-system matter, this isotope has

solar abundance of ^{60}Ni $= 1.29 \times 10^4$ per million silicon atoms.

That is, it is 1.3% as abundant as the element silicon, revealing ^{60}Ni as one of the abundant nuclear species. It is 29th most abundant of all nuclei.

Nucleosynthesis origin

Along with ^{58}Ni, stable ^{60}Ni is also synthesized by equilibrium and quasiequilibrium processes during supernova nuclear burning. They are synthesized together in both explosive *silicon burning* (see **Glossary**) and by the *alpha-rich freezeout* (see **Glossary**). Roughly 0.004–0.01 solar masses of ^{60}Ni is created by the thermonuclear explosions of white dwarfs (Type Ia supernovae). Only about 1/5th as much is created as radioactive progenitors ^{60}Zn and ^{60}Cu by each core-collapse supernova (Type II), but they are about 4 to 5 times more frequent than Type Ia explosions. This means that roughly equal proportions of the galactic total of ^{60}Ni are synthesized by Ia explosions and by core-collapse Type II.

A significant portion of ^{60}Ni is also made as itself by the *s process* (see **Glossary**), as it occurs in the thermonuclear shells of pre-core-collapse stars. The weak s process occurs during helium burning that makes the large (C, O) core of the presupernova star, and also during the C burning in the presupernova. A portion of the ^{60}Ni is made there and survives the explosion.

Anomalous isotopic abundance

Small excesses of ^{60}Ni are detected in Fe-rich metal from meteorites. These have shown larger ^{60}Ni/Fe ratios in metal having larger Fe/Ni ratios, suggesting that this excess ^{60}Ni derives from radioactive ^{60}Fe that was still alive in the early solar system when these meteorite metals formed. (See ^{60}**Fe**, **Abundance**, Excess ^{60}Ni in *meteorites* for more on this extinct radioactivity.)

^{61}**Ni** | $Z = 28, N = 33$
Spin, parity: $J, \pi = 3/2-$;
Mass excess: $\Delta = -64.217$ MeV; $\qquad S_n = 7.82$ MeV

Abundance

From the isotopic decomposition of normal nickel one finds that the mass-61 isotope, ^{61}Ni, is the second least abundant of nickel isotopes; 1.13% of all Ni. Using the total abundance of elemental Ni $= 4.93 \times 10^4$ per million silicon atoms in solar-system matter, this isotope has

solar abundance of ^{61}Ni $= 557$ per million silicon atoms,

revealing ^{61}Ni to be a moderately abundant nuclear species, 49th most abundant of all.

Nucleosynthesis origin

Along with ^{58}Ni and ^{60}Ni, ^{61}Ni is also synthesized by quasiequilibrium processes during supernova nuclear burning. All three nuclei are synthesized together in both explosive *silicon burning* (see **Glossary**) and by *alpha-rich freezeout* (see **Glossary**). Roughly 10^{-4} solar masses of ^{61}Ni is created by the thermonuclear explosions of white dwarfs (Type Ia supernovae). About as much is created as radioactive progenitors ^{61}Zn and ^{61}Cu by each core-collapse supernova (Type II), but they are about 4 to 5 times more frequent than Type Ia explosions. This means that about one-fifth of the galactic total of ^{61}Ni is synthesized by Ia explosions and about four-fifths by core-collapse Type II. The alpha-rich freezeout is more effective than the equilibrium processes in this Type II production of ^{61}Ni.

A significant portion of ^{61}Ni is also made as itself by the s process (see **Glossary**), as it occurs in the thermonuclear shells of presupernova stars. The weak s process occurs during helium burning that makes the large (C, O) core of the presupernova star, and also during the C burning in the presupernova structure. Half of ^{61}Ni is made there and survives the explosion.

^{62}Ni | $Z = 28, N = 34$
Spin, parity: $J, \pi = 0+$;
Mass excess: $\Delta = -66.743$ MeV; $\qquad S_n = 10.60$ MeV

Abundance

From the isotopic decomposition of normal nickel one finds that the mass-62 isotope, ^{62}Ni, is the third most abundant of nickel isotopes; 3.59% of all Ni. Using the total abundance of elemental Ni $= 4.93 \times 10^4$ per million silicon atoms in solar-system matter, this isotope has

solar abundance of ^{62}Ni $= 1770$ per million silicon atoms,

revealing ^{62}Ni to be a moderately abundant nuclear species, 42nd most abundant f all.

Nucleosynthesis origin

Along with lighter stable Ni isotopes, ^{62}Ni is synthesized by equilibrium and quasiequilibrium processes during supernova nuclear burning. They are synthesized together in both explosive *silicon burning* (see **Glossary**) and by the *alpha-rich freezeout* (see **Glossary**). But the alpha-rich freezeout is much more effective than the equilibrium processes in this Type II production of ^{62}Ni. Roughly 0.002 solar masses of ^{62}Ni is created by the thermonuclear explosions of white dwarfs (Type Ia supernovae). About one-quarter as much is created as the radioactive progenitor ^{62}Zn by each core-collapse supernova (Type II), but they are about 4 to 5 times more frequent than Type Ia explosions. This means that comparable quantities of the galactic total of ^{62}Ni are synthesized by Ia explosions and by core-collapse Type II.

Much of ^{62}Ni is made as itself by the *s process* (see **Glossary**) in the thermonuclear shells of pre-core-collapse presupernova stars. The weak s process occurs in helium burning as it makes the large (C, O) core of the presupernova star, and also during the C burning in the presupernova structure. Half of ^{62}Ni is made as itself there and survives the explosion.

^{64}Ni | $Z = 28, N = 36$
Spin, parity: $J, \pi = 0+$;
Mass excess: $\Delta = -67.096$ MeV; $\qquad$ $S_n = 9.66$ MeV

This is but the second stable nucleus to have eight excess neutrons, which makes it an especially interesting abundance for nucleosynthesis theory. The first was ^{48}Ca, which is even more neutron-rich (percentagewise) than is ^{64}Ni. They share the magic number 28 for stability, as neutron number N for ^{48}Ca and as proton number Z for ^{64}Ni.

Abundance

From the isotopic decomposition of normal nickel one finds that the mass-64 isotope, ^{64}Ni, is the least abundant of nickel isotopes; 1.13% of all Ni. Using the total abundance of elemental Ni $= 4.93 \times 10^4$ per million silicon atoms in solar-system matter, this isotope has

$\qquad$ solar abundance of ^{64}Ni $= 449$ per million silicon atoms,

revealing ^{64}Ni to be but a moderately abundant nuclear species. Though only the 50th most abundant of all, it is of high diagnostic interest to nucleosynthesis theory.

Nucleosynthesis origin

^{64}Ni is synthesized by equilibrium and quasiequilibrium processes during supernova nuclear burning. Core-collapse Type II supernova explosions are calculated to be relatively ineffective in the synthesis of ^{64}Ni because most of the ejected matter is not anticipated to be sufficiently neutron-rich for production of such a neutron-rich isotope. Exploding white dwarfs (Type Ia) can be much more effective if their masses are near enough to the Chandrasekhar mass limit for electron capture to make the central gas quite neutron-rich – roughly one excess neutron per eight total nucleons in the bulk composition. This situation for ^{64}Ni resembles that of ^{50}Ti and ^{54}Cr, and probably accounts for the interstellar-dust carriers of the *cosmic chemical memory* (see **Glossary**) that produces anomalies in those isotopes within *calcium–aluminum-rich inclusions* (see **Glossary**).

$\qquad$ But much of ^{64}Ni is made by the s process, as it occurs in the thermonuclear shells of pre-core-collapse presupernova stars. The weak s process occurs during helium burning as it makes the large (C, O) core of the presupernova star, and also during the C burning in the presupernova structure. About half of ^{64}Ni is made there and survives the explosion.

29 | Copper (Cu)

All copper nuclei have charge +29 electronic units, so that 29 electrons orbit the nucleus of the neutral atom. Its electronic configuration can be abbreviated as an inner core of inert argon (a noble gas) plus eleven more electrons: $(Ar)3d^{10}4s^1$, which locates it at the top of Group IB of the periodic table (above silver and gold). Copper commonly has two oxidation states designated by valence +1 or +2, so that its oxides are *tenorite* (CuO) and *cuprite* (Cu_2O). The metal is reddish brown, malleable and ductile. It melts at 1083 °C and boils at 2567 °C. Copper is not very common in the Earth's crustal rocks, but it fortunately exists in large ore deposits of its oxides and sulfides (Northern US, Canada, the former Soviet Union, Chile and Zambia). Its name derives from Cyprus, however, which was the leading supplier to the Roman Empire.

Mankind has used copper throughout recorded history. People learned to refine it from copper ore near 5000 BC. It was used for pottery, tools, coins and jewelry. Because of its softness, Cu was not useful for weapons and tools until it was hardened by alloying it with other metals: brass is Cu and zinc; bronze is Cu and tin. Modern alloys are copper–aluminum and copper–nickel. Copper is one of the best conductors of electricity, so that it is widely used commercially for wiring. Its resistance to tarnishing by oxidation makes it a popular but expensive roofing material.

Copper is not a rare element. It ranks 25th in abundance among the 81 stable elements. Both isotopes are produced by a wide variety of nucleosynthesis processes and settings.

Natural isotopes of copper and their solar abundances

A	Solar percent	Solar abundance per 10^6 Si atoms
63	69.2	361
65	30.8	161

^{63}Cu | Z = 29, N = 34
Spin, parity: J, π = 3/2−;
Mass excess: Δ = −65.576 MeV; S_n = 10.84 MeV

Of the nuclei synthesized primarily in stars, ^{63}Cu has a smaller proton separation energy than any lighter stable nucleus. This results from the increasingly large repulsive Coulomb energy owing to its having so many protons.

Abundance

The mass-63 isotope is the more abundant (69.17%) of the two stable isotopes of copper. It has

solar abundance of ^{63}Cu = 361 per million silicon atoms.

Solar and meteoritic abundances are in good agreement.

Metal-poor stars   Spectral-line comparisons of Cu and of Fe in stars show that the Cu/Fe abundance ratio increased by a factor near 5 while the metallicity of the Galaxy grew during the first billion years. The cause lies either in the increasing yield of Cu from massive stars as they become more metal rich, certainly their s-process component, or in the onset of Type Ia delayed detonations after the earlier galactic epoch of Type II supernovae only.

Nucleosynthesis origin

^{63}Cu is synthesized in massive stars by the s process, the r process, and the quasiequilibrium *alpha-rich freezeout* (see **Glossary** for an account of all three processes). The r process probably contributes little, however. From massive supernovae about half of ^{63}Cu originated from the s process, where it is synthesized in its stable form in the He-burning shell, but as the radioactive ^{63}Ni (halflife 100 yr) progenitor in the C-burning shell. Unfortunately the ^{63}Ni decay has no astronomically observable effects that have yet been measurable. In the supernova explosive burning, ^{63}Cu progenitors exist owing to the *quasiequilibrium* (see **Glossary**) processes of stellar nucleosynthesis. It is synthesized during the alpha-rich freezeout not as itself but as a radioactive progenitor, ^{63}Zn, having more nearly equal numbers of protons and neutrons. The mass of ^{63}Zn ejected depends upon the mass cut dividing material that gets out of the supernova core from material that falls back onto the remnant. The greater the mass that falls back, the farther out is the mass cut, and the smaller is the ratio of the ^{63}Zn/ ^{63}Cu ejected. The Cu yield increases by a factor of 5 while the initial metallicity of the supernova increases owing to chemical enrichment of the Galaxy.

The relative importance of Type Ia supernovae for Cu nucleosynthesis remains uncertain because it depends on the nature of the dynamic model chosen to explain the Type Ia explosion, and it is still not known which type, if either, is correct. Deflagration models account for less ^{63}Zn than Type II supernovae; but delayed detonation models account for more.

^{65}Cu | $Z = 29$, $N = 36$
Spin, parity: J, $\pi = 3/2-$;
Mass excess: $\Delta = -67.260$ MeV; $\quad S_n = 9.91$ MeV

^{65}Cu is the lightest stable isotope having seven excess neutrons. This results from the increasingly large repulsive Coulomb energy owing to its having so many protons.

Abundance
The mass-65 isotope is the lesser abundant (30.83%) of the two stable isotopes of copper. It has

solar abundance of ^{65}Cu = 161 per million silicon atoms.

Solar and meteoritic abundances are in good agreement.

Metal-poor stars Spectral-line comparisons of Cu and of Fe in stars show that the Cu/Fe abundance ratio increased by a factor near 5 while the metallicity of the Galaxy grew during the first billion years. The cause lies either in the increasing yield of Cu from massive stars as they become more metal rich, certainly their s-process component, or in the onset of Type Ia delayed detonations after the earlier galactic epoch of Type II supernovae only.

Nucleosynthesis origin
^{65}Cu is synthesized in massive stars by the *s process*, the *r process*, and the quasiequilibrium *alpha-rich freezeout* (see **Glossary** for an account of all three processes). The r process probably contributes little, however. Most of ^{65}Cu comes from the s process, where it is synthesized as its stable form in massive stars in the He-burning shell, but some also emerges from the C-burning shell. In the supernova explosive burning, ^{65}Cu progenitors exist owing to the *quasiequilibrium* (see **Glossary**) processes of stellar nucleosynthesis. ^{65}Zn is synthesized during explosive silicon burning and during the alpha-rich freezeout. The mass of ^{65}Zn ejected depends upon the mass cut dividing material that gets out of the supernova core from material that falls back onto the remnant. The greater the mass that falls back, the farther out is the mass cut, and the smaller is the ratio of the ^{65}Zn/^{65}Cu ejected. The Cu yield increases by a factor of 5 while the initial metallicity of the supernova increases owing to chemical enrichment of the Galaxy.

The relative importance of Type Ia supernovae for Cu nucleosynthesis remains uncertain because it depends on the type of dynamic model chosen to explain the Type Ia explosion, and it is still not known which type, if either, is correct. Deflagration models account for less ^{65}Cu than Type II supernovae; but delayed detonation models account for more.

30 | Zinc (Zn)

All zinc nuclei have charge $+30$ electronic units, so that 30 electrons orbit the nucleus of the neutral atom. Its electronic configuration can be abbreviated as an inner core of inert argon (a noble gas) plus twelve more electrons: $(Ar)3d^{10}4s^2$, which locates it at the top of Group IIB metals of the periodic table (above cadmium and mercury). Zinc's oxidation state is designated by valence $+2$, so that its oxide is ZnO, a famous white compound used for white pigment in paints. Zinc dichromate ($ZnCr_2O_7$) and zinc chromate ($ZnCrO_4$) also find use as pigments, orange-red and brilliant yellow, respectively. The pure metal is hard and blue-white. Because it does not corrode easily, it is used to coat other metals (called "galvanized iron" when electroplated onto iron). Alloyed with copper it makes brass, a historically very significant alloy. Zinc metal melts at 420 °C and boils at 907 °C. Today zinc is produced commercially by electrolysis of zinc sulfate ($ZnSO_4$).

Zinc is not a rare element. It ranks 23rd in abundance among the 81 stable elements. Its five isotopes are produced by a wide variety of nucleosynthesis processes and settings. Two of its isotopes can be abundantly produced in rare Type Ia explosions having a large excess of neutrons. Many contributing processes mean that zinc is a rich candidate for isotopic anomalies, either by cosmic chemical memory or within presolar grains (see **Glossary**). But this requires a sensitivity not yet achieved except for a single "FUN" Ca–Al-rich inclusion (FUN – showing Fractionation and Unknown Nuclear anomalies).

Natural isotopes of zinc and their solar abundances

A	Solar percent	Solar abundance per 10^6 Si atoms
64	48.6	613
66	27.9	352
67	4.10	51.7
68	18.75	236
70	0.62	7.8

^{64}Zn $Z = 30, N = 34$
Spin, parity: $J, \pi = 0+$;
Mass excess: $\Delta = -66.000$ MeV; $\qquad S_n = 11.86$ MeV

> ^{64}*Zn is the first stable nucleus after ^{9}Be to have an alpha-particle binding energy less than 4 MeV. This is a manifestation of the general decline in binding energies above the iron abundance peak.*

Abundance

From the isotopic decomposition of normal zinc one finds that the mass-64 isotope, ^{64}Zn, is the most abundant of zinc isotopes; 48.6% of all Zn. Using the total abundance of elemental Zn = 1260 per million silicon atoms, making Zn the 23rd most abundant element in solar-system matter, this isotope has

solar abundance of ^{64}Zn = 613 per million silicon atoms,

revealing ^{64}Zn to be a moderately abundant nuclear species, 47th most abundant of all.

Nucleosynthesis origin

^{64}Zn is synthesized in stars by the *s process* and the quasiequilibrium *alpha-rich freezeout* (see **Glossary**). That ^{64}Zn from the s process is synthesized directly in massive stars in the He-burning shell. The AGB stars do not contribute significantly, although they produce the main s process at larger atomic weights. In the supernova explosive burning, ^{64}Zn progenitors exist owing to the *quasiequilibrium* (see **Glossary**) processes of stellar nucleosynthesis. It is synthesized during explosive silicon burning and during the alpha-rich freezeout not as itself but as a radioactive progenitor, ^{64}Ge, having more nearly equal numbers of protons and neutrons. The mass of ^{64}Ge ejected depends upon the mass cut dividing material that gets out of the supernova core from material that falls back onto the remnant. The greater the mass that falls back, the farther out is the mass cut, and the smaller is the ratio of the ^{64}Ge/^{64}Zn ejected. In the neutron-rich quasiequilibrium alpha-rich freezeout the ^{64}Zn yield is large; but the frequency of such natural events is unknown. The astrophysical uncertainties mean that the nucleosynthesis of this important nucleus can not yet be determined.

^{66}Zn | $Z = 30$, N $= 36$
Spin, parity: J, $\pi = $ 0+;
Mass excess: $\Delta = -68.897$ MeV; $\qquad S_n = 11.06$ MeV

*In neutron-rich quasiequilibria ^{66}Zn is a very abundant final product with special meanings. It is actually synthesized as ^{66}Ni in settings with about 15% more neutrons than protons. Its pairing with ^{48}Ca reveals a consequential difference between nuclear thermal (statistical) equilibrium and quasiequilibrium (see **Glossary**). In the former, the final ^{66}Zn will be too abundant for ^{48}Ca synthesis to be possible, whereas in the latter, ^{48}Ca becomes the more overabundant of the two.*

Abundance

From the isotopic decomposition of normal zinc one finds that the mass-66 isotope, ^{66}Zn, is the second most abundant of zinc isotopes; it is 27.9% of all Zn. Using the total abundance of elemental Zn = 1260 per million silicon atoms, making Zn the 23rd most abundant element in solar-system matter, this isotope has

solar abundance of ^{66}Zn = 352 per million silicon atoms,

revealing ^{66}Zn to be a moderately abundant nuclear species, 53rd most abundant of all.

Nucleosynthesis origin

^{66}Zn is synthesized in stars by the *s process* and the *quasiequilibrium alpha-rich freezeout* (see **Glossary**). Neutron-rich alpha-rich freezeout is especially effective and is related to ^{48}Ca nucleosynthesis, which requires a rare class of Type Ia supernovae that produce somewhat too many nuclei for nuclear statistical equilibrium to be achieved. In the supernova explosive burning, ^{66}Zn progenitors exist owing to the quasiequilibrium processes of stellar nucleosynthesis. In the neutron-rich quasiequilibrium alpha-rich freezeout the ^{66}Zn yield can be large; but the frequency of such natural events is unknown. That proportion of ^{66}Zn resulting from the s process is synthesized directly in massive stars in their He-burning shell. The AGB stars produce only 1% of ^{66}Zn, although they produce the main s process at larger atomic weights. The astrophysical uncertainties mean that the nucleosynthesis of this important nucleus can not yet be determined.

Anomalous isotopic abundance

The isotopes of Zn will, when more accurately measurable, provide a rich spectrum of isotopic deviations from the solar ratios in presolar materials and in solar materials prepared retaining the *cosmic chemical memory* (see **Glossary**) of presolar materials.

Calcium–aluminum-rich inclusions (CAIs) After removing the progressive mass-dependent fractionation that occurs in the measuring process and in the formation process for the samples, isotopic anomalies for ^{66}Zn are observed in certain types of CAIs (the "FUN" inclusions). Only one detection exists to date, an excess of 1.67 parts per thousand for ^{66}Zn in one FUN inclusion. A deficit of ^{70}Zn exists in that same CAI. Some form of cosmic chemical memory is probably involved.

^{67}Zn | $Z = 30$, $N = 37$
Spin, parity: $J, \pi = 5/2-$;
Mass excess: $\Delta = -67.877$ MeV; $\qquad S_n = 7.05$ MeV

> *^{67}Zn has the lowest neutron binding energy of any stable nucleus after ^{21}Ne. This is a manifestation of the general decline in binding energies above the iron abundance peak.*

Abundance

From the isotopic decomposition of normal zinc one finds that the mass-67 isotope, ^{67}Zn, is the fourth most abundant of zinc isotopes; 4.10% of all Zn. Using the total abundance of elemental Zn = 1260 per million silicon atoms, this isotope has

solar abundance of ^{67}Zn = 51.7 per million silicon atoms,

revealing ^{67}Zn to be a moderately abundant nuclear species, 67th most abundant of all.

Nucleosynthesis origin

^{67}Zn is synthesized in stars by the s process and, to some poorly known degree, by the quasiequilibrium alpha-rich freezeout. About half of ^{67}Zn from the s process is synthesized directly in massive stars in their He-burning shell. The AGB stars produce only 2% of ^{67}Zn, although they produce the main s process at larger atomic weights.

^{68}Zn | $Z = 30$, $N = 38$
Spin, parity: $J, \pi = 0+$;
Mass excess: $\Delta = -70.004$ MeV; $\qquad S_n = 10.20$ MeV

Abundance

From the isotopic decomposition of normal zinc one finds that the mass-68 isotope, ^{68}Zn, is the third most abundant of zinc isotopes; 18.75% of Zn. Using the total abundance of elemental Zn = 1260 per million silicon atoms, this isotope has

solar abundance of ^{68}Zn = 236 per million silicon atoms,

revealing ^{68}Zn to be a moderately abundant nuclear species, 57th most abundant of all.

Nucleosynthesis origin

^{68}Zn is synthesized in stars by the s process, to some poorly known degree by the quasiequilibrium alpha-rich freezeout, and, perhaps, by the r process. About 1/2 to 2/3 of ^{68}Zn from the s process is synthesized directly in massive stars in their He-burning shell. The AGB stars produce only 3% of ^{68}Zn, although they produce the main s process at larger atomic weights.

^{70}Zn | $Z = 30, N = 40$
Spin, parity: $J, \pi = 0+$;
Mass excess: $\Delta = -69.560$ MeV; $\qquad S_n = 9.22$ MeV

> ^{70}Zn is actually radioactive, with halflife greater than 5×10^{14} yr. In neutron-rich quasiequilibria (see **Glossary**) ^{70}Zn is a very abundant final product with special meanings. It is actually synthesized as ^{70}Ni in settings with about 17–20% more neutrons than protons. The synthesis of ^{66}Zn and ^{48}Ca requires neutron excesses less than 17% in order that the final abundance of ^{70}Zn not be too great.

Abundance

From the isotopic decomposition of normal zinc one finds that the mass-70 isotope, ^{70}Zn, is the least abundant of zinc isotopes; 0.62% of all Zn. Using the total abundance of elemental Zn = 1260 per million silicon atoms, this isotope has

solar abundance of ^{70}Zn = 7.8 per million silicon atoms,

revealing ^{70}Zn to be a moderately rare nuclear species, 83rd most abundant of all.

Nucleosynthesis origin

^{70}Zn may be synthesized in stars by the s process in sufficiently intense neutron fluxes to pass through ^{69}Zn (56 min), and, to some poorly known degree, by the quasiequilibrium alpha-rich freezeout having 17% or more excess neutrons, and, perhaps, by the r process if it contributes at such small atomic weights. The AGB stars do not contribute significantly, although they produce the main s process at larger atomic weights.

Anomalous isotopic abundance

Calcium–aluminum-rich inclusions (CAIs) After removing the progressive mass-dependent fractionation that occurs in the measuring process and in the formation process for the samples, isotopic anomalies for ^{70}Zn are observed in certain types of CAIs ("FUN" inclusions). Only one detection exists to date, a deficit of 2 parts per thousand for ^{70}Zn in one FUN inclusion. An excess of 1.7 parts per thousand for ^{66}Zn exists in that same CAI. Some form of *cosmic chemical memory* (see **Glossary**) is probably involved.

31 | Gallium (Ga)

All gallium nuclei have charge $+31$ electronic units, so that 31 electrons orbit the nucleus of the neutral atom. Its electronic configuration can be abbreviated as an inner core of inert argon (a noble gas) plus thirteen more electrons: $(Ar)3d^{10}4s^24p^1$, which locates it in Group IIIA of the periodic table. Other elements in this group are boron, aluminum, indium, and thallium. The existence of gallium was actually predicted in 1871 by Dmitri Mendeleyev on the basis of his new chart of the elements. He named the undiscovered element "ekaaluminum" because it should fill the space beneath Al in his chart. Chemically Ga is indeed most like Al, and is in fact mined as a byproduct of Al mining. Gallium has oxidation states designated by valence $+3$, so that its oxide is, like Al's, Ga_2O_3. The metal is blue-white, and so soft it can be cut with a knife. It melts at $29.8\,°C$ and boils at $2403\,°C$. Its name derives from the Latin *Gallia*, "France," however, where it was first isolated.

Gallium is not a rare element. It ranks 31st in abundance among the 81 stable elements. Its two isotopes are produced by a wide variety of nucleosynthesis processes and settings. It was rescued from obscurity in the 1970s when *gallium arsenide* (GaAs) light-emitting diodes, or LEDs, became widely used for electronic displays. The physics description is that the metal becomes a semiconductor when doped with arsenic (As).

One additional application thrills all who love isotopes in the universe. During the 1970s an appreciable fraction of gallium production was purchased for experiments in Germany/Italy and in the (then) Soviet Union to measure the rate of arrival of neutrinos from the center of the Sun, the so-called "solar-neutrino experiments" (see ^{71}**Ga**). The German/Italian experiment, directed by Till Kirsten from the Max Planck Institute for Nuclear Physics in Heidelberg, was named "GALLEX" for "Gallium Experiment" and used 30 thousand kilograms of Ga in the Gran Sasso tunnel near Rome. SAGE the "Soviet–American Gallium Experiment" in the Caucasus region used sixty thousand kilograms of gallium. These experiments found a deficiency of almost a factor of 2 in the expected number of neutrinos from the Sun. This was a key step to the discovery in 2001 that neutrinos oscillate into other *flavors* during their journey from Sun to Earth, forever changing mankind's view of neutrinos. "Flavor" is a technical physics term for classifying elementary particles in the grand-unified theory.

Natural isotopes of gallium and their solar abundances

A	Solar percent	Solar abundance per 10^6 Si atoms
69	60.1	22.7
71	39.9	15.1

^{69}Ga $\quad$ Z = 31, N = 38

Spin, parity: $J, \pi = 3/2-$;

Mass excess: $\Delta = -69.321$ MeV; $\qquad S_n = 10.32$ MeV

Abundance

The mass-69 isotope, ^{69}Ga, is the more abundant (60.11%) of the two stable isotopes of gallium. It has

solar abundance of ^{69}Ga $= 22.7$ per million silicon atoms.

This is comparable in abundance to the element boron, ranking 75th of 283 stable nuclei.

Nucleosynthesis origin

^{69}Ga is synthesized in stars by the s *process*, the r *process*, and the *quasiequilibrium alpha-rich freezeout* (see **Glossary** for an explanation of these terms). The r process may contribute up to one-third of ^{69}Ga, although it is still uncertain that the r process proper contributes in this mass range. Perhaps it should be called the neutron-rich quasiequilibrium alpha-rich freezeout. About one-half to two-thirds of ^{69}Ga comes from the s process in massive stars, where it is synthesized in the He-burning and C-burning shells. Only about 5% comes from the AGB stars that provide the main component of the s process.

^{71}Ga | $Z = 31, N = 40$
Spin, parity: $J, \pi = 3/2-$;
Mass excess: $\Delta = -70.135$ MeV; $\qquad S_n = 9.31$ MeV

The ^{71}Ga nucleus can absorb a neutrino, changing to radioactive ^{71}Ge (halflife 11.43 d), if the neutrino energy is greater than 0.233 MeV. This fact forms the basis for the solar neutrino experiment with gallium. This is done by counting the number of radioactive ^{71}Ge atoms produced in the tank. The exciting aspect of this low threshold energy is that it makes it possible to absorb the neutrinos created in the pp reaction ($H + H \rightarrow D + $ neutrino $+ e^+$) in the solar center, a reaction fundamental to nuclear fusion power for the Sun. Those pp neutrinos range up in energy to 0.42 MeV. The original solar neutrino experiment based on chlorine absorption is insensitive to pp neutrinos.

Abundance

The mass-71 isotope, ^{71}Ga, is the lesser abundant (39.89%) of the two stable isotopes of gallium. It has

solar abundance of ^{71}Ga $= 15.1$ per million silicon atoms.

This is comparable in abundance to the element boron, ranking 78th of 283 stable nuclei.

Nucleosynthesis origin

^{71}Ga is synthesized in stars by the s process, the r process, and the quasiequilibrium alpha-rich freezeout. The r process may contribute up to one-third of ^{71}Ga, although it is still uncertain that the r process proper contributes in this mass range. Perhaps it should be called the neutron-rich quasiequilibrium alpha-rich freezeout. About two-thirds of ^{71}Ga comes from the s process in massive stars, where it is synthesized in the He-burning and C-burning shells. Only about 5% comes from the AGB stars that provide the main component of the s process.

Glossary

This Glossary amounts to small essays on topics that are central to the subject of isotopes as they appear in the cosmos. Italicised words indicate other entries to be found here. The same format within the text of the chapters also refers to Glossary entries.

Alpha-rich freezeout

This description refers to the inner shells of a core-collapse *supernova*. When the neutron star forms at its very core, it presents a hard dense obstacle for infalling matter to crash into. This sends a strong shock wave through that infalling matter. After being shocked the matter is very hot, perhaps 6–10 billion degrees, so that nuclei are decomposed into free nucleons and tightly bound helium nuclei (which are called "alpha particles"). This is the hottest matter to be ejected into space since the *Big-Bang* start to the universe.

 Most of this matter remains on the neutron star, bound to it by strong gravity, but some of it (outside what is called the "mass cut") is ejected from the star as the innermost portions of the supernova. As it expands it cools rapidly owing to the great work done by the high pressure; and this cooling allows neutrons and protons to recombine into more alpha particles. As the expansion and cooling continue, alpha particles are reassembled into heavier nuclei, beginning with the same triple-alpha reaction that begins *helium burning* in stars. When this occurs, the abundances of free neutrons and protons fall to a very low level because heavy nuclei are, unlike alpha particles, able to absorb and retain free nucleons into bound states of new isotopes. The temperature remains quite high, about 4–5 billion degrees, as the new nuclei are being assembled from alpha particles; moreover, that high temperature maintains the small free density of nucleons by photoejection processes from the nuclei. This steady state between absorption and reemission of nucleons defines what is called a "nuclear *quasiequilibrium*" because it has many resemblances to complete nuclear *thermal equilibrium*, wherein all absorption and emission processes occur at equal, balancing rates. New nuclei of ^{12}C assembled by the triple-alpha reaction capture many more alpha particles and nucleons as they become heavier. This growth stops with isotopes of nickel, at which time the gas is composed primarily of nickel and alpha particles, with many other coexisting species of lesser abundance.

 As expansion and cooling proceeds further, if all of the alpha particles can be assembled into heavier nuclei before the temperature falls below 3.5 billion degrees, the matter approaches thermal and nuclear statistical equilibrium; however, if the density is not high enough for all alpha particles to reassemble, the expansion cools below 3.5 billion degrees with large numbers of excess alpha particles remaining in the gas. That state is called the "alpha-rich freezeout." The "freezeout" refers to the temperature falling below 3.5 billion degrees, which is not hot enough to reassemble the remainder in the seconds available before the gas is much too cool for nuclear reactions. So the *alpha-rich freezeout* is hot expanding material that cannot fuse its alpha particles into heavy nuclei in the short time available for the expansion. It "freezes" with alphas left over.

The alpha-rich freezeout is a very important part of the nucleosynthesis of elements in supernova explosions. Because most of the assembled nuclei are maintained in radioactive isotopes of the element Ni, their radioactive transmutation into isotopes of iron heats the gas, as in a nuclear reactor. Eventually some of the gamma rays that accomplish this escape the thinning supernova remnant, giving substance to the astronomy of gamma-ray lines in young supernovae. The prediction of this in 1969 was selected as one of the most significant astrophysics papers of the century and reprinted in the Centennial Volume of *Astrophysical Journal*. These gamma-ray lines were seen for the first time in supernova 1987A, which ejected such alpha-rich freezeout material. The daughter products become Fe, "radiogenic Fe," so called because it is the decay product of radioactive Ni. In supernova 1987A, a mass equal to 7% of the Sun's mass was ejected as pure radioactive isotopes of nickel, prompting the description of that exciting event as "the largest nuclear accident of all time." Issues surrounding the nucleosynthesis of Fe in this way provided one of the most important building blocks of fundamental nucleosynthesis theory. (See **26 Iron (Fe)** for more details of this – or the commentary in the Centennial Volume of *Astrophysical Journal*.) The alpha-rich freezeout also produces much radioactive ^{44}Ti, about 1/10 000th of the Sun's mass of that isotope alone; and its gamma-ray emission lines have also been detected in the three-centuries-old supernova remnnant Cassiopeia A because it lives long enough to still be partly alive today.

Abundance per nucleon, Y

Several different abundance scales are used in astrophysics. One of these compares the number of atoms of any given isotope $N(^ZA)$ with the total number of nucleons contained in all nuclei present, usually in the solar mixture of the elements. That ratio is designated by $Y(^ZA)$. It is the "*abundance per nucleon*" of that isotope. One imagines for this purpose a representative sample of the Sun and calculates the ratio $Y(^ZA)$ of the numbers of nuclei of each isotope in that sample to the total number of nucleons in the sample. The latter is numerically equal to the total mass of the sample when expressed in atomic mass units. The total number of nucleons is dominated by the two light gases, H and He, which together constitute about 98% of the nucleons in solar gas. Thus Y (^{Z}A) will be a small number for a single isotope of a heavy element. Its smallness renders it somewhat remote.

To converse with more reasonable numbers it is common to instead express the abundance of an isotope N (^{Z}A) as its number per million atoms of silicon in the solar system. One imagines for this purpose a sample of the Sun containing in its total mix one million silicon atoms, and counts the numbers of nuclei of each isotope.

AGB stars

AGB stands for "asymptotic giant branch" and refers to red-giant stars that are in a later stage of evolution than the common red giants that are fusing helium to carbon in their cores. The AGB stars have already done that. They now have a core of C and O as a result of the exhaustion of He. This core is destined to grow by the fusion of He into C in a burning shell atop that core, which will remain a degenerate white-dwarf star after the remainder of the mass has been lost as a wind from the star. This configuration evolves upward in the Hertzsprung–Russell diagram that plots stellar luminosities against their surface temperatures. The track of the star in the Hertzsprung–Russell diagram parallels the track it had made earlier when ascending the red-giant branch for the first time. Hence it is called the asymptotic giant branch. Stars having initial masses between 2 and 8 times the mass of the Sun do this. In the end they eject the final mass remaining above the C and O core as a glowing round nebula known for historic reasons as "planetary nebula." These are favorites of stellar photographers. The inert core can no longer sustain nuclear fusion and cools off in a terminal state known as a white dwarf, a dead star.

Although the AGB stars live but 2 to 10 million years, depending on their mass, they play a very important role in isotopic astronomy. Their complicated structures support several nuclear burning situations that alter the isotopic composition of the envelope soon to be ejected. During their earlier evolution as ordinary red giants, the convection of envelope gases between the surface and the base of the convection zone dredged up matter to the surface; matter that had been altered by nuclear reactions with hot hydrogen. This enriched the envelope in ^{13}C and in ^{14}N, and it partly destroyed isotopes of Li, Be, and B. But the most exciting action begins when the AGB phase arrives. The giant star develops two burning shells, a lower one in which He fuses to C and O, depositing the C and O ashes onto the growing white-dwarf core, and an upper one in which H fuses to He via the CN cycle. These burning shells, surrounded by the outer gasecus envelope, are alternately ignited and extinguished out of phase with each other. When the He burns, it does so with a runaway "flash," expanding the star and quenching the H burning in the upper shell. The surface convection gets so deep during this expansion that it reaches from the envelope surface downward to the top of the extinguished H-burning shell. Matter in this convective cycle undergoes further H reactions called "hot-bottom burning." After a lengthy period of this burning the extinguished He shell reignites in another near-explosive flash. This convects hot matter within the "helium flash" in a boiling motion that grows to the top of the He shell and on into the lower part of the envelope, where the subsequent deepening surface convection will dredge it up to the surface. This so-called third dredgeup brings many new nuclear products to the surface. The double-shell action constitutes a heat engine that transports new nuclei to the surface via the alternately growing and contracting convection zones. Meanwhile the surface matter is being steadily lost from the star as a stellar wind that carries the new nuclear products away to join the interstellar gas and dust.

New carbon nuclei fused from He in the He shell make the surface increasingly C-rich through repeated dredgeups, and eventually the AGB star becomes recognized as a "carbon star" when the surface C abundance exceeds that of O. That transition is very noticeable owing to the abrupt change of the optical spectra of atoms and molecules in its surface. The hot-bottom burning can turn much of the new C into N, thereby enriching the interstellar gas in both C and N. Roughly half of the interstellar supply of both elements results from such stars rather than from *supernovae*. Furthermore, free neutrons are liberated in the *He-burning* shell, mostly from the $^{13}C + {}^4He \rightarrow {}^{16}O + n$ reaction, where the ^{13}C is itself made cyclically when new C reacts with hot protons. This is the primary site for *s-process* nucleosynthesis by neutron capture of elements heavier than Fe. New ^7Li is also made when the ^3He produced by the pre-AGB evolution encounters hot He during the hot-bottom burning. The AGB star is a complicated object.

To top off the action, dust grains condense within the cooling surface gas as it leaves the star as a dense wind. They are sprayed into the interstellar matter. The "mainstream SiC grains" have this origin. They record the unusual isotopes that were present in the surface gas when it cooled. These and other *presolar grains* are found within *meteorites*. They have inspired a new astronomy of AGB stars based on the isotopic analysis of the grains. They record isotopic information about the star with far greater precision than can be obtained by other astronomical means.

Big Bang

The Big Bang is a paradigm for understanding the universe. Almost all astrophysicists adhere to it as the "best bet" for how our universe began and evolved. Its name was coined by British cosmologist Fred Hoyle to make fun of the idea; but the idea has passed every test that has been asked of it and is today the accepted term for the very early state of the universe. "Big Bang" calls to mind an explosion; but it is not an explosion in the sense one normally takes that word. There exists no center for the explosion from which all matter emanates, because the standard Big-Bang picture is of a universe that is the same everywhere. There exists no vacuum for matter to explode into. The only similarity with an explosion lies in the increasingly high values for the density of matter, for the temperature, and for the pressure that must have existed as one regards the universe at earlier and earlier times. According to Einstein's much confirmed theory of gravity, a dense hot state everywhere is not stable, but will expand. The same expansion occurs at all neighboring points, so it naively appears that the expansion has no place to go. More careful interpretation shows that space itself expands in every direction, so the distance between two parcels of matter increases with time. The density therefore declines everywhere, and the temperature falls as it does so. The "observable universe" corresponds to matter in a sufficiently small volume that light signals have had time to move from one parcel of matter to another during the age of the universe. The "horizon" separates that observable universe from the unobservable matter lying exterior to it. The term Big Bang is usually not used to describe the universe today but rather the early violent state that has expanded to the present conditions. Steven Weinberg's brilliant book regarded it as "the first three minutes," the title he gave to his book; and that duration makes a useful though arbitrary sense of the time involved in what one usually means by the Big Bang. Excellent books exist that describe this theory at differing levels of sophistication and that also describe the many observations of the universe that agree with this interpretation.

The importance of the Big Bang for the isotopes derives from its natural explanation of the abundances of the isotopes of hydrogen and helium. They were created during the Big Bang and inherited by the later universe.

Four major landmark discoveries, coupled with Einstein's relativistic theory of gravity, have guided mankind to this picture. They are the four great facts that are explained by the Big Bang but that are decidedly difficult for other interpretations of the universe. In chronological order, these decisive discoveries are:

1 About 1925 a young Englishwoman, Cecilia Payne, demonstrated that hydrogen atoms constituted 90% of the atoms in the universe. This great shock was not at first believed by bewildered astronomers who had regarded hydrogen as a rarity, as it certainly is on Earth, and as it appeared naively to be in the Sun. A similar shock followed when it was learned that most of the other atoms were helium, a very rare element on Earth, and apparently also in the Sun. This huge error hid the fact that the Sun is actually 99% H and He. That 99% of the atoms of the universe are H and He constitutes the first decisive fact. This fact is accommodated naturally by Big-Bang cosmology, because that is what the expansion of the first three minutes leaves as residue. Prior to the Big-Bang picture the dominance of H and He was regarded by most as a "just so" story; or the work of the Creator.

2 In 1929 Edwin Hubble published the famous law that today bears his name; namely, all galaxies recede from each other, and they do so in such a way that the greater the distance between a pair, the greater their velocity of separation. Hubble observed this from the Earth's point of view; namely, the more distant

a galaxy from Earth, the faster it recedes from Earth. The expansion of a homogeneous unending universe, the gravitational consequence of the Big-Bang picture, naturally provides a symmetric situation for all galaxies and for all observors within them.

3   In 1965 Arno Penzias and Robert Wilson discovered that our local universe is permeated by microwave radiation, exactly as if the universe were inside a microwave oven generating a temperature of 2.7 K for the microwave photons that fill it. Their distribution is exactly thermal, coinciding with that deduced by Max Planck. That radiation arrives equally from every direction. This is the thermal isotropic cosmic microwave background. It can occur naturally in the Big Bang, because the earlier much hotter photon gas is cooled by the stretching of its wavelengths as space expands. The universe of photons retains its thermal distribution as space expands, but all wavelengths increase by a common factor that increases as time progesses. It is a miracle of physics that occurs with disarming simplicity in mankind's picture based on the Big Bang. This temperature is not guaranteed to be the observed value, 2.7 K, but that result follows naturally from the observed ratio of photons to matter. One hopes that someday that ratio may also be predicted from theory.

4   Recently, as if to mark the beginning of the new millenium, it was shown that at much earlier times, as can be seen in very distant galaxies owing to the great travel times for light, that microwave radiation was indeed much hotter than today, just as the Big-Bang theory demands. This discovery was published in the final week of 1999 by three European astronomers. This observation was accomplished by measuring the temperature of isolated carbon atoms in distant galaxies (therefore long ago), as reflected by their distribution of fine-structure states. This discovery occurred after the Big Bang was already almost universally accepted by leading scientists; but it is important in its elimination of other pictures.

Although the Big-Bang picture has successfully interpreted many other observations, it is these four facts that pin the human race into a conceptual corner. It is these four facts that the Big Bang yields so naturally as to have the smell of scientific correctness, and that other theories must strive with severe difficulty to accommodate.

Calcium–aluminum-rich inclusions (CAIs)

These millimeter-sized stones that formed from interstellar gas and dust in the early solar system's planetary disk carry many anomalous isotopic ratios. These stones are found within the bulk of a *meteorite*, which may itself be thought of as a collection of rocky debris from the solar system's disk. Apparently these stones (CAIs) formed during the very earliest events in that disk. They have the oldest documented ages of any solar-system object, 4.56 billion years. Their name derives from the concentration of the elements Ca and Al into these stones (in the form of the oxides of these elements). They are Ca-and-Al-rich stones, included (inclusions) within a larger groundmass of the large meteoritic rock. Thus they are known as Ca–Al-rich inclusions, abbreviated by the acronym CAIs. To become rich in Ca and Al it appears that the precursor dust or rock to the CAIs themselves must have been heated to temperatures so high that they almost completely evaporated. Oxides of Al and Ca are among the last minerals to evaporate if such a setting occurs. The minerals found today in the CAIs seem to instead be condensates from a gas, probably a gas prepared locally by evaporation of dust in a dust-enhanced region, which then recondensed.

One of the famous aspects of the CAIs is that isotopic anomalies were discovered in their minerals. These anomalies are of two types: (1) intrinsic isotopic differences of the reservoir from which the CAIs formed, which show that reservoir to have differed isotopically from that of the bulk solar system – an early but isotopically different portion of it; (2) matter that appears to be of the solar-system reservoir but which has been systematically enhanced in the heaviest isotopes by a thermal chemical process that preferentially retains the heavier isotopes. The discovery three decades ago of excess ^{16}O isotope of oxygen in "spinels" (gem-like crystals of $MgAl_2O_4$) and other refractory minerals within CAIs set off a historic search for presolar matter of more highly anomalous constitution. These have been hard to find; and it now seems likely that the minerals of the CAIs are condensates from gaseous vapor. To account for the ^{16}O-richness that gas must have been ^{16}O-rich. That happened when pockets of matter having a high ratio of dust to gas were almost totally vaporized by heat, giving ^{16}O-rich gas from the initial collection of ^{16}O-rich dust (see ^{16}**O** for more details of that historic discovery). Some entire inclusions containing isotopically fractionated Mg and Si, however, and consisting of many different minerals, are uniformly anomalous in the isotopes ^{50}Ti and ^{48}Ca and other neutron-rich isotopes of the heavy elements. That subgroup of the CAIs is designated by the acronym FUN (for Fractionated and Unknown Nuclear anomalies). In these FUN CAIs the elements O, Mg, and Si are each progressively enriched (fractionated) in their heavier isotopes by an unknown process that depends on the atomic mass. Strangely, it is these fractionated inclusions that contain the large nuclear isotopic anomalies at ^{48}Ca and ^{50}Ti. These nuclear anomalies are apparently inherited from the initial materials from which these CAIs were assembled (see *Cosmic chemical memory*). It is not known why two groups of CAIs having such differing isotopic properties exist. The CAIs are thus viewed as some kind of "Rosetta Stones," capable of revealing important history of the isotopes and of the early solar system; but no consensus yet exists on exactly how they were formed or what their information conveys.

Carbon burning

Carbon burning is the name of the process by which carbon nuclei are fused by nuclear reactions into heavier elements. The need for this occurs naturally during stellar evolution when the helium-burning core of red-giant stars has completed its fusion of helium into carbon and oxygen. The core then ceases to provide enough nuclear power to keep the interior hot. Subsequent heating is accomplished instead by contraction to a higher density and this contraction ceases when the temperature becomes high enough for carbon burning to begin. Thermonuclear power from the fusion of carbon nuclei into Ne, Na, Mg, and Al heats the core thereafter for a moderately long time (until the carbon is exhausted). Unlike He burning, this fusion can utilize binary collisions between two ^{12}C nuclei. The primary initial nuclear reactions are:

$$^{12}\text{C} + {}^{12}\text{C} \rightarrow {}^{20}\text{Ne} + {}^{4}\text{He, and } {}^{12}\text{C} + {}^{12}\text{C} \rightarrow {}^{23}\text{Na} + {}^{1}\text{H,}$$

but a variety of secondary reactions make ^{24}Mg another abundant final product. This stage in a star's life is a final portion of its giant stage, which can be either red or blue depending on the star's mass.

"Explosive carbon burning" occurs in supernova explosions when the carbon in a shell surrounding the core of the star is suddenly heated to temperatures well in excess of its stable burning temperature and therefore burns quite rapidly (in seconds). In Type Ia supernovae this happens when a runaway nuclear burning of a carbon white dwarf causes the temperature to overshoot stable burning, leading to a *quasiequilibrium* that is quite close to *thermal equilibrium*; in Type II supernovae it occurs when the rebounding shock wave reaches the carbon shell of the presupernova star and heats it suddenly. This is nature's major source for ^{20}Ne, ^{24}Mg, ^{25}Mg, ^{26}Mg, ^{23}Na, and ^{27}Al.

Chemical evolution of the Galaxy

This term describes the process by which the new elements created in stars enrich the interstellar gas as the stellar material is ejected from the star into the gas. This is a monotonic process, because the Galaxy began with only hydrogen and helium, remnants of the light-element synthesis that occurred during the Big-Bang expansion at the beginning of the universe. Much later, perhaps one billion years later, that gas was cool and had sufficient time to contract to make large gravitationally bound clumps of gas. These turned into galaxies. When the first stars were born in the galaxies, they contained only H and He and traces of ^{7}Li. Therefore the first stars can manufacture only those nuclei that are naturally made from H and He during the thermonuclear evolution of the stars. These are called "primary nuclei."

The massive stars, those with masses between 12 and 60 times the mass of the Sun, are especially good thermonuclear nucleosynthesis machines. So when they explode as supernovae at the end of their lives (typically 10 million years) they eject large abundances of the primary nuclei into the interstellar gas. The concentrations of primary nuclei (*e.g.* ^{12}C, ^{16}O, ^{28}Si, ^{40}Ca, and ^{56}Fe just to sample a few) in the supernova ejecta are about 15 times greater than the concentrations today of those elements in the Sun. These ejecta are mixed into the interstellar gas, from which new stars continue to form. The gas from which these new stars form now has some abundance of the primary nuclei from the earlier supernovae admixed within it. The abundances of primary nuclei therefore grow with time, so that astronomers frequently select easily observable iron and use the Fe/H ratio observed in a star's spectrum as a marker for the time of its birth. These stars too undergo their natural evolution, live their natural lives. At their ends they also explode, ejecting not only more primary nuclei into the interstellar gas but also ejecting new isotopes that can be made only from the primary nuclei within the stars. These isotopes are called *secondary nuclei*. They cannot be made from H and He by stellar evolution, but require the initial presence in new stars of a concentration of primary nuclei from which seed they grow the secondary nuclei. Examples are ^{13}C, ^{17,18}O, ^{29}Si, to name but a few. The secondary nuclei are neutron-rich isotopes of common primary elements.

This process of star birth, evolution, ejection of new nuclei synthesized by nuclear reactions in the stars, mixing with interstellar gas, and formation of yet more new stars, continues up till the present day. This process is called the chemical evolution of galaxies. It is reflected by a continuing growth with time of the abundances of the elements in the interstellar gas, and therefore in the stars born later rather than earlier. Thus chemical evolution of galaxies might better be called "abundance evolution of galaxies," because it is not really chemistry that evolves but the abundances of the chemical elements. Nonetheless, for traditional reasons that are not uncommon in science, the initial terms stuck. It may be evident that the secondary nuclei do not grow as rapidly as the primary. The abundance of secondary nuclei ejected is proportional to the abundances of primary nuclei in the initial star. Because the primary abundances grow in time, the productivity for making secondary nuclei from them also grows in time. So in the oldest metal-poor stars surviving today, low-mass stars whose abundances reflect those they initially inherited from the interstellar gas when the stars formed, have not only smaller ratios of C/H, O/H, Si/H, etc., but also smaller ratios of abundances like ^{17,18}O/^{16}O. An entire astronomical science surrounds these issues. Its data consists primarily of the numbers of old stars as a function of their ages and the initial abundances from which they formed. The old stars are fossils that preserved for viewing today the abundances of the elements at the time they were born.

Cosmic chemical memory

This theoretical picture attempts to explain why solid objects in the solar system contain isotope ratios that differ from average solar-system isotope ratios, a phenomenon that was called "isotopic anomalies" when it was first discovered in the 1970s. Three basic discoveries prompted both it and competing theoretical pictures. They were an unusual neon gas (called Ne-E) that could be demonstrated with the *three-isotope plot* to be common in carbonaceous meteorites, excess ^{16}O that could be demonstrated with the three-isotope plot to be common in *calcium–aluminum-rich inclusions (CAIs)* within carbonaceous *meteorites*, and small but significant anomalous isotopic ratios in many elements within the minerals of the CAIs. None of these samples was itself "presolar;" they were organized into their present forms by chemistry within the early solar system. Chemical memory is their isotopic memory of the earlier materials from which they were assembled.

At the time of these dicoveries (1970s) the solar system was widely regarded as having initially been a well-mixed atomic gas. That assumption agreed with the general constancy of isotope ratios found in Earth, Moon, Sun, and meteorites. The early solar system was thought to have been sufficiently heated to erase any preexisting solids, an incorrect view that was supported by a long list of leading meteorite chemists until the manifestly *presolar grains* were discovered in 1987. During the 1970s and 1980s the most common picture held that into that well-mixed solar system was injected a modest portion of its matter from a nearby supernova, presumably the same supernova that was regarded to have injected the *extinct radioactivities* into the solar cloud. Variable admixtures of that injected supernova matter were presumed to have resulted in the small isotopic anomalies found in small meteorite rocks. Papers from the meteorite chemists advanced that view.

The cosmic chemical memory interpretation was advanced by the writer as a superior way to think of these isotopic anomalies. This picture argued that the early solar system was not hot enough to vaporize the entirety of most solids but only their volatile parts and portions of their refractory Ca-and-Al-rich minerals. The refractory parts had survived to that time from their earlier condensation as stardust and were fused into the CAI assemblages found today in the meteorites. That fusion occurred while the gas that was vaporized from a dust-rich presolar mixture recondensed as the main minerals of the CAIs. The refractory cores, being stardust that had condensed even earlier within individual stars and supernovae, contain the isotope ratios from those distinct sources. When these cores were fused into the CAIs found today, the chemistry "remembered" the isotope ratios of the source presolar grains, so that solar-system rocks (CAIs) remembered their isotopic parentage. Hence the name "cosmic chemical memory." See ^{16}O for a fuller account of the historical role played by the experimental discovery of ^{16}O-rich minerals within the CAIs, and of how the memory of ^{16}O-richness was saved.

Some of these presolar grains did survive. When incontestably presolar grains were discovered in 1987 the cosmic-chemical-memory picture was suddenly taken seriously. Indeed, the earlier predictions of the existence of refractory isotopically anomalous presolar grains endowed this picture with experimental credibility.

Today it is an accepted part of isotope astronomy and is seldom mentioned by name in isotopic studies of meteorites.

Cosmic rays

Cosmic rays represent one of the three directly accessible samples of matter from outside our solar system. Another is interstellar dust that drifts into the solar system. The third is the *presolar grains* that have survived the formation of the solar system and are found in *meteorites*.

Cosmic rays are ordinary nuclei of atoms, but moving through space at high speeds. Their speeds are such that their kinetic energy (energy of motion) is comparable to their rest-mass energy, ($E = mc^2$). The cosmic-ray flux (the number arriving per second per unit area) falls off rapidly with increasing energy, so the more abundant ones that have been suitable for measurement of isotopic abundances normally range in kinetic energy from 100 MeV to 10 000 MeV per nucleon. The rest-mass energy, E, of a single nucleon is 931 MeV. The cosmic rays are not measurable at Earth below 100 MeV because such low-energy cosmic rays cannot penetrate the magnetic bubble surrounding the solar system. That magnetic bubble is carried away from the Sun as an ionized magnetized gas (plasma) that leaves the solar system carrying its own magnetic field along with it. The particle identity (charge, mass, energy) of a cosmic ray is measured in specialized experiments, originally performed in high-altitude balloons, but now done much better from satellites launched into orbit around the Earth and, preferably, into their own orbit around the Sun. These experiments resemble those of Earth-based high-energy particle physics. Today they are quite successful, and cosmic-ray data has moved to one of the forefronts of astrophysics, where it was in the days of the first discoveries of cosmic rays.

The most abundant cosmic rays are H nuclei and He nuclei, just as these elements are the two most abundant in the universe. But H and He are not as dominant as they are in the gas from which stars form. The other abundant elements from carbon ($Z = 6$) through nickel ($Z = 28$) resemble the distribution of the elements in the solar system. Thus the cosmic rays are not manifestly exotic. They may even be atoms of interstellar gas accelerated to high speeds by shock waves blasting through the *interstellar medium*. Such shock waves are launched by *supernova* explosions. Or they may be atoms from supernovae directly, new nucleosynthesis products that are accelerated to high energy by interacting with the shock wave launched by that supernova. These are the two favorite ideas for the origin of the matter that is contained in the cosmic rays. It is the subject of lively contemporary debate.

Examination of the abundances in the cosmic-ray nuclei shows that they have not moved in a vacuum. There is ample evidence that many have suffered collisions with interstellar atoms along their paths, and these collisions have broken off nuclear fragments of new nuclei created by the cosmic rays themselves. These collisions are called spallation reactions, for they are nuclear reactions between the cosmic ray and an interstellar atom that spall a fragment from one or both of them. The clearest evidence of this is the very high abundance of the three rare light elements Li (Z = 3), Be (Z = 4), and B (Z = 5). Their relative numbers are many thousands of times more abundant (relative to C and O, say) in the cosmic rays than they are in interstellar gas or in supernova ejecta. This means that several percent of C and O nuclei (among others) have been broken apart by such collisions during their journey. From this and from the probability for such nuclear collisions to occur (measured in the laboratory) it can be concluded that the cosmic rays have, on average, traveled through a thickness of gas that amounts to many grams per square centimeter. (Reminder: 1 g/cm² is equivalent to a sheet of water 1 cm thick). They appear to have traversed a continuous spectrum of path lengths since the time of their acceleration, with mean path length equivalent to a thickness of gas in the range 6–9 g/cm². For the low density of interstellar gas this means that the cosmic rays have on average traveled about 10 million years from their sources to our detectors. The question is, "What sources?"

Another isotopic manifestation is the so-called "Cosmogenic Radioactivity" which is produced in meteorites owing to reactions of their stable atoms with the cosmic rays that bombard the meteorites. Owing to this, meteorites have many live short-lived radioactivities within them when they fall to ground, and measuring their amounts determines when the meteorite fell and how deep within it was the location of the sample prior to the collision that released it into space (see *Extinct radioactivity* and *Meteorites*).

Their large energies have always raised the question of where the cosmic rays originate. They clearly arrive from outside the solar system. In that sense they are a sample of matter from elsewhere in the Galaxy. As new waves of NASA space missions improve this data base rapidly, scientists are looking to the day when its precision will enable one to state the source of the cosmic rays and even to do new science that is made possible by knowing what they represent and by having very accurate isotopic ratios. By computer calculation the cosmic-ray composition can be computed backward along their paths (like viewing a movie in reverse) to get at the "cosmic ray composition at the source." This is the composition of the particles before they had suffered so many spallation reactions along their paths. When this is reliably done, science discoveries based on the isotopic differences from solar matter will occur. Some already have occurred, and the reader will meet many such cases among the isotopes in this book. However, one topic that will not be described is the nuclear one – the many interesting questions of nuclear interaction that occur along these cosmic-ray paths. This is a science field in itself, but a rather technical one based on nuclear reaction probabilities, whereas this book emphasizes the science ideas that are fueled by the abundance differences that remain when the nuclear effects have been removed – the "cosmic ray composition at the source."

Extinct radioactivity

An extinct radioactive nucleus, or simply an "extinct radioactivity," is a radioactive nucleus that was once noticeably abundant in the early solar system but which now has undetectable abundance. It is not surprising that such nuclei once existed. The processes of nucleosynthesis produce all manner of radioactive nuclei, especially during explosions of stars (*supernovae*). They are ejected from those stars along with the more common stable isotopes. As a consequence, the interstellar matter contains some of those radioactive nuclei that are produced in the supernova explosions. Broadly speaking they may be divided into three categories along lines of their detectability in solar-system solids: (1) long-lived radioactivities (such as the well-known isotopes of uranium) that still exist on Earth; (2) extinct radioactive nuclei that once existed in measurable quantities in early-solar-system solids but which can not now be found within them; (3) short-lived radioactive nuclei that are characterized by such short lifetimes for their decay (their halflives) that they did not live long enough after their ejection from the stars to ever be incorporated into chemical solids. If class (3) exists, the nuclide production must have been from collisions in the early solar system between energetic particles such as solar cosmic rays.

Because Earth and other solar-system bodies are 4.56 billion years old, those nuclei that can still be found live on Earth or in the *meteorites* must have long halflives. In practice this requires $t_{1/2}$ to be greater than 100 million years. Nuclei having longer halflives (*e.g.* uranium) can still be found on Earth, although their abundances may be small. The extinct radioactive nuclei are not themselves still present in the solar system, but because they were still in abundance when the solid objects took form, their presence is revealed by excess numbers of their "daughter isotopes." The daughter isotope is the isotope to which a radioactive nucleus decays. The radioactive parent's previous presence in the solid can be noticed because the daughter has an excess abundance in a solid in which the element of the parent nucleus is abundant. An example is radioactive ^{26}Al, whose halflife is 700 000 years. The daughter nucleus to which it decays (beta decay) is ^{26}Mg. So, excess ^{26}Mg is found in aluminum-rich solids that formed in the early solar system. At that time significant ^{26}Al abundance still existed, so the aluminum-loving chemistry that creates aluminum-rich solids will have taken the live ^{26}Al into the solids along with the stable ^{27}Al isotope. Later, after the ^{26}Al has decayed, that solid will have excess numbers of daughter ^{26}Mg isotopes. Solids having much more aluminum than magnesium heighten the effect.

Another interesting distinction exists. If the solids did form in the early solar system, the radioactive nucleus must have been alive then and there. This is the most common picture. It was once believed that all solids examined on Earth did form in the solar system from the previous interstellar gas. But in the past decade solids older than the Earth have been found in meteorites. They were once part of the interstellar matter and were incorporated into the meteorites when they assembled from dust and gas in the early solar system. Those presolar solids appear to have been condensed (like soot) from cooling gases while they were being ejected from the stars. In this case the extinct radioactivity need only live long enough for the small grains to have condensed as they were leaving the star. This typically requires times of months to years. The excess daughter nuclei in such cases could be called "fossil" daughter nuclei, because they are fossils of decay that occurred before the Earth and planets even existed. This phenomenon was predicted in 1975 and was the subject of intense controversy. But following the detection of *presolar grains* in meteorites in 1987, those predictions of fossil abundances were confirmed and helped identify the sources of the presolar dust.

The parent nuclei listed in the accompanying table have been found to have observable initial abundances in the early solar system from the excesses of their daughter abundances, primarily within meteorites. The table identifies them, their halflives, given in millions of years, and their initial abundance, given as a fraction of that of a stable reference abundance to which it is compared. The daughter nucleus to which each decays is also listed. Finding a natural astrophysical history that rationalizes the relative amounts of these extinct radioactivities is the main goal of their study. They carry a puzzling fingerprint of events associated with the origin of our solar system.

Initial abundances of now-extinct radioactive nuclei in the solar system

Radioactive nucleus	Daughter	Halflife (Myr)	Initial abundance
^{7}Be	^{7}Li	53 days	2.2×10^{-1} ^{9}Be
^{10}Be	^{10}B	1.51	9×10^{-4} ^{9}Be
^{26}Al	^{26}Mg	0.72	5×10^{-5} ^{27}Al
^{36}Cl	^{36}Ar	0.30	1.4×10^{-6} ^{35}Cl
^{41}Ca	^{41}K	0.10	1.5×10^{-8} ^{40}Ca
^{53}Mn	^{53}Cr	3.7	6×10^{-6} ^{55}Mn
^{60}Fe	^{60}Ni	1.5	1.5×10^{-6} ^{56}Fe
^{92}Nb	^{92}Zr	35	1.1×10^{-3} ^{93}Nb
^{97}Tc	^{97}Mo	2.6	5×10^{-5} ^{92}Mo
^{107}Pd	^{107}Ag	6.5	2×10^{-5} ^{108}Pd
^{129}I	^{129}Xe	15.7	1×10^{-4} ^{127}I
^{146}Sm	^{142}Nd	103	5×10^{-3} ^{144}Sm
^{182}Hf	^{182}W	9	1×10^{-4} ^{180}Hf
^{244}Pu	^{232}Th	81	7×10^{-3} ^{238}U

Because the Sun and planets are believed to have resulted from the gravitational contraction and collapse of a dense cloud core within a molecular cloud, one asks how these radioactivities existed in that cloud core. Supernovae do not explode within the cloud cores, and rarely explode within the parent molecular clouds. When the latter does occur, it disrupts a large portion of the cloud, which is heated and changed from cold molecular gas into hot atomic gas. The ejecta from supernovae begin as hot gas as well, although those ejecta do contain grains condensed within the supernova (SUperNOva CONdensates, each called SUNOCON (*sue-no-con*) from that acronym). How and when does that radioactivity initially in a hot-gas phase get into a cold core of a molecular cloud?

The answer to the foregoing question involves the cycling of the atoms of the *interstellar medium* among the distinct physical phases of interstellar gas: hot diffuse ionized medium; neutral atomic gas near 8 000 K temperature (HI phase, or warm neutral medium); individual small cold atomic clouds; and massive cold molecular clouds, wherein hydrogen atoms become molecular H_2, near 10–100 K. It is within the latter, higher-density molecular clouds that stars and their planetary systems are born. It appears that atoms of the interstellar medium exist within local gas that is transformed by cooling and heating from one phase to another. Individual atoms thus spend their interstellar history in all phases, doing so cyclically. The mean time for one phase to transform into another appears to be about 50 Myr, with considerable variations about that mean value. Thus, on average, one might expect 50 Myr to elapse between ejection of radioactivity by a supernova and that same radioactivity finding itself within a star-forming molecular cloud. But that 50 Myr is just an average; some radioactivity is continuously being cooled and admixed into clouds, while clouds are also continuously being heated to return to diffuse gas. Some radioactivity may be admixed within a few Myr, and some may be delayed by several hundred Myr. It is possible to define an average concentration within the star-forming clouds based on a steady cycling model; but individual clouds, including that from which our solar system was born, may have differed significantly from that average.

A related question is this: "Which types of stars or supernova explosions produce the extinct radioactive nuclei that are found in the solar system?" Here the reader is referred to the entry for each specific isotope. But this much must be appreciated first. Some radioactivities appear to be made primarily by the thermonuclear explosions of white-dwarf stars, called Type Ia supernovae. Others are created primarily in massive stars whose cores collapse to become neutron stars to initiate an explosive ejection (Type II supernovae). Type II supernovae occur three to five times more frequently than do Type Ia supernovae. Some radioactive nuclei are made within differing portions of each event, some prior to the ejection, but some during the heat of the ejection process. And still other radioactive nuclei are created within evolved stars that do not become supernovae (red giants). This diversity of origin renders uncertain the identity of those extinct radioactivities that are to be attributed specifically to that supernova that is supposed to have triggered the formation of the solar system. In recent scientific studies of these complications, the following few general results seem to be emerging.

1   Most of the shortest-lived nuclei within the table seem to have been admixed just prior to the solar birth from the outer envelope of a nearby exploding massive star.
2   Those nuclei arising from several distributed explosions of Type Ia prior to event 1 seem to have been mixed into the Sun's birthing molecular cloud with average mixing times near 50 Myr, as expected from general astronomical studies.
3   Those nuclei arising from several distributed explosions of Type II prior to event 1 seem to have been mixed less efficiently into cloud cores, with longer mixing times near 300 Myr on average.

A subset of these radioactivities (^{26}Al, ^{41}Ca, ^{60}Fe, and ^{107}Pd) may have been synthesized in *AGB stars* rather than in supernovae; that is, to put it more precisely, the reason for their existence in the early solar cloud may have been a nearby AGB star rather than a supernova. This extra degree of freedom gives yet another dimension to the topic.

Another dimension recognizes that the extinct radioactivities (or a subset of short-lived ones) may have been present owing not to stellar nucleosynthesis but rather to creation by events in the early solar system. The extinct nucleus ^{7}Be is one such example, because it does not live long enough to have been injected from any star into the early solar system. Events suggested involve the acceleration of nuclei to cosmic-ray energies, after which they collide with the dominant unaccelerated solar matter and create radioactive nuclei in the nuclear collisions that ensue. In principle, this irradiation may have occurred in the molecular cloud, in the solar accretion disk gas, on the first solids in that accretion disk, or at specific critical points within the accretion disk through which material is cycled. The last type of model appears the most promising. Solar high-energy particles are accelerated by the very active early Sun and its associated magnetic fields. These particles irradiate especially matter that is thrown upward at the X point in the solar disk, near to the Sun, and then settles back onto the disk further out, in the vicinity of planet formation. This so-called "X-wind model" has been raised in plausibility by observations made with the *Chandra X-Ray Observatory*, which show that almost all young "suns" in Orion have much more frequent flares than does the Sun today. Extinct ^{60}Fe, and perhaps also ^{182}Hf, are important counterexamples in that they can not be created in this way by cosmic-ray collisions but instead require production in stars and quite rapid admixture into the forming solar system.

Age dating   The extinct radioactivities can in many cases be used as clocks to determine when demonstrably once-molten meteorite samples solidified. Before this can be done one must know the initial amount of that radioactivity, which is usually taken to be the amount known to have existed in the oldest sample, the one to have solidified earliest. This appears today as a measurable excess of the daughter nucleus in samples rich in the parent element. For example, excess ^{53}Cr as determined by the sample having a larger than normal ^{53}Cr/^{52}Cr ratio, is attributed to live ^{53}Mn in the Cr-bearing mineral when it solidified. A second sample solidifying later may have a smaller ^{53}Cr/^{52}Cr ratio because there existed less ^{53}Mn when it solidified, or it may have a larger ^{53}Cr/^{52}Cr ratio because the sample solidified with larger elemental Mn/Cr abundance ratio. Similar applications are utilized in cosmochemistry for each of the extinct radioactivities listed above. Comparisons of this type determine the relative time of solidification of very primitive samples that each formed early in the history of our solar system.

Cosmogenic radioactivities Extinct radioactivities from the early solar system, 4.6 billion years ago, are now extinct owing to that great age; but those nuclei may be found alive today in samples of *meteorites* that have been exposed to bombardment from *cosmic rays* on their recent journey to Earth. Meteorites that are seen to fall to Earth are obviously known to have been quite recently in space. The cosmic-ray interactions break apart the stable nuclei within the meteorites into radioactive fragments. Their rates of decay after their fall to Earth show the rate at which they were being created by cosmic rays during the travel to Earth from the asteroid belt. These are usually called "cosmogenic radioactivity" among professionals, with "cosmogenic" meaning a genesis within space by cosmic rays. Prominent among those measured in meteorites are many of those listed in the table above as alive in the early solar system in much higher abundance: namely, ^{10}Be, ^{26}Al, ^{36}Cl, ^{41}Ca, ^{53}Mn, ^{59}Ni, and ^{60}Fe. Unlike the more abundant extinct radioactivity from the early solar system, the abundances of these cosmogenic nuclei are measured directly by counting their radioactive decays within the meteorite sample after its fall to Earth.

These several aspects make the study of the extinct radioactivities a problem with many facets, and an active area of astrophysical research. The true origins of these extinct radioactive nuclei are still largely unknown.

Helium burning

Helium burning is the name of the process by which helium nuclei are fused by nuclear reactions into heavier elements. The need for this occurs naturally during stellar evolution when the hydrogen-burning core has completed its fusion of hydrogen into helium. The core then ceases to provide nuclear power to keep the interior hot. Subsequent heating is accomplished instead by contraction to higher density and this contraction ceases only when the temperature is high enough for helium burning to begin. Thermonuclear power from the fusion of helium into carbon heats the core thereafter for a very long time (until the helium is exhausted). This stage in a star's life is called the red-giant stage, because the outer layers expanded and cooled (turned more red) while the core was contracting to become hot enough to begin helium burning. This explanation of the red giants, first advanced by Hoyle and Schwarzschild was a landmark in the history of understanding the stars (see also *Stellar structure*).

Grave difficulties for the thermonuclear evolution of matter were presented by the fact that no stable nuclei exist at mass $A = 5$ or at mass $A = 8$. Because of that, no mixture of hydrogen and helium is able to fuse into heavier nuclei by two-particle reactions. Helium burning utilizes the transient existence of ^{8}Be, the fusion of two alpha particles. Though rare, owing to their very short lifetime before breaking apart again into two alpha particles, the ^{8}Be nuclei can sometimes capture a third alpha particle, because the complex of three alpha particles can be stabilized by emission of a photon, resulting in the ^{12}C nucleus (for more details, see ^{12}C, **Nucleosynthesis origin**). Following production of ^{12}C by the triple-alpha reaction, the ^{12}C nuclei can capture yet another alpha particle, thereby producing ^{16}O. When helium is finally exhausted, ^{12}C and ^{16}O dominate the matter. Further nuclear evolution does not occur until the stellar core contracts more and becomes sufficiently hot to begin *carbon burning*.

Interstellar medium (ISM)

The interstellar medium is the medium that fills the space between the stars. This space is far from empty. It includes magnetic fields, gas composed of atoms and ions at several different temperatures and densities, cosmic rays, and dust particles. The material content of the ISM changes with time owing to the formation of new stars from it and the ejection of matter from stars into it. The latter include the new nuclei that have just been assembled by nucleosynthesis in the stars. The state of this medium is turbulent, driven by the shock waves from exploding supernovae. Dust comprises about one percent of the mass of the interstellar matter. It is measured by its infrared radiation and by its obscuration and reddening of starlight.

The gaseous component is far from uniform, but rather exists in differing physical phases: a hot (million degree) gas of high ionization and very low density filling half of the volume; a moderately hot (8000 K) gas in which H is ionized, so that recombination lines of H are emitted by this phase, which is also about half of the volume; a cooler (100 K) gas filling 5% of the volume in which H exists as neutral atoms, allowing the kinematics of the ISM to be measured by the 21-cm-wavelength emission from atomic H; and molecular clouds, cold (50 K) dense gas in which H exists as H_2 molecules. The molecular clouds' densities range from 100 to 10^6 H per cubic centimeter, very large for interstellar gas but still a near vacuum by terrestrial standards. The molecular clouds comprise less than 1% of the volume of the ISM but almost half the mass radially interior to the orbit of the Sun's galactic radius. Rotational lines of the CO molecule are the main astronomical tracers of the molecular clouds, within which stars are born. The molecular clouds cause so much extinction of the light from stars behind them that those regions of the sky appear to the eye as dark holes in the pattern of stars. Over a hundred different molecules have been observed in these clouds by radio astronomers. Importantly for this book, those molecules include differing isotopomers of otherwise identical molecules. An isotopomer is the same molecule but having differing isotopes of its elements. For example, ^{12}CH and ^{13}CH are isotopomers of the CH molecule. Equally important is the depletion of gaseous atoms by condensation onto dust grains, which is observed in a diffuse cloud by the relative strengths of the absorption lines of the elements as starlight from a star behind the cloud is absorbed during its transit of the cloud. These show that some elements (sulfur, for example) remain more gaseous than other more chemically reactive elements (aluminum, for example).

Because of the dynamic state of the ISM, these phases continuously transmute from one to another. Hot stars heat and disperse molecular clouds in which they were born, and diffuse gas cools and forms new molecular clouds. During these cycles the elements ejected from stars, especially from supernovae, are mixed with the pool of existing atoms, and much important chemistry occurs, termed cosmochemistry. In particular, when ISM matter cools to become molecular, the free condensible atoms are captured onto the surfaces of preexisting cold dust grains in the cloud. Energy-dependent differences in chemical reaction rates slowly fractionate one isotope of an element from its isotopic brothers, creating isotopic anomalies. And some of the dust grains are ab initio isotopically very anomalous because they condensed within the matter of a single star as the matter left that star (see *Presolar grains*). Complex chemical structures are generated by these cycles, giving birth to *cosmic chemical memory*. It is a goal of the theory of cosmic chemical memory to explain the chemical and isotopic state of matter from which our solar system formed. The science of meteoritics has assumed great significance as a marker of cosmic chemical memory, from the ISM to the heating and subsequent chemistry of interstellar matter that occurred in the solar disk as planetary bodies were forming. Superimposed on these cycles of interstellar phases, the bulk contents of the ISM become richer in heavy elements with the passage of time. This result derives from nucleosynthesis in stars, because the ejecta from stars are richer in heavy elements than was the interstellar gas from which those stars had been born. This gives rise to *chemical evolution of the Galaxy*.

The physics and chemistry of the ISM is perhaps the richest of all areas of study for astronomy and astrophysics, because it is so varied and because there exists so much complex information relating to the huge chemical machine that is the interstellar medium.

Mass excess Δ; Neutron separation energy S_n

Atomic mass unit   The unit of atomic mass, symbol m_u, is defined historically as 1/12th of the mass of the ^{12}C atom. Therefore, the atomic mass unit is given by

$$m_u = m(^{12}C)/12.$$

So the mass of the ^{12}C atom is exactly 12 atomic mass units, by virtue of the definition.

Atomic mass excess   Listed in the headings for each isotope in this book is the "mass excess" for that atom (isotope) designated by the symbol Δ (Greek upper-case delta). The quantity Δ represents the mass of that isotope in a special way; namely, the mass excess Δ is the difference between the actual mass of the neutral atom and the mass of a fictitious related atom having the same integer number A of nucleons but an atomic mass that is a factor A times greater than the atomic mass unit.

For a given nucleus, having nuclear charge number (atomic number) Z representing its number of nuclear protons and nucleon number (mass number) A representing its total number of nucleons (neutrons + protons), the mass excess of atom (Z, A) is defined by

$$m(Z, A) = Am_u + \Delta(Z, A),$$

that is, Δ is the difference between the true mass of the atom and an integer number of atomic mass units. This definition honors the historical observation that the true mass $m(Z, A)$ of this atom is approximately given by the first term on the right-hand side of the equation, Am_u, the product of the nucleon number and the atomic mass unit. This is a "first guess" at the mass of the atom. That first term is typically thousands of times greater than the second term, the mass excess. If one knows Δ one also knows the mass of the atom (including the electrons) by adding Am_u. The quantity Δ is that listed in each isotopic heading in the book, as is conventional in nuclear physics. Distinct isotopes containing equal numbers of nucleons are called *isobars*. For example, ^7Li and ^7Be are isobars of A = 7.

In this book the value of Δ is given in energy units rather than units of mass, which is possible because of the equivalence of mass and energy expressed in Einstein's famous formula

$$E = mc^2.$$

Indeed, the energy and mass relations in the nuclei constitute one of the earliest proofs of that amazing relationship. The energy units commonly used in nuclear physics to express mass are millions of electron-volts, MeV. One MeV is defined as the energy picked up by one electron when it falls through an electric potential of one million volts. It is a convenient unit for nuclear masses, because it is also a convenient unit of measure for the quantity of energy required to pluck one nucleon from its binding within a nucleus and set it free. That separation energy is usually several MeV, typically 8 MeV.

Nuclear binding energy   This is the energy needed to disassemble a nucleus into its component nucleons, and is an important demonstration of Einstein's formula. It can best be understood by considering the nucleus ^{12}C. The atomic mass unit, by being defined as 1/12th of the ^{12}C atomic mass, causes Δ for ^{12}C to be identically equal to zero; that is, $\Delta(^{12}$C$) = 0$. But the ^{12}C atom is assembled from six H atoms (six protons with their six electrons) and six neutrons, each of which is more massive than m_u. Their atomic mass excesses are

$$\Delta \text{ (H atom)} = 7.289 \text{ MeV} \qquad \Delta \text{ (neutron)} = 8.071 \text{ MeV}$$

and yet when assembled in the ^{12}C atom their average Δ is zero. The sum of the 12 partial masses is greater than the mass of ^{12}C. The sum weighs less than the sum of its parts! The resolution of that apparent paradox is that the nuclear force binding the ^{12}C together gives a negative energy to ^{12}C. Binding energies are negative energies because positive energy must be added to separate the nucleus into its component nucleons. By Einstein's law the mass of ^{12}C is reduced by the negative binding energy, in an amount given by the difference between its atomic mass and the summed masses of its constituents:

$$\text{nuclear binding energy} = \text{(difference of mass)} \times c^2,$$

and since the binding energy is a negative energy, the difference in mass is negative. The isotope ^{12}C actually weighs less than the sum of the weights of the six H atoms and six neutrons from which it is assembled by $6(7.289 \text{ MeV}) + 6(8.071 \text{ MeV}) = 92.16 \text{ MeV}$. The total binding energy of the ^{12}C nucleus is, therefore, 92.16 MeV. That much energy is required to break it apart into six H atoms plus six neutrons. Dividing that total by 12 nucleons fixes the average binding energy per nucleon of the ^{12}C nucleus as 7.68 MeV. For other nuclei the binding energy can be computed from Δ in the same way.

The atomic mass excess Δ is one of the important characteristics of the nuclei for the understanding of their abundances. This is because Δ signifies how tightly the nucleons are bound together in each nucleus. Nuclei differ in that regard; for example, the nucleons in the ^{28}Si nucleus are very tightly bound together by the nuclear force and the quantum laws that organize the structure of nuclei, whereas the nucleus ^{2}H (heavy hydrogen, or D) has very weak binding between its two nucleons.

Separation energy   This is defined as the energy to remove one nucleon from a nucleus. Consider an example of a neutron removed from a nucleus, taking that nucleus to be ^{29}Si just for purposes of illustration. The ^{29}Si nucleus is assembled from 14 protons and 15 neutrons. Removing one neutron can be represented by the nuclear-identity equation

$$^{29}\text{Si} + \text{energy} = {}^{28}\text{Si} + \text{neutron},$$

giving equal numbers of both protons and neutrons on both sides of the equation. Thus, in calculating the mass (energy) changes involved, the "first-guess" terms involving 29 nucleons (times m_u) are the same on both sides of this reaction, and the energy required is given by the increase in mass (in energy units):

$$\text{separation energy for that neutron} = \Delta\,(^{28}\text{Si}) + \Delta\,(\text{neutron}) - \Delta\,(^{29}\text{Si}),$$

which the curious reader may confirm from the entries for these isotopes to be

$$\text{separation energy of neutron from } ^{29}\text{Si} = S_n(^{29}\text{Si}) = 8.473\,\text{MeV}.$$

This calculation from the Δ values alone shows that 8.473 MeV of energy is needed to remove one neutron from the ^{29}Si nucleus. The value of that energy is important to nucleosynthesis processes in which nuclei are heated to high temperatures (see p *process* for example). Such energies determine at high temperature which species survive, and which will be destroyed in the battle to synthesize the nuclei in stars. Because it is so important in the scheme of nucleosynthesis, the neutron separation energy, S_n, for each specific isotope is also listed throughout the book.

A similar calculation reveals whether a reaction in stars between two nuclei (A + B) releases energy (is "exothermic") when it transmutes to a final product C + D. If the combined atomic masses of A and B are greater than those of the products C and D, the reaction will liberate energy. It may therefore occur in stars more easily than if the masses of C and D are greater, in which case the reaction must absorb energy (be "endothermic"), usually requiring explosively hot conditions to have a chance of occurring. Exothermic reactions dominate what occurs in stars during their evolution.

These subtle meanings are hidden within the otherwise routine-appearing entry, for each isotope, its atomic mass excess Δ.

Metal-poor stars

The oldest stars bear testimony to the Galaxy having begun with only hydrogen and helium (and a little Li), remnants of the light-element synthesis that occurred during the *Big-Bang* expansion at the beginning of the universe (in the accepted picture of our day). Much later, perhaps one billion years later, that gas was cool and had sufficient time to contract to denser clouds that became large gravitationally bound clumps of gas. These turned into galaxies. When the first stars were born in the galaxies, they contained only H and He and traces of ^{7}Li. Therefore the first stars can manufacture only those nuclei that are naturally made from H and He during the thermonuclear evolution of the stars. These are called "primary nuclei."

Along with the massive stars, regarded as those with masses between 12 and 60 times the mass of the Sun, which are especially good thermonuclear nucleosynthesis machines, a large number of low-mass stars were born. Stars having mass less than that of the Sun still exist today, because such low-mass "dwarfs" take so long to evolve to red giants that this has not yet happened. These stars are still visible today, although they are hard to find. More useful for stellar spectroscopy are those low-mass stars that are just today turning into red giants, because they are much more luminous and therefore easier to find and measure. Within them the abundances of the elements generally are very low in comparison with the Sun. Especially good measures of this low metallicity are the abundances of oxygen and iron. Because these low-mass stars do not synthesize elements, their outer gaseous envelopes retain their initial compositions; the elements that went to produce our Galaxy. The most metal-deficient of the observed stars have abundance ratios of Fe/H that are 1/10 000th those in today's Sun and stars! These are precious stars to astronomers, because they record something very significant about the conditions when our Galaxy was born. They also bear testimony to subsequent nucleosynthesis, those more massive stars that ended their lives long ago and enriched the *interstellar medium* with new elements.

When the most massive first stars explode as *supernovae* at the end of their lives (typically 10 million years) they eject large abundances of the primary nuclei into the interstellar gas. The concentrations of primary nuclei (*e.g.* ^{12}C, ^{16}O, ^{28}Si, and ^{40}Ca, just to sample a few) in the supernova ejecta are about 15 times greater than the concentrations today of those elements in the Sun. These ejecta are mixed into the interstellar gas, from which new stars form. The gas from which these new stars form now contains some abundance of the primary nuclei synthesized from the earlier supernovae (see *Chemical evolution of the Galaxy*). In this way the element concentration of the interstellar gas has increased since those early days. These elements do not all grow together, however; O grows most rapidly (see ^{16}O, **Astronomical measurements**), but Fe and C grow noticeably more slowly at first. That is, stars very deficient in Fe are not so deficient in O; this is evidence that more slowly evolving stars (Type Ia supernovae) created the major portion of the iron, but starting more slowly because of the time required to prepare those stars (see ^{56}Fe, **Astronomical measurements**).

In modern times it has been possible to measure the abundances of the heavy elements in the metal-poor stars. Apparently the r *process* occurred earlier in galactic history than did the s *process*, and even earlier than the synthesis of iron. As an example, osmium (Os) is an important part of that pattern, and because its solar abundance is dominated by the r process, the ratio of its abundance in these metal-poor stars relative to an element whose solar abundance has resulted primarily from the s process is observed to be much larger than in the solar system. For example, the Os/Ba abundance ratio is 5 to 10 times larger in these old stars than in the Sun. Such observations are evidence that not only does nucleosynthesis occur in different processes which are later mixed together, but that some of those processes began earlier than others; namely, the r process began earlier than the s process. Of similar significance is the observation that the ratio Os/Fe is larger in these stars than in the Sun, meaning that the supernovae that promptly began to produce the r-process isotopes were not the major source of iron in the galactic history. A later, more slowly prepared source of Fe was the onset of Type Ia supernova explosions. Similar r-process overabundances (relative to Fe) in old metal-poor stars have been recorded for the elements Sm, Eu, Gd, Tb, Dy, Ho, Er, Tm, Yb, Hf, and Pt, each of which is primarily an r-process product in the solar abundance mixture. But in these oldest metal-poor stars, these element abundances stand in proportion to their r-process abundances, rather than to their total abundances in the Sun. This discovery has provided exciting insight into the nucleosynthesis history of the heavy isotopes.

Meteorites

Meteorites are rocky aggregates that fall from the sky. Some few are <u>seen</u> to fall ("Falls"), but most ("Finds") fall unobserved and are found on the surface of the Earth or, latterly, the Moon. They are pieces of small planetary bodies that were broken off during a collision in the asteroid belt around a million years ago. Some meteorites ("irons" and "stones") are fragments of the cores of small planets which have differentiated by melting into inner metallic portions and outer stony mantles. More important for isotopic astronomy is another class of meteorites, the geochemically primitive collection of debris known as chondrites. They are so named for the many millimeter-sized spherules (chondrules) that are distributed abundantly among the chondrites. The chondrites are rubble piles, formed on the surfaces of small asteroids that collected all manner of primitive debris in the early solar disk. As time progressed they were fused into a rock-like structure; but they were never very hot and retain a fine record of early materials. Those chondrites rich in carbon, the carbonaceous chondrites, are the most significant, for it is they that contain in the debris collection large numbers of *presolar grains*. The huge isotopic anomalies within the presolar grains have spawned a new field in astronomy.

Extinct radioactivities from the early solar system, 4.6 billion years ago, are now extinct owing to that great age; but those radioactive nuclei may be found alive today in samples of meteorites that have been exposed to bombardment from *cosmic rays* on their recent journey to Earth. Meteorites that are seen to fall to Earth are especially known to have been quite recently in space. The cosmic-ray interactions break apart the stable nuclei within the meteorites into radioactive fragments. Their rates of decay after the fall show the rate at which the were being created by cosmic rays during the travel to Earth from the asteroid belt. These are usually called "cosmogenic radioactivity" among professionals, cosmogenic meaning a genesis within space by cosmic rays.

Neutron-burst nucleosynthesis

Neutron-burst nucleosynthesis occurs in *supernovae* interiors at the time of the explosion. A strong pressure wave, called a shock wave (which one may think of like a sonic boom, a sound wave that travels faster than the speed of sound), travels outward through the hot supernova gases, making them suddenly hotter after the shock passes. When the gas is suddenly heated, nuclear reactions that were already occurring proceed much more rapidly. Although the temperature may be increased only by 50%, say, it is a property of thermonuclear reactions that they proceed very much more rapidly with increasing temperature.

Some of these reactions release free neutrons. The reaction $^{22}Ne + {}^4He \rightarrow$ $^{25}Mg + n$ is a leading example. The other nuclei capture those free neutrons as quickly as they can, depending on their capturing probabilities. Within seconds the gas, now much hotter and therefore at higher pressure, expands to relieve its excess pressure, and, also within seconds, the cooling effect turns off the speeded-up reactions. The process can be thought of as a burst of free neutrons, liberated in seconds and consumed by capture within a few more seconds. Those captures move the isotopes of stable elements toward heavier isotopes. The isotopic composition is altered to favor the heavier isotopes by this. Not every portion of the supernova has neutron sources ready to yield neutrons, however. This process happens only in special thin shells of material wherein the composition is just right to enable the neutron burst.

Evidence of neutron-burst nucleosynthesis has been found most convincingly from the isotopic study of *presolar grains*. Many that were made within supernovae contain, in addition to their other telltale compositions, the evidence of the rapid burst of free neutrons. Interestingly, this is what occurs in an atomic bomb. The testing of these weapons gave the first inkling of something that is common in the universe.

Neutron-capture cross section

The neutron-capture cross section of any isotope represents the probability with which it is able to capture free neutrons passing by it. This quantity is important for the *s process* of nucleosynthesis. This process was named "s" owing to the need to patiently make the s isotopes of the heavy elements by the "slow" capture of free neutrons, neutrons liberated by other nuclear reactions within the gas in stellar interiors. It was one of the first nucleosynthesis processes identified historically. The capture of a free neutron by a nucleus increases its mass number A by one unit. As the captures continue, each nucleus in the gas is rendered heavier, little by little, capture by capture. When an isotope of mass number A of an element with atomic number Z captures a neutron, the compound nucleus formed from their union becomes an isotope of the same element but having mass number greater by one unit (i.e. A + 1).

The value of the neutron-capture cross section is conceived as follows. Imagine neutrons randomly bombarding an area containing the isotope in question. The number of neutrons striking each unit area per second is called the flux, F_n, of neutrons. The neutron-capture cross section is the probability during each second that the isotope will capture a neutron divided by the flux F_n of neutrons. That is, it is that probability for a flux equal to one neutron per unit area each second. The neutron-capture cross section has units of area because, as the fraction of neutrons captured from the flux F_n striking the unit area, it may be regarded as the fraction of the area that the isotope occupies.

The neutron-capture cross section is tiny, about the size of the nucleus itself. In the early days of neutron physics a very large cross section was excitedly called "a barn, because it is as big as a barn!" Actually the barn is 10^{-24} cm², a square 10^{-12} cm on a side. This is comparable to the size of the largest known atomic nucleus. Smaller cross sections are usually measured in thousandths of a barn, or "millibarns." The data used changes annually, and the cross sections vary with the temperature of the gas into which the neutrons are freed.

One sees that the probability for an isotope to capture a given neutron (one out of many) is really quite small, typically 100 millibarns (10^{-25} cm²). But this small size must be seen in the perspective of the s process. In the helium-burning shell of an AGB *star* the flux of free neutrons is typically $F_n = 10^{16}$ neutrons per cm² each second – and that condition may last 300 years, which is duration 10^{10} seconds. During that time a huge 10^{26} neutrons strike each cm². That is sufficient for ten neutron captures during that interval by the typical isotope. The s process is patient, like a river.

Oxygen burning

Oxygen burning is the name of the process by which oxygen nuclei are fused by nuclear reactions into heavier elements. The need for this occurs naturally during stellar evolution when the carbon-burning core of a giant star has completed its fusion of carbon into neon and magnesium. A huge mass of oxygen also remains at that time from the prior ashes of helium burning. The carbon core then ceases to provide nuclear power to keep the interior hot. Contraction to higher density is stopped only when the temperature is high enough for oxygen burning to begin. Thermonuclear power from the fusion of oxygen nuclei into Si, Mg, and S heats the core thereafter for a short time (until the oxygen is exhausted). As in carbon burning, oxygen fusion also utilizes binary collisions, now between two ^{16}O nuclei. The initial primary nuclear reactions are

$$^{16}O + {}^{16}O \rightarrow {}^{28}Si + {}^{4}He, \text{ and } {}^{16}O + {}^{16}O \rightarrow {}^{31}P + {}^{1}H;$$

but a variety of secondary reactions make ^{28}Si the most abundant final product.

Explosive oxygen burning occurs in supernova explosions when the oxygen is suddenly heated to temperatures well in excess of its stable burning temperature and therefore burns quite rapidly (in seconds). In Type Ia supernovae this happens when a runaway nuclear burning causes the temperature to overshoot stable burning, leading to a *quasiequilibrium* that is quite close to *thermal equilibrium*; in Type II supernovae it occurs when the rebounding shock wave reaches the oxygen shell of the presupernova star and heats it suddenly, so that the oxygen undergoes a thermonuclear explosion. This burning process produces ^{28}Si, ^{32}S, ^{36}Ar, and ^{40}Ca together in solar proportions, linking the nucleosynthesis of those dominant "alpha" isotopes of Si, S, Ar, and Ca.

p process

One of the processes of nucleosynthesis of the elements heavier than iron, this process was named "p" owing to the need to make the most proton-rich (neutron-poor) isotopes of the heavy elements. These isotopes cannot be produced through reactions induced by irradiating the elements with neutrons (the *s process* and the *r process*). Such neutron-shielded isotopes are called p isotopes. Either adding protons or removing neutrons from other nuclei is required to synthesize the p nuclei. Examples are the isotopes of xenon, ^{124}Xe and ^{126}Xe, which illustrate the sense in which they are shielded; namely, ^{124}Xe and ^{126}Xe cannot be produced by placing other elements within a flux of free neutrons (such as in a nuclear reactor or in a stellar interior that is liberating neutrons by thermonuclear reactions). Atomic numbers $Z = 52$ and 53 immediately proceed Xe ($Z = 54$). When tellurium, Te ($Z = 52$) captures neutrons, there result new stable isotopes of Te from $A = 122$ to $A = 126$. Only at ^{127}Te does beta decay occur, and that to a stable isotope of iodine, ^{127}I. After it in turn captures a neutron, its radioactive ^{128}I decays to ^{128}Xe, thereby sending some abundance flow into the element Xe at the earliest possible branch. But the two light Xe isotopes, ^{124}Xe and ^{126}Xe, were bypassed in this flow. Thus they are called p isotopes.

In heavy elements the lightest isotopes, or p isotopes, of an element are often much less abundant than the isotopes that were created by the capture of neutrons by lighter nuclei. The p nuclei are synthesized by the stripping down of heavier isotopes by the action of the abundant high-energy photons that exist in the thermal bath of a supernova explosion. The temperature reaches 3–4 billion degrees in the explosive oxygen and neon shells of supernovae. There the photons strip neutrons off the heavier isotopes by photoneutron reactions (gamma, n). For example, ^{126}Xe can be created by removing neutrons in this way from ^{128}Xe and then from ^{127}Xe. This photon process is what is meant by "the p process" today, although historically it was imagined that the p isotopes might be created instead by proton addition. The gamma-ray stripping is controlled by the neutron-separation energies S_n, which reveal two things. The lightest isotope will have the largest p abundance in any element; and the very heavy nuclei (*e.g.* osmium) will be eroded faster than will the medium-heavy nuclei (*e.g.* tin). This conflict makes it difficult to create the very heavy and the intermediate heavy nuclei at the same time.

Although it is quite efficient when it occurs, the p process acts on only a small amount of mass in stars. This causes the yields of the p process to be much less than those for the neutron-capture processes, which are restricted to lower temperatures. But those neutron irradiations at lower temperatures either occur in a much larger stellar mass region (as for the s process) or produce a much higher concentration (as for the r process), in either case giving a more abundant yield of nuclei than does the p process.

Presolar grains

Presolar grains are small mineralized, organized assemblies of atoms. Those found in *meteorites* to date might be described as tiny "gems." Interstellar dust has long been known by astronomers to exist, but today the hard gem component is actually found within meteorites, from which they are extracted and subjected to a battery of microcharacterizations in laboratories on Earth. They are typically 1/10 000th of a centimeter in size, but with a wide range. They are therefore not visible without a microscope. The more familiar terrestrial rocks of all types are assembled from grains of varying sizes, so geochemists are familiar with them. Presolar grains are such grains, but ones that existed prior to the Earth's existence. The vast majority of grains in meteorites were instead created from well-mixed atoms by chemistry within our solar system; but the presolar grains already existed in the molecular cloud of gas and dust that would collapse to form our Sun and planets. They are thus older than the solar system.

Holding objects in our hands, figuratively speaking, that predate our solar system came as a shock to many scientists, as it does also to mankind in general. It was long thought that man could not lay his hands on matter older than the solar system, that all which we have was created in its present chemical form by chemical reactions among prior constituents during the evolution of the solar system. But presolar grains were successfully extracted from the meteorites in 1987. They are identified by having relative abundances of isotopes of the elements that differ from the normal ratios with which we were familiar. These (anomalous) isotopic ratios are in many cases so wildly different from the usual ratio that the grains themselves can be identified as having formed in the matter from some other (unidentified) star that died before our Sun was born. Those grains would then have been present in the solar gases predating solar birth. The presolar grains enable a cosmic archaeology, the studying of things that went before the solar system and after the Big Bang. Amazingly, they still exist. They survived intact within the meteorites, sheltered from chemistry and heat for the 4560 million years of the solar system's existence. Thousands have been found and studied in laboratories on Earth.

The presolar grains constitute a gift of "astronomical" proportions to the world of nuclear astrophysics, that science community that concerns itself with the origin of the elements and their isotopes. Their isotopic constitutions enable study of nuclear processes directly, unmixed with other cosmic sources. One has long regarded the goal of nucleosynthesis theory to be the understanding of the origin of the isotopes and their relative abundances. In a practical sense this meant of solar abundances, for the abundances within our solar system were the only ones that had been amenable to study with detailed precision. Since the 1950s it has been recognized that many differing nuclear processes, occurring in different types of stars, are necessary for the full production of the rich mix of isotopes that mankind has inherited. These different nuclear processes produced specialized isotopic abundance patterns that were admixed in the interstellar gases to give the full pattern. Understanding of the solar abundances had long been seen as a problem of decomposition, the act of separating the solar abundances into their several formation processes, whose relative contributions to the solar gases reproduce the solar abundance. But the individual processes previously had to be taken on faith. And it is that faith that the presolar grains have justified. Within them isotopes exist in much different ratios than those of our common experience. They reveal individual processes of nucleosynthesis directly. As a concrete example, the presolar grains of SiC contain within them trapped xenon (Xe) atoms, whose isotopic abundance ratios are those obtained when matter is exposed to a lengthy irradiation by free neutrons (see *s process*). In stars the corresponding process is called the *s process*. On a personal note, the s process was formulated mathematically in the writer's Ph.D. thesis in 1961. And if astronomers long knew that it was one of the contributors to our solar soup, few dared believe that its isotopic pattern could be found isolated from all others, naked. This clear view is what the presolar grains have provided. It is, along with the astronomies of gamma-ray lines and of the X-ray maps of supernova ejecta, the greatest advance in nuclear astrophysics since the 1960s and 1970s when the early sketches of the nucleosynthesis processes were formulated in great detail. Those decades were a high water mark in nuclear astrophysics, but they have been equalled by these new discoveries during the last 15 years of the 20th century.

Carbon grains   *Meteorites* yielded first the presolar grains that are based on the chemistry of carbon. These grains were of three types: diamonds, a distinct mineral form of pure carbon; graphite, another mineral form of pure carbon; and silicon carbide. This was facilitated by the fact that these grains do not dissolve in strong acids, whereas all of the common rocks within meteorites do. Thus the meteorite can be dissolved, leaving a residue of carbonaceous material. This process has been likened to "burning down the haystack to find the needle!" Within this carbonaceous residue, itself quite complex, could be found the individual presolar carbon grains. They were isolated, characterized, identified, and measured in their isotopic abundances by a device called a sputtering mass spectrometer. Those first mass spectrometric measurements rocked the foundations of our natural world. They revealed a naturalist's wonderworld, where wildly differing abundance ratios exist. From measuring isotopic anomalies in parts per thousand deviation from solar the world leaped to measuring them in factors of tens to hundreds times the solar value! For grounding in a common example, the abundance ratio of carbon's two stable isotopes, ^{13}C and ^{12}C, which was always found to be very close to the value $^{12}C/^{13}C = 89$ in solar-system material, was measured to range between the value 3 and 10 000 in presolar graphite grains. This let a new genie from the bottle.

Presolar oxide grains   Meteorites contain far more oxide grains than carbon grains. In this they are similar to the Earth. But the searches for presolar grains of carbon were easier than those for presolar oxide grains. The presolar oxide grains are a small number within a much larger population of oxide grains made by chemistry during the early solar system. The challenge was to identify presolar oxides in the face of that large background of normal grain material. The only sure way to identify them is by their unusual isotopic ratios. These mark a particle's origin as being at another time and in another place, when the isotopes had different relative abundances. These sites are the stars themselves, or more exactly the gaseous matter that they lose late in their lives. That outflowing matter cools toward the temperatures when dust can form, like soot, but having the isotopic and chemical composition of that star's matter. Painstakingly, oxide grains have been isolated from meteorites and placed in a sputtering beam of ions that can be followed by automated isotopic counting of the sputtered ions. When these are anomalous, the grains have been selected out of the lineup for more careful study. Especially exciting have been the isotopic analyses of oxygen itself in those grains. The measurements taught us new facts about stars and about nuclear reactions within them.

Quasiequilibrium

In *thermal equilibrium* the rate of every reaction is balanced by the rate of the inverse reaction. For example, the rate of ^{44}Ti(alpha, p)^{47}V owing to free alpha particles is exactly equal to the rate of ^{47}V(p, alpha)^{44}Ti owing to the density of free protons in thermal equilibrium. This equality sets the abundance ratio ^{44}Ti/^{47}V. Moreover, this is true for every reaction, not just most. In nuclear quasiequilibrium almost all reactions, but not all, are balanced by their inverses; within quasiequilibrium one link (at least) is not in balance. In nucleosynthesis the link that is too slow is that for converting free nucleons into heavy nuclei and vice versa. This means that the total number of heavy nuclei differs from what would be reached given sufficient time to reach thermal equilibrium. The quasiequilibrium rate of ^{44}Ti(alpha, p)^{47}V owing to free alpha particles is still almost exactly equal to the rate of ^{47}V(p, alpha)^{44}Ti owing to the density of free protons; but there may exist either too many or too few free nucleons for thermal equilibrium. This means that the densities of free alpha particles and free protons are also not at their equilibrium values, even though their densities are also steady in time owing to equal rates of absorption and of emission of them. The quasiequilibrium state is achieved in at least three major variations in *supernovae*, where most nucleosynthesis of heavy elements occurs. It was first discovered in *silicon burning* in presupernova cores, where the number of heavy nuclei is almost twice as great as thermal equilibrium would demand (while silicon is transforming into nickel); it was next found in the *alpha-rich freezeout* reassembly of nucleons and alpha particles, where insufficient time existed for producing enough heavy nuclei; and lastly it was discovered in the explosions of white-dwarf stars (Type Ia supernovae) where, at high density, too many heavy nuclei are first assembled in the explosion. These are subtle details of nucleosynthesis; but quite significant for the frontiers of the subject today, in which one attempts to truly understand the explosive events and the isotopes produced.

r process

One of the neutron-capture processes of nucleosynthesis of the elements heavier than iron (the *s process* being the other), this process was named "r" owing to the need to make the r isotopes of the heavy elements by the "rapid" capture of free neutrons. The captures must be more rapid than the beta decays. Each capture increases the mass number A by one unit. (If an isotope of mass number A of an element with atomic number Z captures a neutron, the compound nucleus is an isotope of the same element Z but now having mass number $(A + 1)$. That new nucleus may be a stable isotope of element Z, or it may be radioactive (usually capable of emitting an electron and an antineutrino, called "beta decay," in transforming to the isotope of mass number $(A + 1)$ belonging to the next element $(Z + 1)$). The adjective "rapid" describes the pace of successive captures in the r process, calling to mind the short wait between successive neutron captures. The neutron flux is so high that the time to capture a neutron is generally less than one-thousandth of a second. Beta decays cannot occur so rapidly, so an element continues to capture neutrons until the nuclear structure can hold no more. This in effect stops the succession of neutron captures only when very neutron-rich beta-unstable isotopes have been reached. When the r-process event terminates, these neutron-rich beta-unstable isotopes then have sufficient time to beta decay toward the stable isotopes. Depending upon the element in question, four to twenty decays may subsequently occur before a stable isotope is reached. The final nucleus reached by that series of beta decays will of necessity be a neutron-rich stable isotope of its own element. The r-process isotopes are those that can be reached in this way, by the beta decay of radioactive neutron-rich isotopes. Such beta-decay-produced isotopes are called r isotopes.

Like the s isotopes, the r isotopes are produced by neutron irradiation, but the flux must be sufficiently intense that the neutron captures would be "rapid" compared to the beta-decay lifetimes of the radioactive isotopes. For the s-process path through stable isotopes these beta-decay times are typically minutes to hours; but on very neutron-rich paths encountered in the r-process, the corresponding times may be but tiny fractions of a second. In s-process circumstances (10 to 100 years between captures), decay of radioactive isotopes will occur first and the r-process isotopes are not reached. By contrast, the much more rapid r-process will last but a few seconds, with the neutron captures occurring in milliseconds and the beta decays in fractions of a second. So the r-process truly is "rapid." Its rapidity places it within explosive stellar events, probably supernovae.

The series of neutron captures can not continue indefinitely, no matter how fast. Eventually the captured neutron is quickly liberated again by the energetic photons at high temperature. The energy required to separate the added neutron (the "neutron separation energy") declines as the isotope mass increases. After significant captures, the separation energy required to remove the next neutron after it is added is small, comparable to a significant fraction of the photon energies that also exist in the gas at its temperature T. These photons become able to remove the added neutron when its separation energy S_n is below 2–3 MeV (depending upon the temperature and neutron density). Consider the removal of a neutron from ^{137}Xe:

$$separation\ energy\ for\ that\ neutron\ = \Delta\,(^{136}\text{Xe}) + \Delta\,(\text{neutron}) - \Delta\,(^{137}\text{Xe}),$$

where $\Delta\,(^{137}\text{Xe})$ is the *mass excess* of ^{137}Xe. Using the mass excess for the neutron $\Delta\,(\text{neutron}) = 8.071$ MeV, this example becomes

$$S_n(^{137}\text{Xe}) = \Delta\,(^{136}\text{Xe}) + 8.071\ \text{MeV} - \Delta\,(^{137}\text{Xe}) = 4.025\ \text{MeV}.$$

Although $S_n = 7$–9 MeV for most stable nuclei, its value falls to $S_n = 2$–3 MeV after 10 to 20 neutron captures, depending on the element. When it does so, the captures effectively cease, because the ejection of the next neutron by photons is even faster than the rate of neutron captures. This location is called a "waiting point" because the series of captures must then wait until beta decay has been able to transmute the nucleus to the next higher element, for which the neutrons will have greater binding. The waiting point concept was important to the development of the theory of the r process.

The nine stable isotopes of xenon illustrate these r-process systematics. The lightest two isotopes, ^{124}Xe and ^{126}Xe, can not be produced by placing other elements within a flux of free neutrons (such as in a nuclear reactor or in a stellar interior liberating neutrons by thermonuclear reactions), and are therefore "proton-rich" p isotopes. Each of the seven heavier Xe isotopes can be produced by captures of free neutrons. The s-process chain of increasing mass number produced by capturing a neutron passes through five stable isotopes of Xe. ^{128}Xe is the lightest s-process isotope of xenon and the heaviest is ^{132}Xe. Radioactive ^{133}Xe decay causes the s-process abundance flow to leave xenon at that point. This leaves xenon's two heaviest isotopes, ^{134}Xe and ^{136}Xe to be produced by the r process. They are not reached under s-process circumstances because ^{135}Xe and ^{137}Xe beta decay in times of 9.1 hr and 3.8 min, respectively, both much shorter than 10 to 100 years. The neutron-rich radioactive nuclei created by the rapid neutron captures decay after the event, not only to ^{134}Xe and ^{136}Xe but also to ^{131,130,129}Xe. Those last three isotopes are thus produced by both the r process and by the s process. Their nucleosynthesis is shared by the two neutron-capture processes. One frequently hears that ^{131,130,129}Xe are both s and r isotopes, with the solar abundances of each being the sum of their astrophysical abundances from each process. The s and r isotopes are made in separate events, distinct kinds of star for the most part, and only mixed together later, during their long residence within the interstellar gas. Similar distinctions occur for the other heavy elements.

The large neutron flux required by the r process does not arise naturally during the normal evolution of stars. It requires perhaps 10^{20} to 10^{24} free neutrons per cm^3, at temperatures somewhat in excess of 10^9 degrees. This corresponds to free neutrons having a density several percent that of normal water! The total mass density of all nuclear species is much, much higher, however, requiring a dense and hot explosive situation. Two exotic possibilities have therefore dominated astrophysical thinking on the r process: (1) supernova explosions, or (2) close encounters between neutron stars. In the supernova, the core becomes so compressed and hot that it decomposes naturally into neutrons plus helium nuclei (alpha particles). During its subsequent expansion the alpha particles partially reassemble into heavier nuclei, which then find themselves within an intense flux of the free neutrons. Rapid captures of those neutrons ensue. This can lead to the r process; but the astrophysics is not straightforward. Very high entropy is required within the expanding matter to achieve the r process. High entropy means that the pressure of the photons exceeds that of the particles during the expansion. "Light" exerts more pressure than "gas;" and it has to date proven difficult to find a cause for the entropy being sufficiently high. Nonetheless this remains the r process site of choice, although perhaps it is an act of faith to assume that the conditions will be right. During collisions of neutron stars, on the other hand, chunks of neutron-rich matter can be expelled from one of the neutron stars. As that highly compressed chunk expands, set free of the enormous pressure of the neutron star, it also undergoes a sequence of states suitable for the r process. Unfortunately, it cannot be alleged with confidence which site is truly responsible, or indeed if it is not instead an unsuspected third alternative.

When the first stars were formed, the r-process nuclei were among the first nuclei to be produced by nucleosynthesis in their explosions. This fact is deduced from the oldest *metal-poor stars*, whose surface abundances showed that they contain proportionately more r-process atoms than s-process atoms (even though the absolute amount of each element is much less than in the Sun and the newest stars). There may even be two distinct r process sets of nuclei, because those heavier than $A = 130$ stand in just the same ratio as do the same r process nuclei in the Sun, whereas those lighter than $A = 130$ do not. So the matter containing the heavier group seemed always to produce the same relative numbers, whereas the matter containing the lighter group was more variable in its yield.

The mathematical solution of the r process series of neutron captures was presented in 1965 by Phillip A. Seeger and the writer, following conceptual formulations by Hoyle, Fowler and Cameron. Since that time effort has concentrated on the many details of nuclear physics and of the natural events that have conspired to produce the r-process abundances that are found in nature. Most favored modern scenarios involve the formation of a neutron star by the collapsing core of a Type II *supernova*. One model utilizes a fast wind of neutron-rich matter from the neutron-rich newly born neutron star. The other utilizes bipolar jets of neutron-rich matter ejected when a highly compressed disk of neutron-rich material establishes an accretion disk of matter to be added to the neutron star. Even after 40 years the correct natural site has not been convincingly modeled.

s process

One of the processes of nucleosynthesis of the elements heavier than iron, this process was named "s" owing to the need to make the s isotopes of the heavy elements by the "slow" capture of free neutrons, neutrons liberated by other nuclear reactions within the gas in stellar interiors. It was one of the first nucleosynthesis processes identified historically. The capture of free neutrons increases the mass number A by one unit at a time. They render each nucleus in the gas heavier, little by little, capture by capture.

When an isotope of mass number A of an element with atomic number Z captures a neutron, the compound nucleus formed from their union becomes an isotope of the same element Z, but having mass number greater by approximately one unit (i.e. A + 1). That new nucleus may be a stable (nonradioactive) isotope of element Z, or it may be radioactive (usually emitting an electron and an antineutrino, called "beta decay," in transforming to the isotope of mass number (A + 1) belonging to the next element (Z + 1)). The adjective "slow" describing the captures in the s process refers to the long wait between successive neutron captures. In stars it may typically be about 10 to 100 years. The s-process isotopes are those that can be reached through neutron reactions induced by irradiating the elements with neutrons, in contrast to the p isotopes which cannot be produced in this way. Such neutron-produced isotopes are called s isotopes. Not all isotopes will be reached in this way, because if a radioactive isotope is created, it usually decays to the next heavier element. This decay prevents the series of captures from reaching those heavier stable isotopes of the lighter element. Those heavier stable isotopes, heavier than the isotope that beta decays, are called r isotopes. The r isotopes are also distinguished from the s isotopes in another phenomenological way. The r isotopes can be produced by neutron irradiation, but the flux would have to be sufficiently intense that the neutron captures would be "fast" compared to the beta-decay lifetimes of the radioactive isotopes (typically minutes to hours). In s-process circumstances (i.e. those stellar interiors within which the number density of free neutrons is such as to require 10 to 100 years between captures), beta decay will occur first and the r-process isotopes will not be produced.

Isotopes of xenon can illustrate the systematics. Xenon has nine stable isotopes. The middle five are s isotopes, the lightest two are p isotopes, and the heaviest two are r isotopes. The lightest two, ^{124}Xe and ^{126}Xe cannot be produced by placing any element within a flux of free neutrons (such as in a nuclear reactor or in a stellar interior liberating neutrons by thermonuclear reactions), and are therefore p isotopes. The other heavier Xe isotopes can be produced via neutron-capture reactions. For clear identification it helps to think also of the two neighboring lighter elements, tellurium and iodine. The chain of increasing mass number produced by capturing a neutron passes first through Te and I just before reaching Xe. Atomic numbers Z = 52, Te, and Z = 53, I, immediately proceed Xe (Z = 54). When Te (Z = 52) captures neutrons, there result new stable isotopes of Te from A = 122 to A = 126. Only at radioactive ^{127}Te does beta decay occur in the time available between captures, and that to stable ^{127}I. After it in turn captures a neutron, its radioactive isotope produced thereby, ^{128}I, decays to ^{128}Xe, thereby sending some abundance flow into the element Xe at that mass number 128. From there successive captures produce the A = 129, 130, 131, and 132 isotopes of Xe. The isotope ^{128}Xe is the lightest s-process isotope of xenon; the heaviest is probably ^{132}Xe. This leaves xenon's two even heavier isotopes, ^{134}Xe and ^{136}Xe to be produced by the r process. They are not reached under s-process circumstances because ^{133}Xe and ^{135}Xe beta decay in times of 5.24 d and 9.14 hr, respectively, both much shorter than 10 to 100 years. Similar distinctions occur for the other heavy elements.

Within stars the s process occurs at two main sites. In the asymptotic-giant-branch red giants, called *AGB stars*, the thermonuclear *helium burning* pulses in the shells of those stars are mildly explosive, causing neutrons to be emitted by (alpha, n) reactions on ^{13}C and ^{22}Ne, both of which are present. Portions of this matter are dredged up to the surface of the red-giant star by convection following some pulses, and this enriches the atmosphere of the star both in carbon (which is the main product of the He-burning pulses) and in s-process isotopes. The observed s-process isotopes in atmospheres of those AGB stars are so abundant that they can only have been generated from iron seed nuclei. The fluence of neutrons is large in these pulses. These stellar masses are about 2–3 times the Sun in mass. The star is eventually turned into a "carbon star" by the dredgeups of the new carbon, and astronomers study them intently. The large abundances of the s-process chain beginning with iron account for most of the Galaxy's s-process nuclei having mass number A > 87. These isotopes are placed in the *interstellar medium* (gas and dust) when a wind flows from the star, carrying its mass away, typically at speeds of 10 km/s.

The other main site for the s process occurs in massive stars, more than twelve times more massive than the Sun. They burn their helium in a large convective ("boiling") core, usually comprising about 40% of the entire mass of the star. A portion of the ^{22}Ne reacts there with He nuclei and liberates neutrons by the same (alpha, n) reaction that occurs in *AGB stars*. There results a weaker fluence of neutrons liberated here, so that nuclei more massive than $A = 87$ are difficult to create with an efficiency that competes with that of the AGB stars. But those of smaller mass number ($A < 87$) are produced effectively in these massive stars, which will eventually eject their shells as a supernova, placing the new s nuclei in the interstellar gas. An excellent marker for this s-process site has been ^{80}Kr, which is exclusively the product of this s-process site. This s process during core helium burning in massive stars has come to be called the "weak s process," because the total fluence of neutrons is too weak to produce the nuclei heavier than $A = 90$ by it. But so many iron nuclei are irradiated by this "weak s process" that it creates significantly more of those isotopes having $A < 88$ than does the "main s process" in the AGB stars. Considerable uncertainty remains over just how many of the nuclei between ^{58}Fe and ^{70}Ge are in fact due to this process rather than to explosive burning in supernova explosions. These are frontier issues in the science today.

On a personal note, the mathematical description of this series of neutron captures called the s process was solved by the writer in his Ph.D. thesis, and was published in 1961. This provided me the opportunity to assist in small ways with the development of programs of the measurement of neutron-capture cross sections for the s process, first at Oak Ridge and later at Karlsruhe. These were great privileges in a fortunate life.

Secondary nuclei

This is a technical term from the nucleosynthesis of the elements, and has a very specific definition to distinguish "secondary nuclei" from "primary nuclei." The "primary nuclei" are those that can be abundantly manufactured during the evolution of a star that initially contains only those abundances inherited from the Big-Bang origin of the universe. In effect this means nuclei that can be assembled from hydrogen and helium (^{1}H and ^{4}He). Leading examples of such nuclei are ^{12}C and ^{16}O, which are fused from helium, or ^{24}Mg which is the main product when the ^{12}C fuses with itself. Stars of pure ^{1}H and ^{4}He produce them abundantly. A secondary nucleus is one that cannot be synthesized from a star made initially of ^{1}H and ^{4}He. Leading examples are found in the two heavy isotopes of oxygen or of silicon and in the s-process nuclei. Nature's production of ^{17}O, for example, utilizes ^{16}O from the initial composition of that star (see **^{17}O**); but that initial ^{16}O must itself have been produced from hydrogen and helium (^{1}H and ^{4}He) during the prior evolution and death of an earlier star in order that the ^{16}O nuclei can have already been present in the gas from which the ^{17}O-producing star formed.

Secondary nuclei tend to be neutron-rich isotopes of the alpha elements (O, Ne, Mg, Si, S, Ar, and Ca). To make neutron-rich isotopes in stars requires a source of neutrons, and this in turn requires that the gas in bulk contain more neutrons than protons by a small amount. This requires that the star contain an abundant neutron-rich isotope. This is not easily done in stars formed from only H and He, the earliest stars of the Galaxy. But stars that are born containing initial C and O abundances produce neutron-rich nuclei during *helium burning* from the ^{14}N (not n-rich) that is the dominant ash of nuclear hydrogen burning. The neutron-rich isotope ^{18}O is synthesized during helium burning by the nuclear reaction $^{14}N + {}^4He \rightarrow {}^{18}F$, followed by positron emission to leave ^{18}O; and ^{18}O is subsequently destroyed during helium burning by the nuclear reaction $^{18}O + {}^4He \rightarrow {}^{22}Ne$. That ^{22}Ne becomes abundant then in later-generation stars, and it serves both as a source for free neutrons when it interacts with hot He and as a seed nucleus for the construction of secondary n-rich isotopes of Mg, Si, S, Ar, and Ca. The competition between the various secondary nuclei for the free neutrons liberated evokes again the ecological metaphor. The nuclei from which the secondaries are born may be thought of as the parents; *e.g.* production of ^{17}O utilizes parent ^{16}O. The free neutron that must be captured by ^{16}O to do this might be likened to food. Other nuclei compete for that same food, and it is the *neutron-capture cross sections* of the parents that measure their fitness for the competition.

As the interstellar gas of the Galaxy becomes more and more enriched in the carbon and oxygen products of the stars, the new stars being born from that enriched gas become increasingly able to manufacture the secondary nuclei. In the ecological metaphor, the ^{16}O must first evolve before the ^{17}O can evolve from it. This concept is very important for understanding the abundance evolution of the Galaxy (see *Chemical evolution of the Galaxy*).

Silicon burning

Silicon burning is the name of the process by which silicon nuclei are transformed by nuclear reactions into heavier elements, primarily nickel. The need for this occurs naturally during stellar evolution when the *oxygen-burning* core of a giant star has completed its fusion of oxygen into silicon. A huge mass of oxygen had remained from the prior ashes of *helium burning*. The oxygen-burning primary nuclear reactions are

$$^{16}O + {}^{16}O \rightarrow {}^{28}Si + {}^4He, \text{ and } {}^{16}O + {}^{16}O \rightarrow {}^{31}P + {}^1H;$$

but a variety of secondary reactions make ^{28}Si the most abundant final product. Statistical equilibrium demands that if the silicon is maintained hot for a sufficiently long time it will transmute into nickel. But how this is accomplished was unclear for more than a decade after the problem was recognized. In the late 1960s it became clear that the transmutation occurred by a silicon nucleus getting hot enough that fragments are photoejected from it, whereupon the capture of these fragments by other Si nuclei slowly reorganizes two Si nuclei into one Ni nucleus. This process is not fusion, but is a nuclear rearrangement that occurs when silicon becomes too hot to be stable (about 3.5 billion degrees). It has been called "nuclear melting" in analogy with the calling of nuclear fusion "nuclear burning." The process is well described as a *quasiequilibrium* within which reactions are balanced by their inverse reactions except for a slow leak that enables two heavy nuclei to be replaced by one nucleus. During silicon burning the "slow leak" is the slow photodisintegration of silicon. In effect the transmutation $^{28}Si + ^{28}Si \rightarrow ^{56}Ni$ occurs via the quasiequilibrium bath of photons, protons, neutrons and alpha particles.

Explosive *silicon burning* occurs in supernova explosions when the silicon is suddenly heated to temperatures well in excess of its quiescent burning temperature and therefore burns quite rapidly (in seconds). In Type Ia *supernovae* this happens when a runaway nuclear burning causes the temperature to overshoot stable burning, leading to a *quasiequilibrium* that is quite close to *thermal equilibrium*; in Type II supernovae it occurs when the rebounding shock wave blasts the silicon shell of the presupernova star and heats it suddenly, so that the silicon undergoes rapid photodestruction and quasiequilibrium transmutation.

SIMS (Secondary-ion mass spectrometry)

The secondary-ion mass spectrometer is a laboratory instrument for sorting ions into their masses. The sorting is called spectrometry. The acronym SIMS refers to either the process or the instrument. It is one form of mass spectrometry, a lab science that has been very important in the history of science for determining the masses of atoms. "Secondary-ion" refers to the source of the ions whose masses are to be measured. Secondary ions are those caused by primary ions striking a solid target. Each primary ion (usually an accelerator beam of a specific ion type, O^- or Cs^+, for negative or positive primary ions) causes several ions to be ejected when it strikes and imbeds within a solid target. This process is called "sputtering." The secondary ions can be weighed by determining their motion in combined electric and magnetic fields that determine the ratio of charge to mass.

The importance of SIMS is that it is the technique for determining the relative numbers of ions of differing mass (different isotopes) of a sputtered target element from a very small sample. It has provided the isotopic analysis of *presolar grains*. Each grain is isolated and mounted on a plate for bombardment by primary ions. These grains are but 1/10 000th of a centimeter in size and contain only about one million atoms. Though that number sounds large to the uninitiated, it is too few atoms to be sorted into isotopes of several elements by any process other than SIMS. This has made SIMS a kind of "telescope" for the observation of stars by way of their grains. That science has rejuvenated the meaning of isotopes in astronomy, because the isotope ratios are measured with enormously greater precision in the laboratory than astronomers can measure in the wavelength spectrum of light from stars. Both the science of stellar nucleosynthesis and the science of the internal structure of stars have been revolutionized by this new detailed knowledge.

Stellar structure

Although it was a mystery to men but two centuries ago, a star is a ball of hot gas. It is held against dispersal by force of its own weight, just as the Earth's atmosphere is held in place by its weight. For the Earth's atmosphere, the retaining force is the considerable gravity produced by the Earth's mass, attracting all other masses toward its center. The star has no solid core, but the principle that each spherical layer of gas is attracted gravitationally by the mass of gas within that sphere remains the same. In both cases the pressure must increase with depth, so that it becomes at each depth sufficient to support the overlying weight. Were those forces not to balance, pressure against weight, the star would either shrink further or expand. This principle is called "hydrostatic equilibrium," the prefix perhaps deriving from its first applications to bodies of water on Earth, where it was discovered that the presssure must increase with depth in just the right balance to support the weight of overlying water. This is the first principle of stellar structure.

The pressure in a gas is proportional to both the density of the gas (the numbers of particles per unit volume) and to its temperature. For stars composed of ordinary gases (e.g. the Sun) the increase in pressure with depth is achieved in part by the increase in the gas density and in part by an increase of temperature with depth. Without the increasing temperature with increasing depth, the force of gravity would become too strong and the star would shrink. Indeed, it must have shrunk at the beginning when it condensed out of more dispersed gas in the interstellar clouds. Thermodynamic science requires that the very act of compression heats the gas, as in the compression phase of an automobile piston. Because the central gas is the most compressed, it is also the hottest. The values in stars quickly exceed human experience; for example, the center of the Sun must heat to near 14 million degrees to provide the Sun's obvious hydrostatic equilibrium. This is the second principle of stellar structure.

Even children quickly learn that heat energy flows from hot to cold rather than the other way round. It is therefore evident that heat must flow outward in the Sun, from the hot center toward the cooler atmosphere. It is that atmosphere that we see as the Sun. Its temperature is near 5800 degrees – hot, but not outside human experience. From that hot surface the Sun radiates heat as surely as a red-hot bowling ball in space would radiate heat. That heat radiation is what we call sunlight, or, when emanating from some other star, starlight. The eye cannot see the greater temperatures at increasingly deeper layers because the dense atmosphere becomes opaque. Application of this third principle of stellar structure determines the power radiated by the Sun and stars. How rapidly the heat flows to the surface where it is radiated into space is determined by how steeply the temperature rises with increasing depth and by the opacity of the solar gases.

One paradox in all this troubled mankind greatly until about the time of the outbreak of World War II. If heat energy leaks from an object, it cools. Children learn this with ease as well. As a red-hot ball in space radiates heat energy, it cools off. This same principle argues that the Sun is cooling off. But no evidence of that can be found. Indeed, life fossils on Earth exceeding 3 billion years in age argue that the Sun has been shining at least that long. The problem comes with calculating how fast the Sun should cool owing to its huge rate of expenditure of heat energy into space. The Sun could remain hot for only about 10 million years in this way. Though this seems long to a man, it is several hundred times too short to reach back to life's fossils on Earth. That is, the Sun has been shining hundreds of times longer than its available thermal supply could provide. "What keeps the Sun hot?" was a harder question to solve than "Why is a star hot?"

The answer is that thermonuclear reactions near the solar center liberate fresh heat energy. The process is the fusion of hydrogen into helium. Mankind hopes to establish similar thermonuclear fusion processes within fusion-power reactors on Earth. Laboratory studies reveal that temperatures of a few million degrees are needed to achieve the fusion of hydrogen. The structural principles summarized above argue that the solar center should be near 14 million degrees, which is adequate to drive the fusion reactions there. Calculations reveal that thermal balance is achieved in the Sun; namely, the thermonuclear power liberated at the solar center exactly replenishes the radiative power lost from its surface. This must be true in other normal stars as well or they would be unable to shine steadily for so long a time. The famous physicist, Hans Bethe, first described during 1939–40 the exact nuclear reactions that achieve this at the solar center, for which his 1967 Nobel Prize was awarded.

Supernovae

Supernovae are super-bright explosions of stars, in which the light output at the peak of the display equals that from the other 100 billion galactic stars combined. These are rare in our Galaxy, about 2–3 per century, but by looking at many other galaxies astronomers see them occurring frequently. And in our own Galaxy they are evident as the bubbles of hot gas that they leave for a few millions years in the *interstellar medium*.

Supernovae are of fundamentally two types. Their names, Type I and Type II, were historically bestowed to distinguish between those that do not reveal H lines in their spectra and those that do, respectively. But a more fundamental distinction between them is the mechanism by which they explode and the state of the presupernova star. Type Ia supernovae are thermonuclear explosions of white-dwarf stars that have become too heavy because of accreting gas from a neighboring star. They are blown apart and leave no remnant behind in a giant thermonuclear explosion. Before explosion the white-dwarf stars were balls of C and O that generate no more nuclear reactions and might live indefinitely were it not for their increasingly severe overweight problem.

Type II supernovae are massive stars, ones that progress in their nuclear fuels well past the fusion of carbon and the fusion of oxygen at their centers. When their cores run out of nuclear fuel, those central regions collapse to form a neutron star, or in some cases a black hole. The incredibly intense emission of neutrinos from the newly born neutron star so heats the overlying layers, aided by an outward moving shock wave of pressure, that those layers partly explode and are ejected. The last of these that was visible to the naked eye occurred in 1987, and demonstrated for the first time the correctness of the intense neutrino burst that is their main energy output.

Supernovae of both types are prolific contributors to the origins of the chemical elements, or nucleosynthesis. Some isotopes, like ^{28}Si and ^{56}Fe, are abundantly made by both types; but other isotopes (^{22}Ne or ^{48}Ca, for example) are made in only one of the two types (Type II or Type I in these examples, respectively).

Thermal equilibrium (statistical equilibrium)

If a reactive system is held at constant temperature and density for a long time it reaches thermal equilibrium. In that condition each abundance assumes a value that remains constant in time because the rates of creation and destruction of each species are in balance. Since nuclei are the objects in question, in "nuclear thermal equilibrium" destructions and creations balance, so the nuclear abundance is unchanging. Even more restrictively, in thermal equilibrium the rate of every reaction is balanced by the rate of the inverse reaction. For example, the rate of ^{44}Ti(alpha, p)^{47}V owing to free alpha particles at thermal equilibrium is exactly equal to the rate of ^{47}V(p, alpha)^{44}Ti owing to the density of free protons. Moreover, this is true for every reaction, not just for most. This means that the densities of free alpha particles and free protons are also maintained at their equilibrium values, so that their densities are also steady in time owing to equal rates of absorption and of emission of them. When these conditions occur, the steady abundance is also described as a "statistical equilibrium," meaning that each abundance can be predicted from thermal statistics, without knowledge of nuclear reaction rates.

Stars actually do not provide enough time at any temperature for this state to be reached. But it is so nearly reached that thermal equilibrium may be a very accurate approximation. Reactions among nuclei are so slow at normal stellar temperatures that this state is not even approximately approached. The best approximation is found within *supernovae*, despite the fact that the explosive environment changes in mere seconds. But nuclear reactions are so fast at T = 5 billion degrees that a fair approximation may be achieved in much less than one second. Even then, the state achieved is *quasiequilibrium*, rather than equilibrium.

The approximation of thermal equilibrium played an important role in the initial formulation of the theory of nucleosynthesis in stars. Fred Hoyle showed in 1946 that stellar centers may achieve temperatures in which thermal equilibrium can be achieved (or nearly so). Hoyle showed that the equilibrium abundances rather naturally created the large abundance peak at iron, specifically at ^{56}Fe. This equilibrium process was called the e process (e for equilibrium). Although this was regarded as a great triumph and underpinned Hoyle's idea that stars synthesized all of the elements, it proved inadequate in details. The isotope ^{56}Fe is actually synthesized as radioactive ^{56}Ni (see **^{56}Fe**); and within a quasiequilibrium rather than within a statistical thermal equilibrium.

Three-isotope plot

This is a graph constructed from two isotope ratios formed from three isotopes of a single element. It is important for its ability to demonstrate the existence of mixing between two different samples of matter. The element oxygen provides a good example. When the abundances of all three O isotopes ($A = 16$, 17, and 18) are measured in any sample, it is possible to locate that composition as a point on a graph whose y axis is the ratio $^{17}O/^{16}O$ and whose x axis is the ratio $^{18}O/^{16}O$. Such a graph has been named "the three-isotope plot," because three isotopes make only two independent ratios, the two coordinates then locating the isotopic composition. Each bulk sample is represented by its location point on that graph. When the same isotope is used as the denominator in both isotope ratios, the graph has this intuitively appealing property: two different (isotopically) samples of matter plot at two different points A and B on this graph, and any mixture of the two samples lies along the straight line connecting A with B. Furthermore, the distances of the mixture from A and B are inversely proportional to the relative numbers of atoms from the two sources in the mixture. This property has had great utility in the science of natural isotopes. When a suite of natural samples is measured and found to lie along a straight line in the three-isotope plot, one can conclude that the samples can be just a mixture of two distinct sources, each at or beyond the respective ends of the measured correlation line.

Instead of raw ratios as coordinates of the three-isotope plot it is more common to see ratios normalized to the solar isotopic ratio and even deviations of that normalized ratio from unity. In the latter case the coordinates are labeled as delta (δ).